Optical Fiber Sensors for the Next Generation of Rehabilitation Robotics

Optical Fiber Sensors for the Next Generation of Rehabilitation Robotics

Arnaldo Leal-Junior
Mechanical Engineering Department
Federal University of Espirito Santo
Vitória, Brazil

Anselmo Frizera-Neto
Electrical Engineering Department
Federal University of Espirito Santo
Vitória, Brazil

ELSEVIER

ACADEMIC PRESS
An imprint of Elsevier

Academic Press is an imprint of Elsevier
125 London Wall, London EC2Y 5AS, United Kingdom
525 B Street, Suite 1650, San Diego, CA 92101, United States
50 Hampshire Street, 5th Floor, Cambridge, MA 02139, United States
The Boulevard, Langford Lane, Kidlington, Oxford OX5 1GB, United Kingdom

Library of Congress Cataloging-in-Publication Data
A catalog record for this book is available from the Library of Congress

British Library Cataloguing-in-Publication Data
A catalogue record for this book is available from the British Library

ISBN: 978-0-323-85952-3

For information on all Academic Press publications
visit our website at https://www.elsevier.com/books-and-journals

Publisher: Mara Conner
Acquisitions Editor: Sonnini R. Yura
Editorial Project Manager: Isabella C. Silva
Production Project Manager: Sojan P. Pazhayattil
Designer: Victoria Pearson

Typeset by VTeX

Contents

Part II
Introduction to optical fiber sensing

Part III
Optical fiber sensors in rehabilitation systems

Preface

The advances in medicine and physical therapy in conjunction with new developments of mechatronic devices with a higher level of controllability enabled the development of assistive robotic devices, which are explored by many research groups around the world. Concurrently, there is the development and widespread of optical fiber technology, which is increasingly used as sensors devices. The optical fiber sensors characteristics are well aligned with the requirements of robotic instrumentation, especially the ones with electric motors, commonly used in wearable robots: Optical fiber sensors are immune to electromagnetic perturbations offering precise measurements in noise environments. In addition, the flexibility of optical fibers is also aligned with the new trends in soft and flexible robotic systems, where the sensors can be embedded in the robot's structure or they can be placed on wearable devices for patient monitoring. Years ago, all of these advances resulted in a new research direction, where the optical fiber sensors were used on the robots' instrumentation to extend their control capabilities by measuring parameters that were not commonly measured with conventional electromechanical sensors.

The results of years of research in robotics and optical fiber sensors in a joint effort of the Graduate Program in Electrical Engineering and Mechanical Engineering Department of the Federal University of Espirito Santo (UFES) are summarized in this book. The aim of this book is to provide a comprehensive understanding on this new research topic and its underlying theory and principles. This book was proposed and conceived under the assumption that the next generation of wearable robots and devices not only will include the soft structure and compliant actuators, but also the new optical fiber sensors embedded in the robots' structure and actuation units. We divided the book into four parts. In the first part of this book, the developments in wearable robots and assistive devices as well as human-in-the-loop design and the recent developments on soft robotics are discussed. In the second part, the focus is shifted to optical fibers including the presentation of an overview, the main components, and characteristics of an optical fiber-based detection system and the materials commonly used on the development of optical sensors. Moreover, optical fiber sensors approaches are presented. The third part presents the optical fiber-based instrumentation systems in wearable robots and assistive devices, resulting in the combination of the knowledge acquired in the first and second parts of the book. The discussed systems include wearable robots, smart structures in which the sensors are embedded in rigid and/or soft structures of the robots, compliant actuators and smart wearable textiles for patients monitoring. In the last part of the book, different case studies and additional application are presented to provide a broader view of the many possibilities of optical fiber sensors in assistive devices, which include the developments in smart walker's instrumentation, robotic surgery with manipulators, physiological parameters monitoring using multifunctional textiles, and even in biosensors for health assessment.

This book could not be written without the hard work of the contributors, L. Avellar, V. Biazi, W. Coimbra, and L. Vargas-Valencia, all of them from UFES, contributed for some chapters throughout the book. C. Marques from University of Aveiro, a long time contributor in our research group helped us on the biosensors applications using optical fibers. The advances and methods discussed in this book were developed in the framework of different research projects focused on rehabilitation or optical fiber sensing technologies as follows:

- Active transparent orthosis for rehabilitation and movement assistance (CAPES 88887.095626/2015-01);
- Research Center on Photonics and Advanced Sensing (FAPES 84336650);
- Optical fiber sensors network for patients remote monitoring (FAPES 320/2020);
- Optical fiber sensors in oil-water interface measurement in production tanks (Petrobras 2017/00702-6).

We would also like to thank all the support from our colleagues in writing this book.

Arnaldo Leal-Junior
Anselmo Frizera-Neto
Federal University of Espirito Santo, Vitória, Brazil

Part I

Introduction to soft robotics and rehabilitation systems

Chapter 1

Introduction and overview of wearable technologies

1.1 Motivation

Since the early days of human history, there is a continuous increase in the life expectancy, which leads to the population aging. In half of a century, from 1950 to 2000, the elderly population (over 65 years) rose from 131 million in 1950 to 418 million in 2000, more than a threefold increase in 50 years (Rowland, 2009). This increase in longevity reflects the evolution of the society with advances on public health, medicine, economy and social development (United Nations, 2019). All of these advances contribute to the control of diseases (including the eradication of some diseases), injuries prevention, and reduction of premature deaths (especially in newborns). In summary, many health conditions that were deadly in the past (including diseases such as smallpox and polio) nowadays are treatable or curable. According with United Nations (UN) reports, there are four trends in the global population, which are the population growth, urbanization, international migration, and the population aging (Turner, 2009).

Generally, the elderly population is defined as number of people over 65 years, whereas the working ages are defined as the interval between 25 and 64 years. In addition, there are the children (whose ages are 0 to 14 years) and the youth population, ages between 15 and 24 years. Therefore, common metrics to set the scene of population aging are the percentage composition of the population, considering all four groups, i.e., children, youth, working-age adults, and older population. Therefore, common metrics to set the scene of population aging are the percentage composition of the population, considering all four groups, i.e., children, youth, working-age, and older population (United Nations, 2019). Fig. 1.1 shows the population percentage with 65 years or older from 1950 to 2020 and includes statistical projections for the next 80 years (until 2100).

In the analysis of the population aging, some underlying factors such as accessibility to medical care, public health policies, and social development should also be considered. These factors are not uniformly distribution among countries, and thus there are countries with higher proportion (and increase rate) of elderly people. The increase of elderly population across the countries is higher in more developed regions and in higher income countries. In 1950, France was the country with highest proportion of an older population (11.4%). Then, in

https://doi.org/10.1016/B978-0-32-385952-3.00009-3

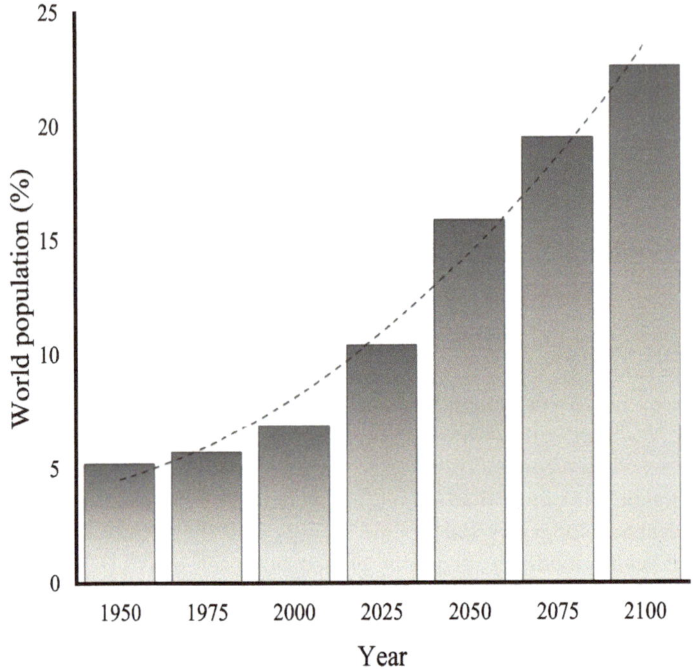

FIGURE 1.1 World population aging throughout the years and predictions for the next 80 years (United Nations, 2019).

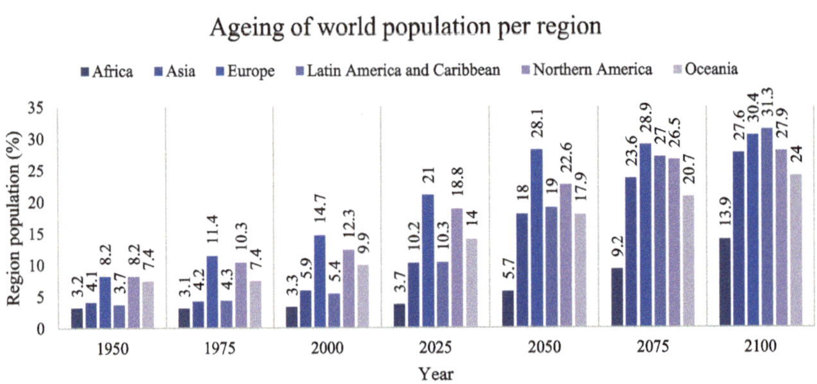

FIGURE 1.2 Population with 65 years or older in each region (United Nations, 2019).

1975, Sweden was the leading country in elderlies with 15.1% of the population over 65 years. The increase of an elderly population continues as Italy had 18.1% in 2000 (Rowland, 2009). This trend continues with Europe and Northern America as the regions with the highest ratio of an elderly population, as shown in Fig. 1.2.

As shown in Fig. 1.2, Northern America, Europe, Australia, and New Zealand are the regions with the highest elderly population proportion. It is also worth noting that the Eastern and Southeastern Asia region is the one with the highest increase rate in the older population, especially after 2010. However, it is possible to observe that almost all regions showed an increase of the elderly population throughout the years. As the older population proportion is the ratio between the population over 65 years and the total population, such increase in the elderly population proportion also is related to a fertility reduction trend in the worldwide population. As depicted in Fig. 1.1, there is no substantial increase on the population between 0–4 years. The so-called age pyramid is the age population distribution across the age groups, as shown in Fig. 1.3. The age pyramid barely resembles a triangular shape nowadays and will continuously change according to statistical projections.

The demographic transition in world population sets new challenges in different areas, in an economical perspective, the increase of an older population increases the demands for pensions, especially when combined with a reduction of the ratio between the elderly and working-age population (Turner, 2009). Another challenge is related to the healthcare of the elderly population that suffers from inherent conditions of normal aging such as immunosenescence, urologic and sensory changes, which include hearing loss, visual acuity, and vestibular function degradation (Jaul and Barron, 2017). Such conditions lead to variation in physical functions, including the reduction of walking speed, mobility disability, difficulty in activities of daily living, and increase of fall risk (Jaul and Barron, 2017). The degradation of physical functions in conjunction with the cognitive reduction can also lead to psychological and social issues (Jaul and Barron, 2017). The population aging also results in an increase of clinical conditions that affect the human health, the so-called chronic age-related diseases and geriatric syndromes (Franceschi et al., 2018). These conditions include osteoarthritis, rheumatoid arthritis, Alzheimer's disease, Parkinson's disease, and weakness of the skeletal muscles. All of these conditions lead to degradation of physical and/or cognitive functions (Franceschi et al., 2018). It is worth noting that strokes, spinal cord injuries, and musculoskeletal injuries can also lead to major locomotor impairments (Huo et al., 2016)

Disabilities and impairments in the world population are increasing due to factors such as population aging and the increase in chronic diseases (Organization, 2011). In 2019, nearly 15% of the world population have at least one of the many types of disabilities, which represent about 1 billion people in the entire world (Organization, 2018). The physical and cognitive disabilities have a major impact in daily life since they impose limitations on work performance, activities of daily living, and hinder the independent development in the community (Allen and Hogan, 2001). If a high-income country such as United States of America (USA) is analyzed, about 26% of adults have some form of disability (Ferneini, 2017). Fig. 1.4 shows the types of functional impairment among the 26% group, which resulted in 61 million people.

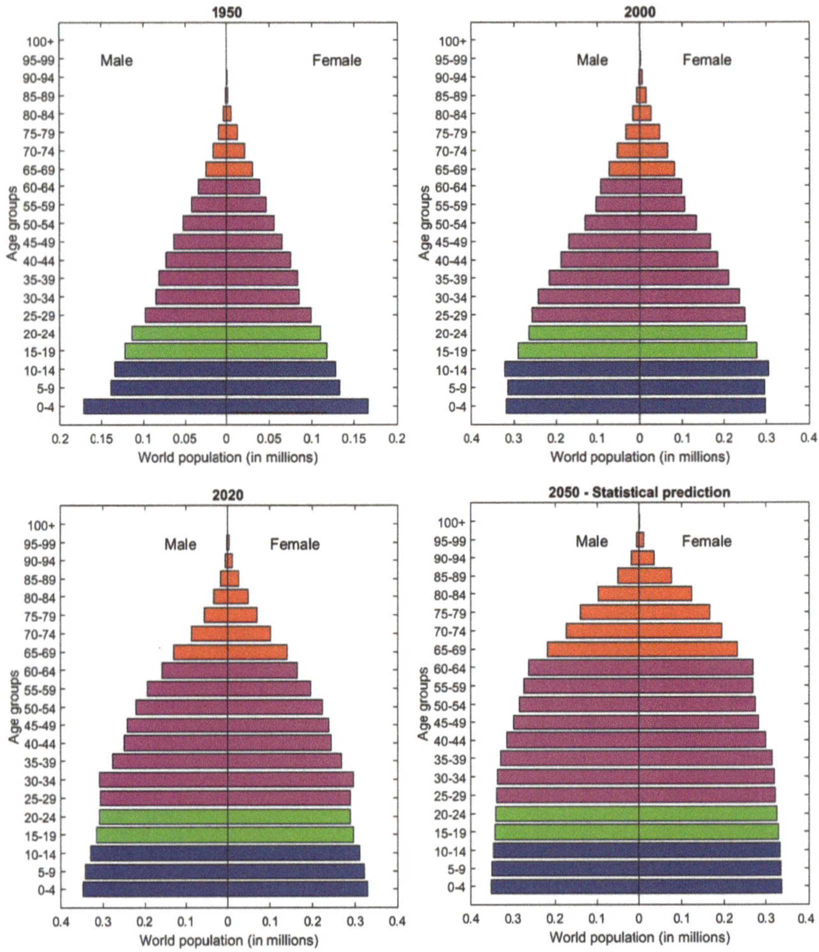

FIGURE 1.3 Age pyramid evolution worldwide from 1950 to 2020, including projections for the next 30 years. Each color represents one age group, i.e., 0–14 years, 15–24 years, 25–64 years, and the ones older than 65 years (United Nations, 2019).

As shown in Fig. 1.4, the most common disability is mobility, caused by locomotor impairment, where different clinical conditions can lead to a multitude of gait disorders, as summarized in Fasano and Bloem (2013). In an attempt of mitigating (or eliminating) the physical impairments, the physical rehabilitation emerges as a feasible option with predefined clinical guides for the rehabilitation of different disorders (Pirker and Katzenschlager, 2017). However, as the population with physical disability increases, many regions report shortage in physiotherapists and rehabilitation personnel (Organization, 2011). Actually, for high-income countries, there is about 5 physiotherapists per 10,000 population and this number is even lower for low-income regions (Organization, 2011).

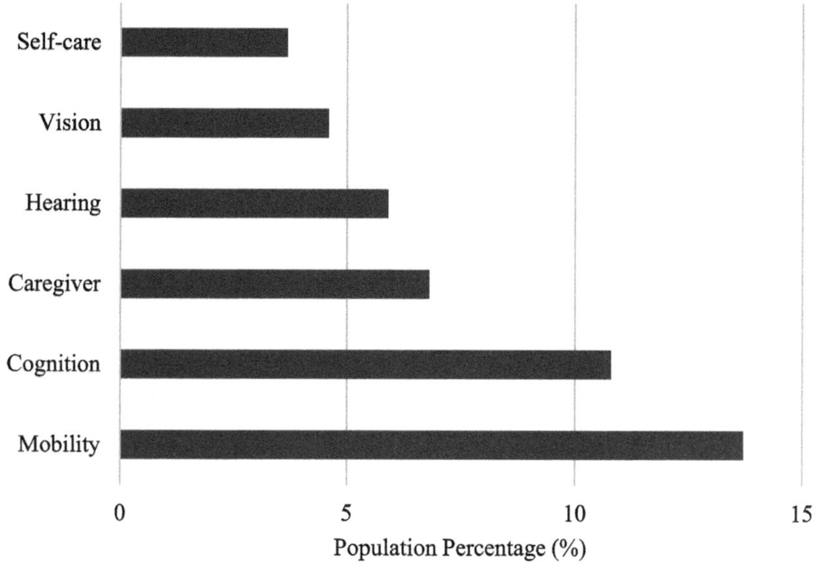

FIGURE 1.4 Types of disabilities in the USA (Ferneini, 2017).

This scenario has pushed the boundaries for novel therapeutic methods and assistance devices for patients with locomotor impairment, which also result in the development of novel devices with the aim of monitoring parameters for human health assessment (Majumder et al., 2017).

In order to offer independence and attenuate the effects of human gait disorders and physical impairments, different assistance devices have been proposed throughout the years, e.g., prostheses (Ha et al., 2011), exoskeletons (Bayon et al., 2016), orthosis (dos Santos et al., 2015) and smart walkers (SWs) (Martins et al., 2012). The latter is generally used as a supporting device in the patients bipedestation, which aids in their balance, and thus, improving the mobility (Martins et al., 2012). SWs present actuators and electronic components aiming to provide a better assistance to the users, where the functionalities of such devices include autonomous control with the possibility of shared or manual control as well, sensorial feedback, higher safety, and the possibility of monitoring the user' state (Martins et al., 2015). Among the wearable robotic devices for rehabilitation, exoskeletons show advantages over conventional rehabilitation therapies related to their higher repeatability in the rehabilitation exercises, possibility of treatment customization, and quantitative feedback of the patient's recovery (Kwakkel et al., 2008). In addition, wearable robots control strategies for human-robot physical and cognitive interactions enable using exoskeletons as assistance devices for daily activities, which include gait assistance (Bueno et al., 2008).

The possibility of monitoring parameters of movement as well as physiological parameters for human health enables novel developments in healthcare in which it is possible to assess the patient's condition for the continuous monitoring of health conditions as well as the possibility of anticipating some diseases and/or disorders. The monitored parameters for human health assessment include foot plantar pressure, which provides important data regarding the human locomotion (Abdul Razak et al., 2012). With the plantar pressure assessment, it is possible to obtain a foot pressure distribution map, which plays an important role on the monitoring of foot ulcerations (of particular importance for diabetes patients). In addition, foot pressure maps enable measurements of foot-function indexes such as arch index, which provide the evaluation of the arch type of each individual that is also related to injuries in runners (Teyhen et al., 2009). Furthermore, the dynamic evaluation of the foot plantar pressure can also aid clinicians on the gait related pathologies diagnosis (Leal-Junior et al., 2018a).

The gait cycle is divided into two main phases: stance and swing, which present many subdivisions (Taborri et al., 2016). The subdivisions of the stance phase can be detected by the plantar pressure variation and it is critical for the control of wearable devices for gait assistance (Villa-Parra et al., 2017). Additionally, the measurement and analysis of joint angles can provide benefits for clinicians and therapists since it is used on the evaluation and quantification of surgical interventions and rehabilitation exercises (Dejnabadi et al., 2005). In addition, such measurements can be applied for training athletes (Hawkins, 2000) and the kinematic data have been employed on the control of neural prostheses (Tong and Granat, 1999).

Furthermore, wearable sensors can be used on healthcare applications (Nag et al., 2017). To that extent, significant advances in sensor technology, wireless communications, and data analysis have enabled a change of scenario, where the health condition assessment is not limited to clinical environments (Korhonen et al., 2003). Thus it is also possible to monitor different physiological parameters for patients at home, which is especially desirable for the elderly population and people with locomotor disabilities (Majumder et al., 2017). Among many important physiological parameters, abnormalities on the heart rate (HR) and breathing rate (BR) are important indicators of some cardiovascular diseases (Böhm et al., 2015), fatigue (Nishyama et al., 2011), apnea (Nishyama et al., 2011), and respiratory abnormalities (Strauß et al., 2014).

These new advances in healthcare technology provide new insights for rehabilitation and therapeutics, where a widespread of wearable technologies has been observed in the last years with an impact in industrial manufacturing for these new products, regulations, and data security (Erdmier et al., 2016). From the user perspective, methods for increasing the patient engagement on the use of such technologies are also proposed (Tran et al., 2019). Furthermore, challenges related to the technology sustainability, failure rates, privacy, and security have been addressed (Bove, 2019). The widespread of wearable assistive technologies in conjunction with the increase on the patient engagement result in a continuous increase on the market of wearable healthcare devices (The European Communities, 2016). Fig. 1.5 shows an overview of the European market on healthcare wearable devices, where a large increase on the market can be seen with the forecast of even higher increase in the coming years. In addition, Fig. 1.5 also shows that almost a half (42%) of the wearable devices are focused on healthcare applications and this value can be even higher if we consider that other healthcare applications are related to monitoring and sensing (16% of all applications).

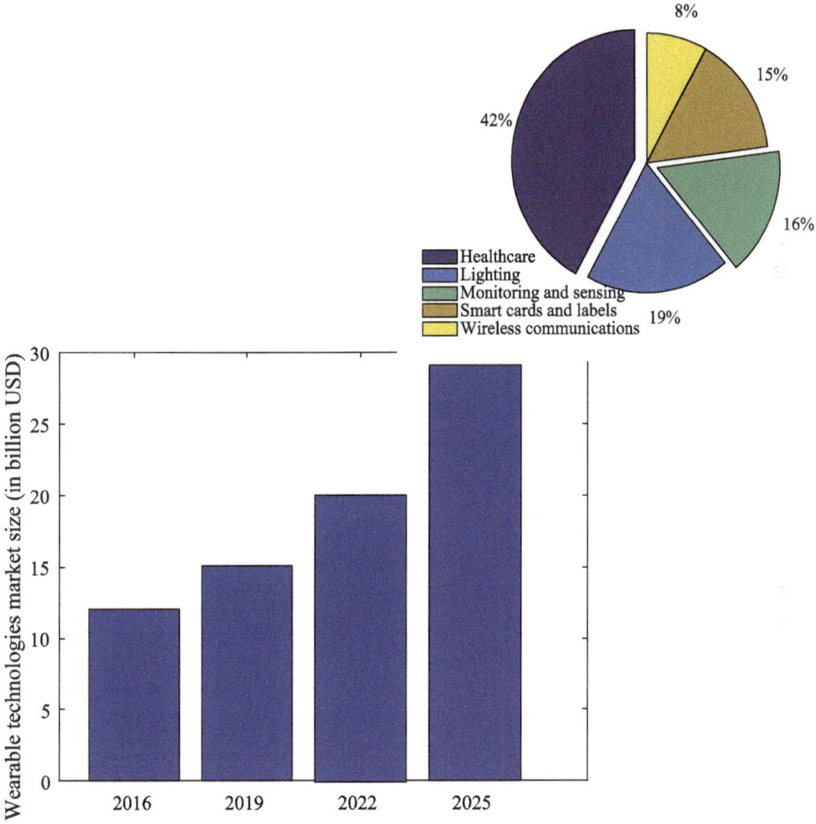

FIGURE 1.5 Wearable devices applications and healthcare market overview (The European Communities, 2016).

The continuous aging of the population as well as the increase on chronic diseases and physical impairments in the world population motivate the development of new smart/robotic devices for human assistance and health condition assessment. Nowadays, such technologies have a large share on the market and are progressively present in our daily life. It is possible to classify such technologies into two major groups: (i) wearable robotics and assistive devices and (ii) wearable sensors and monitoring devices. Both groups are thoroughly discussed in the next sections.

1.2 Wearable robotics and assistive devices

Robots were originally designed to replace humans in repetitive or precise industrial tasks where minimal or no interaction with the operator occurred. Currently, it is usual to notice robots close to the human in an unimaginable set of scenarios, from cleaning robots to rehabilitation and functional compensation devices (Huo et al., 2016). Even in industrial environments, there is human-robot cooperation to develop complex and heavy-duty tasks. In this context, there is a continuous change of the paradigm of robots design and complex (physical and cognitive) human-robot interaction is at the center of technological development (Moreno et al., 2008).

Wearable robots (WR) are defined as those worn by human operators aiming at supplementing or even replacing physical functions (Moreno et al., 2008). Additionally, wearable robots can be used to replace missing limbs, as prosthetic devices, or alongside with human limbs, creating the so-called orthotic devices or exoskeletons. In this context, it is important to define physical human-robot interaction (pHRI) as the generation of supplementary forces to empower and overcome human physical and motor limits deriving from trauma or disease (Alami et al., 2006). Physical human-robot interaction involves a net flux of power between the wearable device and the user. Alternatively, cognitive human-robot interaction (cHRI) implies making the human aware of the possibilities of the robot at the same time that the individual controls the robotic device (Pons, 2010). Considering the context of motor control, cognitive process leads to planning and execution of motor tasks, involving activity from central and peripheral structures. Thus information to decode human intention is gathered from different levels of this process, from central and peripheral nervous systems to human motion, which result in brain-, neural- and movement-controlled exoskeletons (Pons, 2010). Both cHRI and pHRI have direct impact on the usability and dependability of assistive robotic technologies. The concepts of cHRI and pHRI are also translated to other applications of rehabilitation robotics, such as previously proposed in human-robot interaction for locomotion assistance with smart walkers (Cifuentes and Frizera, 2016).

The development of different instances of wearable robots is intricately linked to the applications for which they are proposed. Research and technological developments of WR date from the early 1960s, when the US Department of Defense proposed the concept of powered suits. In parallel, Cornel Aeronautical Laboratories brought to light the concept of human amplifiers as manipulators to enhance the physical capabilities of the operator (Rocon et al., 2008). In fact, according to Moreno et al. (2008), there are different forms of classifying WR. The first one is into prosthetic or orthotic devices. Prosthetic robots are those that substitute lost limbs while orthotic robots operate in parallel with the subject's limbs. A second (and useful) classification is according to the application of use. In this case, applications range from service robots, rehabilitation, and functional compensation devices (also called medical exoskeletons), space applications to devices for military use.

Beyond the potential applications of WR to augment load carrying capacities or to enable the user to work in harsh environments, this book focuses on the rehabilitation and functional compensation wearable devices. Rehabilitation and functional compensation are key in an aging population, where the shortage of caregivers is a reality. While rehabilitation devices can be used for improving lost functions in a large range of applications and disabilities, functional compensation devices are a key to increase independence and performance in daily tasks of individuals with a chronic lesion or permanent dysfunctions (Huo et al., 2016).

Rehabilitation and wearable robots date from the early 1960s. Starting with pioneering work at Case Institute of Technology, a four degree-of-freedom (DoF) externally powered exoskeleton was proposed and, in 1969, the Rancho Golden Arm was presented as a six DoF powered orthosis (Harwin et al., 1995). Another interesting approach that led to the evolution of rehabilitation robots is using industrial robots in combination with interface devices to assist patients. The US Department of Veterans Affairs and Stanford University (VA/SU) robotics program proposed the Robotics Aid Project with the goal of developing a system for people affected by quadriplegia (Van der Loos, 1995). The robot could be voice-controlled to perform preexisting programs. Robots for assisting individuals in Activities of Daily Living (ADLs) were developed by the Clinical Robotics Laboratory at the VA Spinal Cord Injury Center (SCIC). In Europe (Dallaway et al., 1995), the Spartacus Project proposed the use of manipulators to assist individuals with spinal cord injuries. A robot arm was also proposed to assist tetraplegic patients at University of Heidelberg (Germany). The Heidelberg Manipulator used a general-purpose pneumatic end effector was used for manipulation and page turning for which could also be performed by a separately controlled vacuum "finger." For a more detailed historical description of rehabilitation robotics, please refer to Rocon and Pons (2011).

Limitations on the development of WRs were historically related to limitations on power supply, sensor, and actuator technologies. In present days, some of those limitations remain, being one of the main reasons for not finding many WR ambulatory devices. As sensors evolved to miniaturization, with the related advantages in transducing phenomena through different energy domains, the same trend is yet not achieved in power supply technologies and on the development of actuators that are designed to impose a predefined mechanical state on the robotic structure.

In most WR applications, control strategies require force-controlled actuators, which is hardly achieved in most actuator technologies due to impedance, striction, and bandwidth limitations (Pons, 2010). Traditional technologies, such as pneumatic, hydraulic, and electromagnetic actuators are found in several exoskeletal robots (Huo et al., 2016). Direct drive actuators are an interesting manner to achieve close to ideal force sources. However, such systems are power-hungry, bulky, and heavy for exoskeletons, especially those designed to be ambulatory (Duong et al., 2016).

Series elastic actuators (SEAs) are, in this sense, another important approach to achieve a controllable impedance and bandwidth for wearable devices. Electromagnetic actuators are usually set to drive the exoskeleton's joints and to set a controlled force by compressing the elastic element (Blaya and Herr, 2004). SEAs design also enable the possibility of estimating the output torque through spring deflection, which greatly simplify the actuator instrumentation, since only angle sensors can be used (dos Santos et al., 2015).

Another interesting approach is to use the user's muscles as actuators by means of functional electrical stimulation (FES) systems with high selectivity and performance (Springer and Khamis, 2017). It is important to note that the human musculoskeletal system is preserved after some lesions that lead to motor impairments, such as stroke and spinal cord injury. Although such artificial activation of muscles can function as the sole source of actuation, applications of intelligent FES systems in conjunction with other actuators (such as SEA) are also an interesting alternative for increasing user's participation, avoid the decrease of motor function, and at the same time, provide stable locomotion (Seel et al., 2016). Other emerging technologies, such as electroactive polymers (Miriyev et al., 2017), electro- and magneto-rheological fluids (Andrade et al., 2018), and shape memory alloys (Bundhoo et al., 2009) also could be considered promising for WR actuation, but are not as easily found in the literature as the previously mentioned technologies (Moreno et al., 2008).

Considering sensor technology and its close interaction with the scope of this book, the development of compact and energetically efficient sensing devices enable better performance of wearable robotics as more information from the dual physical and cognitive human-robot interactions are gathered, which improves the decision-making process on the robot and the compliance between both intrinsically interfaced agents (human and robot). Sensors allow better feedback for human motor control and are a keystone to monitor the human-robot (and environment) interaction. In this sense, solutions for monitoring bioelectrical activity from the user's neuromuscular system, kinematics (positions, angles, velocities, and accelerations), and the interaction forces and pressures are critical in WR technologies.

Sensors are fundamental to achieve natural interfaces in cognitive interaction. For a better interaction with the user, information should be gathered from different levels, i.e., central nervous system (CNS), peripheral nervous system (PNS), and movement. Considering the CNS, information from the user's brain activity is obtained for the development of brain-controlled exoskeletons (Pons, 2010). In this area, sensors are mainly integrated with brain-machine interfaces and electroencephalogram (EEG) is the most used signal. Advances in wireless, dry and implantable EEG electrodes are also current research and development areas (Xu et al., 2017). Neural control of wearable devices can be achieved by interfacing robots with the human PNS. Surface and implanted electromyography (EMG) electrodes allow a broad range of applications. Although intraneural/implanted interfaces already show promising results, there are important drawbacks that should be considered (which are also found on implanted EEG electrodes), since they suffer from high noise and need direct contact with the measured region; their installation is cumbersome and time-consuming (Moreno et al., 2008). Such sensors also need complex signal processing techniques, and the measured electrical potential is not directly related to the applied force on the human-robot interaction.

The third level of interaction involves the acquisition of kinematic and kinetic information. In this sense, encoders, hall-effect sensors, potentiometers, electrogoniometers, and microelectromechanical systems (MEMS) are already widely used for human and robot joint measurements of parameters such as deformation, angle, torque, and force. Sensors for monitoring the physical interaction between human and robot are also fundamental for the safe operation of the robotic device in close interaction with humans. Beyond the kinematic compatibility between exoskeleton and limb anatomy that should be taken into account during the WR's design, the correct application and monitoring of forces and pressures in the physical interface are necessary for an effective mechanical power transfer between robot and the user. A broad range of technologies, including piezoelectric or capacitive sensors, strain gauges, and piezoresistive polymers can monitor force and pressure interaction between a human and robot (Moreno et al., 2008).

The monitoring comfort and ergonomics play an important role in wearable robots usability and user motivation on the rehabilitation tasks, where suitable monitoring of loads on human tissues (through monitoring force and pressure) and microclimate (temperature and humidity) should be performed in order to avoid pressure ulcers, scars, and other tissue damages. Sensors designed to provide direct measurements of such parameters are essential for achieving the usability and safety requirements in rehabilitation and functional compensation systems. Humidity information can be acquired with different sensor technologies: capacitive, resistive, and thermal conductivity sensors are found. Temperature sensing is also mature for industrial applications, where a broad range of sensitive and precise devices based on thermocouples and semiconductor and resistive sensors are found. Nevertheless, such sensors systems are not usually found in current WR (Huo et al., 2016).

The trend of soft robots as the future of wearable technology brought important constraints and new challenges for the development of flexible technologies for sensors, as the conventional rigid structures for sensors are not suitable for such novel flexible and soft robots. In this sense, new materials for flexible sensors are also the focus of research in several research groups. Soft robots are an emerging field that aims at developing robots that are more adaptable to their surroundings. Such devices are expected to perform different and more autonomous tasks and to mimic the motion and functions of biological systems (Editorial, 2018). Soft and smart materials (such as elastomers and textiles) and fabrication technologies (especially 3D printing (Wallin et al., 2018) and origami folding techniques (Rus and Tolley, 2018)) are important in the design of soft robots.

The technological challenges involved in the development and integration of soft actuators, sensors, control, and power systems to design artificial and intelligent soft robots that can work safely in close interaction with humans are important goals of this field. Applications of soft robots include assistive and wearable devices that can work in space or within human organs. It is important to note that soft robots will not replace traditional rigid (medical) exoskeletons, considering that a rigid structure and powerful actuators are needed for a great number of applications, such as providing the ability to move body parts for patients that are paralyzed below the waist. Instead, soft robots will offer complementary capabilities for applications that require soft systems (Walsh, 2018). The soft material properties provide interesting advantages for assistive robots, especially by minimizing restrictions to the wearer and eliminating the need for aligning robot and biological joints. Additionally, soft technologies can be designed to not interfere with natural movements of the user. Such low inertia features, which are usually hard to achieve in conventional or rigid biomechatronic devices. Fig. 1.6 summarizes the wearable robots technologies discussed in this section and shows the monitored parameters in such devices.

This book addresses a promising and yet mature technology, optical fiber sensors, which represent an evolution on the design and integration of soft sensors in flexible structures, or as part of the fabrication of smart materials. Such technology can be used for developing distributed or quasidistributed sensor systems as will be further explored in the next chapters. Such systems can provide different measurements and parameters to be used in rehabilitation and functional compensation wearable and soft robotic systems.

1.3 Wearable sensors and monitoring devices

The patient monitoring parameters include the biomechanical ones, subdivided into kinematics and kinetics. Such parameters provide important information regarding the human physical condition and are directly related to the efficiency on the daily activity performance as well as the locomotion (Kirtley, 2006). The biomechanics of human movement is the study of the mechanical characteristics and aspects of human movement (Knudson, 2018). As an important feature in human locomotion, the movement analysis includes gait analysis, which comprises the systematic study of human walking, performed by collecting kinematic and kinetic data (Wong et al., 2015). In the kinematics assessment, the description of body motion is performed without considering the causes of motion (Wong et al., 2015). The kinematic parameters include joint angles, center of mass (CoM) displacement velocity, and spatiotemporal gait parameters such as cadence, stride, and step length, among others as discussed in Kirtley (2006). The spatiotemporal gait parameters describe the foot placement, gait events timing, and velocity variables (Kirtley, 2006). The assessment of such

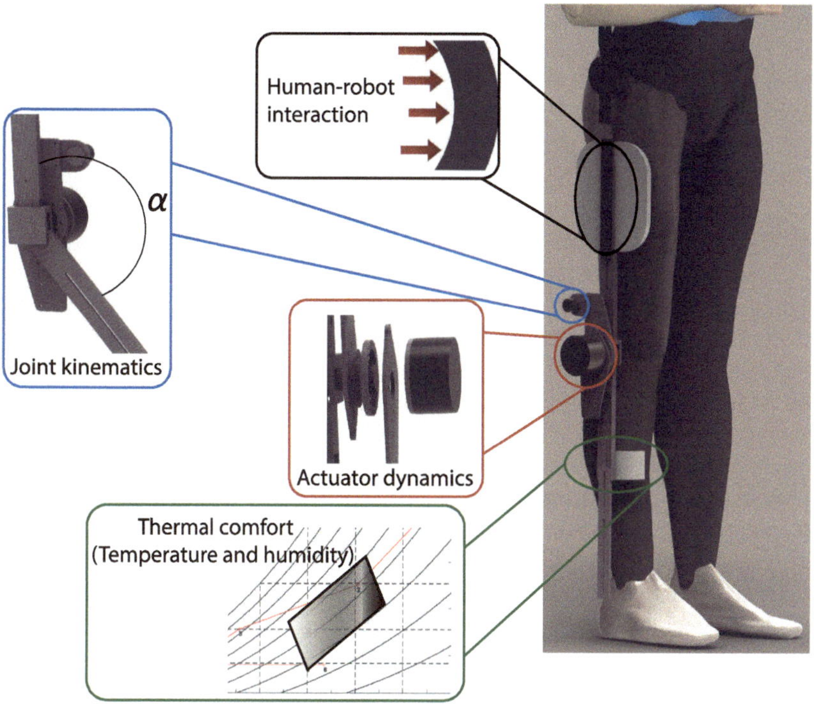

FIGURE 1.6 Schematic representation of the wearable robots (exoskeleton, in this case) and the parameters for their instrumentation.

parameters forms the basis of the gait kinematic analysis as it complements the angular and displacement data in gait. In the movement's kinetics assessment, the forces and torques that initiate the movement are analyzed. Thus, it also considers the forces generated internally in the body that result in human movement (Wong et al., 2015). In general, kinetics parameters include ground reaction forces (GRF), plantar pressure distribution, and joint momentum (Muro-de-la Herran et al., 2014).

The kinematic parameter assessment, especially human joint angles, are applied on rehabilitation, training athletes and the diagnosis of neurological disorders that affects the movement (El-Gohary and McNames, 2012). Moreover, the kinematic data measured can be employed on neural prostheses control and functional electrical stimulation (FES) (Tong and Granat, 1999). Camera-based motion capture systems provide reliable measurements of human kinematics. However, it is a costly and time-consuming technique. As it is limited to laboratory or clinical environment, it cannot be applied on the continuous monitoring of human movement, especially in remote (or in-home) (El-Gohary and McNames, 2012). Therefore, for gait analysis, the motion capture and kinematic measurement are limited to few gait cycles. With the aim of addressing these limitations, another line of research is the development of wearable technologies for joint assessment. Such technologies include electrogoniometers and potentiometers for joint angle assessment in a single axis with the important drawback of large dimensions and weight that not only limit natural patterns of movement, but also are sensitive to misalignments that commonly occur in polycentric joints (Hawkins, 2000). Flexible goniometers, generally based on strain gauges, provide better adaptation to body parts and are not sensitive to misalignments in the polycentric joints' movement. It is also worth noting the development of digital goniometer using encoders to measure joint angles in sagittal plane as an attempt to make a contactless sensor. However, its high sensitivity to misalignment and limitations in speed range discourages its application. Inertial measurement units (IMUs) have experienced a widespread due to advances in MEMS. The combination of a triaxial magnetometer with triaxial accelerometer and triaxial gyroscope is an IMU, which is commonly employed in joint angle assessment in all three planes, where the assessment is commonly performed using sensor fusion algorithms for the extraction of complementary information and interaction between sensors. Despite the wide range of applications of IMUs, they present high sensitivity to magnetic field interferences and need frequent recalibration (El-Gohary and McNames, 2012), which is especially undesired rehabilitation robots or assistive technologies that involve constant activation of electric actuators.

As commonly monitored kinetic parameter in human movement, the plantar pressure and ground reaction forces (GRFs) are commonly assessed through three major monitoring techniques, which include imaging technologies (Mueller et al., 1999), force/pressure distribution platforms (Ballaz et al., 2013), and instrumented insoles (Shu et al., 2010). The imaging technologies generally use expensive equipment and complex signal processing (Abdul Razak et al., 2012). If the analysis is performed in computed tomography machines, there are additional issues due to the inability of performing dynamic analysis of plantar pressure during gait or other dynamic movements (Mueller et al., 1999). As a more affordable option with the possibility of performing dynamic analysis, force platforms are used on the plantar pressure assessment. These platforms generally have a matrix of pressure sensing elements arranged in a rigid and flat platform (Abdul Razak et al., 2012). Even though they provide measurements of the foot plantar pressure and 3D dynamics, they also lack in portability, restricting the tests to laboratory or clinical environments, where there is a limitation on the number of steps per trial. This drawback inhibits the application on wearable robotics, remote and home health monitoring, which is a trend on healthcare applications with the advances in wireless sensor and communication technologies (Mukhopadhyay, 2015). Another drawback of force platforms is the so-called foot targeting effect, where the users alter their natural gait pattern in order to correctly place the foot on the platform, which leads to inaccuracies on the analysis (Sanderson et al., 1993). Consequently, it leads to the necessity of hidden the platform on the ground and repeating the test until a natural gait pattern is obtained with the foot placed within the platform boundaries (Ballaz et al., 2013).

The need of continuous and dynamic assessment of plantar pressure during different dynamic movements motivates the development of instrumented insoles, which throughout the years, became a feasible option to the force platforms. As an important advantage of instrumented insole, there is the possibility of being used inside a shoe, thereby resulting in a portable device to be assessed outside the laboratory environment during daily activities with the natural gait pattern of the users, which enable remote health monitoring and wearable robotics applications (Abdul Razak et al., 2012). Such insoles can present instability on the measurement (with false positives and false negatives) and lack of resistance to the impact loads that commonly occurs in the gait cycle, especially when applying some capacitive or resistive technologies. The number of sensors per insole can also be regarded as an additional drawback in the instrumented insole technology, since the small number of sensor results in a low spatial resolution for the plantar pressure analysis (Abdul Razak et al., 2012). As the human foot has 15 critical pressure areas that support most of the body weight, an ideal sensor system for complete monitoring of the plantar pressure needs at least 15 sensors positioned at these predefined points (Shu et al., 2010). It is worth noting that a higher number of sensors can lead to higher spatial resolution as well as the accuracy in the plantar pressure mapping, which can be achieved with custom fabrics with flexible capacitive sensors. However, they generally present performance limitations due to material features such as low repeatability, hysteresis, creep, and nonlinearities (Abdul Razak et al., 2012).

In addition, parameters, such as heartbeat, oximetry, arterial pressure, glucose, and other physiological indicators are closely related to human health conditions, which can be used as indicators of diseases' evolution. In fact, there are dozens of health indicators and the evolution in healthcare systems is making it possible to obtain quantitative information of many indicators (Organization, 2018). Such parameters can be monitored using several electronic sensors based on different approaches. These approaches include piezoelectric films, dry textile electrodes, flexible capacitive electrodes (among others), with different degrees of success throughout the years, as summarized in some published review works (Majumder et al., 2017). In general, the electromagnetic sensitivity of these electronic-based approaches inhibits their application in magnetic resonance imaging (MRI) (Culshaw, 2013) and can also harm their applications in conjunction with assistive devices in which there is a constant activation of electric motors or in many devices commonly used nowadays with the widespread of portable technologies.

Optical fiber sensors present the intrinsic advantages of lightweight, compactness, chemical stability, immunity to electromagnetic field, and multiplexing capabilities (Peters, 2011), which enable their widespread in different fields. They are an intrinsically safe technology for industrial (Alwis et al., 2016), medical (Mishra et al., 2011), and structural health monitoring (Theodosiou et al.,

2018) applications, also including immunosensors (Guo et al., 2017), biochemical detection (Zhong et al., 2016), and environmental monitoring (Cheng et al., 2018). Regarding to the aforementioned kinematics, kinetics, and physiological parameters, optical fiber sensors are already employed on the measurement of angle (Leal-Junior et al., 2018b), refractive index (Zhong et al., 2015), temperature (Leal-Junior et al., 2018a), humidity (Rajan et al., 2013), acceleration (Stefani et al., 2012), pressure (Vilarinho et al., 2017), breathing rate (Chen et al., 2014), and oxygen saturation (Krehel et al., 2014a). In addition, the geometric and material features of optical fibers enable their embedment in different structures ranging from metals and concrete to fabrics and textiles, which can be used for sensing applications (Krehel et al., 2014b). To that extent, there is also the creation of optical fiber-based textiles, the so-called photonics textiles (Krehel et al., 2014a), which begun as clothing accessory or signaling devices in their first reports (Graham-Rowe, 2007). Recently, the so-called photonics textiles are applied on the body temperature sensing (Li et al., 2012), breath, and heart rates (Ciocchetti et al., 2015). Smart textile approaches offer the advantages of higher transparency between the sensor and the user. Moreover, it does not influence or inhibit the user's natural movements (Quandt et al., 2015). In this context, sensor's compactness and flexibility allow easy installation and removal, which have a positive effect in the system's usability (Majumder et al., 2017). Fig. 1.7 shows a schematic representation and summarizes the physical and physiological parameters measured with wearable technologies. It is worth noting that all these parameters can be measured with optical fiber sensors as will be discussed in the next chapters.

Optical fiber sensors are also well aligned with the requirements of the internet of things (IoT) devices (Islam et al., 2017), which mainly relies on the wireless connectivity of devices with increasing demands toward a constant evolution in wireless systems and their miniaturization as well as low energy consumption. This new paradigm is closely related to remote healthcare (Domingues et al., 2017) applications, which has its widespread motivated by the population aging scenario, as discussed in Section 1.1. Such sensor systems lead to the possibility of continuous monitoring of patients activities, resulting remote assistive services, including diagnosis, early detection of health issues and their transport in case of emergencies (Li, 2019). It is worth noting that there is also an emotional and psychological positive component in monitoring the patient at home instead of hospitals or clinical facilities due to the possibility of performing their daily activities and the sense of independent growth in the community (Na and Streim, 2017). Such smart textiles and optical fiber-embedded smart structures for remote monitoring are also discussed in this book.

1.4 Outline of the book

This book presents the latest developments and applications of optical fiber sensors in healthcare applications, which include the wearable robots and assistive devices instrumentation as well as wearable sensors and human health

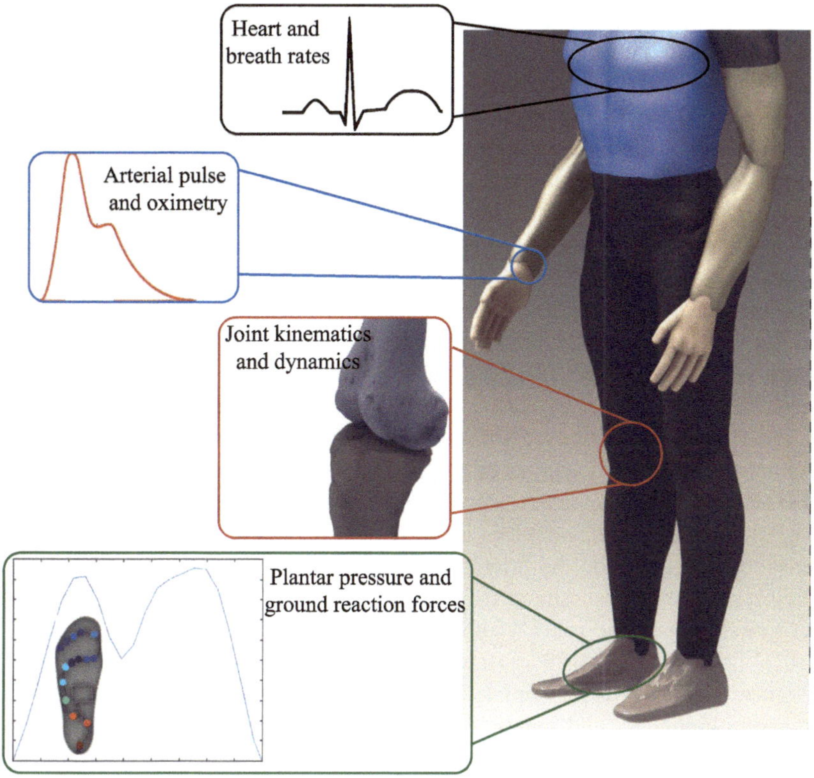

FIGURE 1.7 Kinematics, dynamic (kinect), and physiological parameters for human health assessment.

monitoring systems. In order to comprehensively address all aspects underlying optical fiber instrumentation in healthcare applications, this book is divided into four main sections, namely (i) Introduction to soft robotics and rehabilitation systems, (ii) Introduction to optical fiber sensing, (iii) Optical fiber sensors in rehabilitation systems, and (iv) Case studies and applications. These sections are subdivided into a 13 chapters covering the relevant aspects on optical fiber sensors applications in healthcare.

After setting the scenario in which there is a need of novel technologies for human health assessment and assistance, which include the introduction of wearable and assistive robotics technologies, Chapter 2 presents the current state-of-the-art and new trends on robotic systems. As a prominent new technology on robotics system, Chapter 2 depicts the all relevant aspects on soft robots, which include the different actuators technologies, sensors, and applications as well as the human-in-the-loop design. The discussion includes the control strategies for human-robot interaction as well as the design of compliant structures customized for each user. Then, in Chapter 3, an overview and fundamental aspects of gait analysis are discussed.

The second section of the book presents fundamental aspects of optical fiber technology with emphasis on sensors systems. Thus, Chapter 4 depicts the fundaments, physical principles, and overview of optical fibers in its many variants, which include single and multimode fibers, photonic crystal (microstructured) optical fibers, and other specialty optical fibers. Thereafter, in Chapter 5, the materials used on optical fiber fabrication are discussed, the main classes of materials are glasses and polymers, where their mechanical, thermal, chemical, and optical characterizations are presented. In Chapter 6, approaches regarding the optical fiber sensors, with emphasis and detailed discussion of fiber Bragg gratings, interferometers, intensity variation principle. In addition, Chapter 6 depicts the interrogation (demodulation) techniques for optical fiber sensors in their many variants as well as the compensation techniques for temperature cross-sensitivity reduction.

After the introduction of assistive robotic devices and optical fiber sensor technologies, the third section of this book presents the applications of optical fiber sensor systems in healthcare. Chapter 7 presents the instrumentation of exoskeletons and orthotics devices using optical fiber sensors technologies. The sensor systems include the microclimate assessment, human-robot interaction forces/pressures measurement using different optical fiber-based sensors, and the assessment of the robot's joints position and displacement. In Chapter 8, optical fiber-embedded smart structures and textiles for human movement assessment and analysis are depicted. The discussed systems include the ones for assessing kinematic and kinetic parameters. As the final chapter in this third section, Chapter 9 focuses on novel application of optical fiber sensors systems in soft robotics instrumentation as well as the instrumentation of compliant actuators.

In the last section of the book, case studies and applications are depicted. In contrast with third section, not only wearable technologies are discussed. In Chapter 10, multifunctional smart textiles are discussed, different from the ones previously discussed in third section; the smart textiles discussed in this chapter have multiple functionalities with multiplexed systems that not only are related to the movement analysis, but also on the physiological and health parameters monitoring. The applications discussed in this part of the book also include the development of compliant optical fiber sensors for the instrumentation of smart walkers are presented in Chapter 11. Although it is not a wearable technology, walkers are used by a considerable amount of the elderly population and their robotic counterpart, i.e., smart walkers have a control system for human-robot interaction that enable the navigation in different environment. To achieve these functionalities, sensors for environment mapping, force interaction with the user and odometry are currently employed. In this chapter, optical fiber-based instrumentation of such system is discussed. Another important application is discussed in Chapter 12; the robotic surgeries are closely related to the scope of this book, i.e. optical fiber sensors in robotics and healthcare applications, and the optical fiber sensors systems provide major developments in manipulators and robots for surgeries. In addition, this chapter also discusses the optical fiber embedment in medical instruments used in surgeries and/or medical treatments. As a commonly nonwearable technology with many applications in healthcare, biosensors using optical fiber systems are discussed also in Chapter 12, where the discussion includes glucose sensors, immunosensors, and aptasensors. Finally, the conclusions, final remarks, and future directions in the optical fiber technology for healthcare are drawn in Chapter 13. Fig. 1.8 summarizes the book outline and the topics that will be discussed throughout the book.

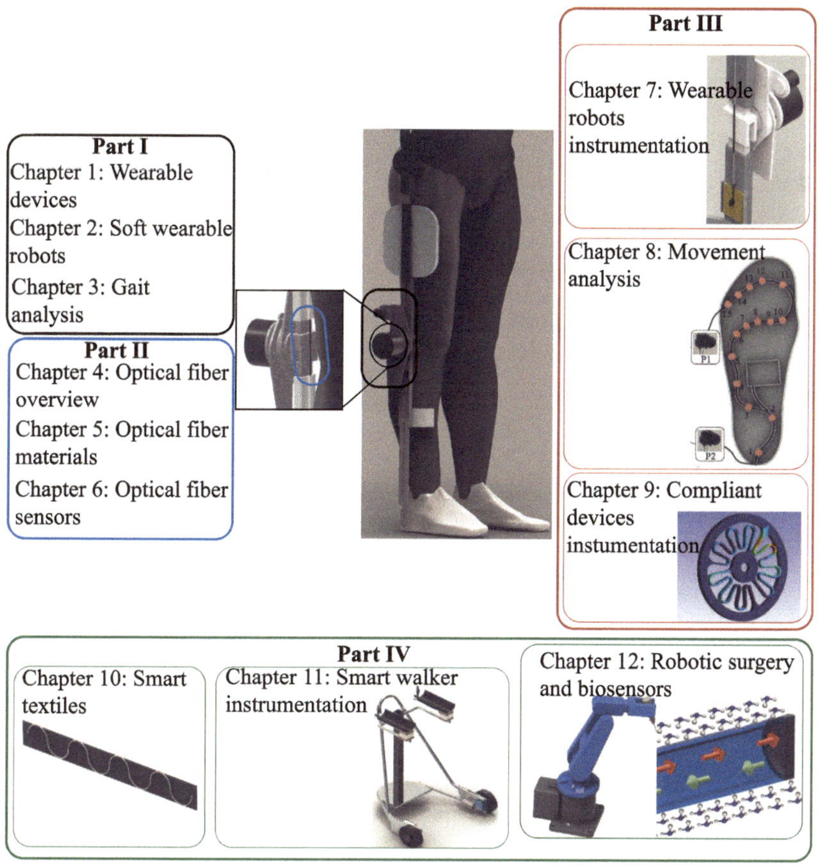

FIGURE 1.8 Schematic representation of the book outline.

References

Abdul Razak, A.H., Zayegh, A., Begg, R.K., Wahab, Y., 2012. Foot plantar pressure measurement system: a review. Sensors 12, 9884–9912. https://doi.org/10.3390/s120709884.

Alami, R., Albu-Schaeffer, A., Bicchi, A., Bischoff, R., Chatila, R., De Luca, A., De Santis, A., Giralt, G., Guiochet, J., Hirzinger, G., Ingrand, F., Lippiello, V., Mattone, R., Powell, D., Sen, S., Siciliano, B., Tonietti, G., Villani, L., 2006. Safe and dependable physical human-robot interaction in anthropic domains. In: State of the Art and Challenges. IEEE International Conference on Intelligent Robots and Systems.

Allen, S., Hogan, D., 2001. Disability, demography of. In: International Encyclopedia of the Social & Behavioral Sciences, pp. 3697–3700.

Alwis, L., Sun, T., Grattan, V.K.T., 2016. Developments in optical fibre sensors for industrial applications. Optics and Laser Technology 78, 62–66. https://doi.org/10.1016/j.optlastec.2015.09.004.

Andrade, R.M., Filho, A.B., Vimieiro, C.B., Pinotti, M., 2018. Optimal design and torque control of an active magnetorheological prosthetic knee. Smart Materials and Structures 27. https://doi.org/10.1088/1361-665X/aadd5c.

Ballaz, L., Raison, M., Detrembleur, C., 2013. Decomposition of the vertical ground reaction forces during gait on a single force plate. Journal of Musculoskeletal Interactions 13, 236–243.

Bayon, C., Lerma, S., Ramírez, O., Serrano, J., Del Castillo, M., Raya, R., Belda-Lois, J., Martínez, I., Rocon, E., 2016. Locomotor training through a novel robotic platform for gait rehabilitation in pediatric population: short report. Journal of NeuroEngineering and Rehabilitation 13, 98. https://doi.org/10.1186/s12984-016-0206-x.

Blaya, J.A., Herr, H., 2004. Adaptive control of a variable-impedance ankle-foot orthosis to assist drop-foot gait. IEEE Transactions on Neural Systems and Rehabilitation Engineering 12, 24–31. https://doi.org/10.1109/TNSRE.2003.823266.

Böhm, M., Reil, J.C., Deedwania, P., Kim, J.B., Borer, J.S., 2015. Resting heart rate: risk indicator and emerging risk factor in cardiovascular disease. American Journal of Medicine 128, 219–228. https://doi.org/10.1016/j.amjmed.2014.09.016.

Bove, L.A., 2019. Increasing patient engagement through the use of wearable technology. Journal for Nurse Practitioners 15, 535–539. https://doi.org/10.1016/j.nurpra.2019.03.018.

Bueno, L., Brunetti, F., Frizera, A., Pons, J.L., 2008. Human – robot cognitive interaction, 1998.

Bundhoo, V., Haslam, E., Birch, B., Park, E.J., 2009. A shape memory alloy-based tendon-driven actuation system for biomimetic artificial fingers, part I: design and evaluation. Robotica 27, 131–146. https://doi.org/10.1017/S026357470800458X.

Chen, Z., Lau, D., Teo, J.T., Ng, S.H., Yang, X., Kei, P.L., 2014. Simultaneous measurement of breathing rate and heart rate using a microbend multimode fiber optic sensor. Journal of Biomedical Optics 19, 057001. https://doi.org/10.1117/1.JBO.19.5.057001.

Cheng, X., Bonefacino, J., Guan, B.O., Tam, H.Y., 2018. All-polymer fiber-optic ph sensor. Optics Express 26, 14610. https://doi.org/10.1364/OE.26.014610.

Cifuentes, C.A., Frizera, A., 2016. Human-Robot Interaction Strategies for Walker-Assisted Locomotion. Springer Tracts in Advanced Robotics, vol. 115. Springer, Cham.

Ciocchetti, M., Massaroni, C., Saccomandi, P., Caponero, M.A., Polimadei, A., Formica, D., Schena, E., 2015. Smart textile based on fiber Bragg grating sensors for respiratory monitoring: design and preliminary trials. Biosensors 5, 602–615. https://doi.org/10.3390/bios5030602.

Culshaw, B., 2013. Optical fibre sensors for industrial applications in safety and security. Optics InfoBase Conference Papers, 1–3.

Dallaway, J.L., Jackson, R.D., Timmers, P.H., 1995. Rehabilitation robotics in Europe. IEEE Transactions on Rehabilitation Engineering 3, 35–45. https://doi.org/10.1109/86.372890.

Dejnabadi, H., Jolles, B.M., Aminian, K., 2005. A new approach to accurate measurement of uniaxial joint angles based on a combination of accelerometers and gyroscopes. IEEE Transactions on Biomedical Engineering 52, 1478–1484. https://doi.org/10.1109/TBME.2005.851475.

Domingues, M.F., Alberto, N., Leitao, C., Tavares, C., de Lima, E.R., Radwan, A., Sucasas, V., Rodriguez, J., Andre, P., Antunes, P., 2017. Insole optical fiber sensor architecture for remote gait analysis – an ehealth solution. IEEE Internet of Things Journal 4662, 207–214. https://doi.org/10.1109/JIOT.2017.2723263.

dos Santos, W.M., Caurin, G.A.P., Siqueira, A.A.G., 2015. Design and control of an active knee orthosis driven by a rotary series elastic actuator. Control Engineering Practice 1–12. https://doi.org/10.1016/j.conengprac.2015.09.008.

Duong, M.K., Cheng, H., Tran, H.T., Jing, Q., 2016. Minimizing human-exoskeleton interaction force using compensation for dynamic uncertainty error with adaptive rbf network. Journal of Intelligent and Robotic Systems: Theory and Applications 82, 413–433. https://doi.org/10.1007/s10846-015-0251-x.

Editorial, 2018. The soft touch of robots. Nature Reviews Materials 3, 71. https://doi.org/10.1038/s41578-018-0017-8.

El-Gohary, M., McNames, J., 2012. Shoulder and elbow joint angle tracking with inertial sensors. IEEE Transactions on Bio-Medical Engineering 59, 2635–2641.

Erdmier, C., Hatcher, J., Lee, M., 2016. Wearable device implications in the healthcare industry. Journal of Medical Engineering and Technology 40, 141–148. https://doi.org/10.3109/03091902.2016.1153738.

Fasano, A., Bloem, B.R., 2013. Gait disorders. Continuum: Lifelong Learning in Neurology 19 (6), 1344–1382. https://doi.org/10.1212/01.CON.0000436159.33447.69.

Ferneini, E.M., 2017. All of us! The American Journal of Cosmetic Surgery 34, 63. https://doi.org/10.1177/0748806817703862.

Franceschi, C., Garagnani, P., Morsiani, C., Conte, M., Santoro, A., Grignolio, A., Monti, D., Capri, M., Salvioli, S., 2018. The continuum of aging and age-related diseases: common mechanisms but different rates. Frontiers in Medicine 5. https://doi.org/10.3389/fmed.2018.00061.

Graham-Rowe, D., 2007. Photonic fabrics take shape. Nature Photonics 1, 6–7. https://doi.org/10.1038/nphoton.2006.35.

Guo, T., Gonzalez-Vila, A., Loyez, M., Caucheteur, C., 2017. Plasmonic optical fiber-grating immunosensing: a review. Sensors 17, 1–20. https://doi.org/10.3390/s17122732.

Ha, K.H., Varol, H.A., Goldfarb, M., 2011. Volitional control of a prosthetic knee using surface electromyography. IEEE Transactions on Biomedical Engineering 58, 144–151. https://doi.org/10.1109/TBME.2010.2070840.

Harwin, W.S., Rahman, T., Foulds, R.A., 1995. A review of design issues in rehabilitation robotics with reference to North American research. IEEE Transactions on Rehabilitation Engineering 3, 3–13. https://doi.org/10.1109/86.372887.

Hawkins, D., 2000. A new instrumentation system for training rowers. Journal of Biomechanics 33, 241–245. https://doi.org/10.1016/S0021-9290(99)00139-6.

Huo, W., Mohammed, S., Moreno, J.C., Amirat, Y., 2016. Lower limb wearable robots for assistance and rehabilitation: a state of the art. IEEE Systems Journal 10, 1068–1081. https://doi.org/10.1109/JSYST.2014.2351491.

Islam, T., Mukhopadhyay, S.C., Suryadevara, N.K., 2017. Smart sensors and internet of things: a postgraduate paper. IEEE Sensors Journal 17, 577–584.

Jaul, E., Barron, J., 2017. Age-related diseases and clinical and public health implications for the 85 years old and over population. Frontiers in Public Health 5, 1–7. https://doi.org/10.3389/fpubh.2017.00335.

Kirtley, C., 2006. Clinical Gait Analysis: Theory and Practice. Elsevier, Philadelphia.

Knudson, D., 2018. The fundamentals of biomechanics. In: Biomechanics of Human Motion, second edition. CRC Press, Taylor & Francis, Boca Raton, Florida, pp. 109–177.

Korhonen, I., Pärkkä, J., Van Gils, M., 2003. Health monitoring in the home of the future. IEEE Engineering in Medicine and Biology Magazine 22, 66–73. https://doi.org/10.1109/MEMB.2003.1213628.

Krehel, M., Schmid, M., Rossi, R.M., Boesel, L.F., Bona, G.L., Scherer, L.J., 2014a. An optical fibre-based sensor for respiratory monitoring. Sensors (Switzerland) 14, 13088–13101. https://doi.org/10.3390/s140713088.

Krehel, M., Wolf, M., Boesel, L.F., Rossi, R.M., Bona, G.L., Scherer, L.J., 2014b. Development of a luminous textile for reflective pulse oximetry measurements. Biomedical Optics Express 5, 2537. https://doi.org/10.1364/BOE.5.002537.

Kwakkel, G., Kollen, B.J., Krebs, H.I., 2008. Effects of robot-assisted therapy on upper limb recovery after stroke: a systematic review. Neurorehabilitation and Neural Repair 22, 111–121. https://doi.org/10.1177/1545968307305457.

Leal-Junior, A.G., Frizera, A., Avellar, L.M., Marques, C., Pontes, M.J., 2018a. Polymer optical fiber for in-shoe monitoring of ground reaction forces during the gait. IEEE Sensors Journal 18, 2362–2368. https://doi.org/10.1109/JSEN.2018.2797363.

Leal-Junior, A.G., Frizera, A., José Pontes, M., 2018b. Sensitive zone parameters and curvature radius evaluation for polymer optical fiber curvature sensors. Optics & Laser Technology 100, 272–281. https://doi.org/10.1016/j.optlastec.2017.10.006.

Li, D., 2019. 5g and intelligence medicine—how the next generation of wireless technology will reconstruct healthcare? Precision Clinical Medicine 2, 205–208. https://doi.org/10.1093/pcmedi/pbz020.

Li, H., Yang, H., Li, E., Liu, Z., Wei, K., 2012. Wearable sensors in intelligent clothing for measuring human body temperature based on optical fiber Bragg grating. Optics Express 20, 11740. https://doi.org/10.1364/OE.20.011740.

Majumder, S., Mondal, T., Deen, M., 2017. Wearable sensors for remote health monitoring. Sensors 17, 130. https://doi.org/10.3390/s17010130.

Martins, M., Santos, C., Frizera, A., Ceres, R., 2015. A review of the functionalities of smart walkers. Medical Engineering and Physics 37, 917–928. https://doi.org/10.1016/j.medengphy.2015.07.006.

Martins, M.M., Santos, C.P., Frizera-Neto, A., Ceres, R., 2012. Assistive mobility devices focusing on smart walkers: classification and review. Robotics and Autonomous Systems 60, 548–562. https://doi.org/10.1016/j.robot.2011.11.015.

Miriyev, A., Stack, K., Lipson, H., 2017. Soft material for soft actuators. Nature Communications 8, 596. https://doi.org/10.1038/s41467-017-00685-3.

Mishra, V., Singh, N., Tiwari, U., Kapur, P., 2011. Fiber grating sensors in medicine: current and emerging applications. Sensors and Actuators. A, Physical 167, 279–290. https://doi.org/10.1016/j.sna.2011.02.045.

Moreno, J.C., Bueno, L., Pons, J.L., Baydal-Bertomeu, J.M., Belda-Lois, J.M., Prat, J.M., Barber, R., 2008. Wearable robot technologies. In: Wearable Robots. John Wiley & Sons, Ltd, Chichester, UK, pp. 165–200.

Mueller, M.J., Smith, K.E., Commean, P.K., Robertson, D.D., Johnson, J.E., 1999. Use of computed tomography and plantar pressure measurement for management of neuropathic ulcers in patients with diabetes. Physical Therapy 79, 296–307.

Mukhopadhyay, S.C., 2015. Wearable sensors for human activity monitoring: a review. IEEE Sensors Journal 15, 1321–1330.

Muro-de-la Herran, A., García-Zapirain, B., Méndez-Zorrilla, A., 2014. Gait analysis methods: an overview of wearable and non-wearable systems, highlighting clinical applications. Sensors 14, 3362–3394. https://doi.org/10.3390/s140203362.

Na, L., Streim, J.E., 2017. Psychosocial well-being associated with activity of daily living stages among community-dwelling older adults. Gerontology and Geriatric Medicine 3. https://doi.org/10.1177/2333721417700011.

Nag, A., Mukhopadhyay, S.C., Kosel, J., 2017. Wearable flexible sensors: a review. IEEE Sensors Journal 17, 3949–3960. https://doi.org/10.1109/JSEN.2017.2705700.

Nishyama, M., Miyamoto, M., Watanabe, K., 2011. Respiration and body movement analysis during sleep in bed using hetero-core fiber optic pressure sensors without constraint to human activity. Journal of Biomedical Optics 16, 017002. https://doi.org/10.1117/1.3528008.

Peters, K., 2011. Polymer optical fiber sensors—a review. Smart Materials and Structures 20, 013002. https://doi.org/10.1088/0964-1726/20/1/013002.

Pirker, W., Katzenschlager, R., 2017. Gait disorders in adults and the elderly: a clinical guide. Wiener Klinische Wochenschrift 129, 81–95. https://doi.org/10.1007/s00508-016-1096-4.

Pons, J.L., 2010. Rehabilitation exoskeletal robotics. IEEE Engineering in Medicine and Biology Magazine 29, 57–63. https://doi.org/10.1109/MEMB.2010.936548.

Quandt, B.M., Scherer, L.J., Boesel, L.F., Wolf, M., Bona, G.L., Rossi, R.M., 2015. Body-monitoring and health supervision by means of optical fiber-based sensing systems in medical textiles. Advanced Healthcare Materials 4, 330–355. https://doi.org/10.1002/adhm.201400463.

Rajan, G., Noor, Y.M., Liu, B., Ambikairaja, E., Webb, D.J., Peng, G.D., 2013. A fast response intrinsic humidity sensor based on an etched singlemode polymer fiber Bragg grating. Sensors and Actuators. A, Physical 203, 107–111. https://doi.org/10.1016/j.sna.2013.08.036.

Rocon, E., Pons, J.L., 2011. Exoskeletons in Rehabilitation Robotics. Springer Tracts in Advanced Robotics., vol. 69. Springer, Berlin, Heidelberg.

Rocon, E., Ruiz, A.F., Raya, R., Schiele, A., Pons, J.L., Belda-Lois, J.M., Poveda, R., Vivas, M.J., Moreno, J.C., 2008. Human–Robot Physical Interaction. John Wiley & Sons, Ltd, Chichester, UK.

Rowland, D.T., 2009. International handbook of population aging. In: International Handbook of Population Aging. https://doi.org/10.1007/978-1-4020-8356-3.

Rus, D., Tolley, M.T., 2018. Design, fabrication and control of origami robots. Nature Reviews Materials 3, 101–112. https://doi.org/10.1038/s41578-018-0009-8.

Sanderson, D.J., Franks, I.M., Elliott, D., 1993. The effects of targeting on the ground reaction forces during level walking. Human Movement Science 12, 327–337. https://doi.org/10.1016/0167-9457(93)90022-H.

Seel, T., Werner, C., Schauer, T., 2016. The adaptive drop foot stimulator – multivariable learning control of foot pitch and roll motion in paretic gait. Medical Engineering and Physics 38, 1205–1213. https://doi.org/10.1016/j.medengphy.2016.06.009.

Shu, L., Hua, T., Wang, Y., Li Qiao, Q., Feng, D.D., Tao, X., 2010. In-shoe plantar pressure measurement and analysis system based on fabric pressure sensing array. IEEE transactions on information technology in biomedicine: a publication of the IEEE Engineering in Medicine and Biology Society 14, 767–775. https://doi.org/10.1109/TITB.2009.2038904.

Springer, S., Khamis, S., 2017. Effects of functional electrical stimulation on gait in people with multiple sclerosis – a systematic review. Multiple Sclerosis and Related Disorders 13, 4–12. https://doi.org/10.1016/j.msard.2017.01.010.

Stefani, A., Andresen, S., Yuan, W., Herholdt-Rasmussen, N., Bang, O., 2012. High sensitivity polymer optical fiber-Bragg-grating-based accelerometer. IEEE Photonics Technology Letters 24, 763–765. https://doi.org/10.1109/LPT.2012.2188024.

Strauß, R., Ewig, S., Richter, K., König, T., Heller, G., Bauer, T.T., 2014. The prognostic significance of respiratory rate in patients with pneumonia. Deutsches Aerzteblatt Online 111, 503–508. https://doi.org/10.3238/arztebl.2014.0503.

Taborri, J., Palermo, E., Rossi, S., Cappa, P., 2016. Gait partitioning methods: a systematic review. Sensors 16, 66. https://doi.org/10.3390/s16010066.

Teyhen, D.S., Stoltenberg, B.E., Collinsworth, K.M., Giesel, C.L., Williams, D.G., Kardouni, C.H., Molloy, J.M., Goffar, S.L., Christie, D.S., McPoil, T., 2009. Dynamic plantar pressure parameters associated with static arch height index during gait. Clinical Biomechanics 24, 391–396. https://doi.org/10.1016/j.clinbiomech.2009.01.006.

The European Communities, T.C., 2016. Smart wearables: reflection and orientation paper. Digital Industry Competitive Electronics Industry 121, 4592–4599. https://doi.org/10.1021/acs.jpcb.7b01309.

Theodosiou, A., Komodromos, M., Kalli, K., 2018. Carbon cantilever beam health inspection using a polymer fiber Bragg grating array. Journal of Lightwave Technology 36, 986–992. https://doi.org/10.1109/JLT.2017.2768414.

Tong, K., Granat, M.H., 1999. A practical gait analysis system using gyroscopes.pdf. Medical Engineering and Physics 21, 87–94.

Tran, V.T., Riveros, C., Ravaud, P., 2019. Patients' views of wearable devices and ai in healthcare: findings from the compare e-cohort. npj Digital Medicine 2, 1–8. https://doi.org/10.1038/s41746-019-0132-y.

Turner, A., 2009. Population ageing: what should we worry about? Philosophical Transactions of the Royal Society B: Biological Sciences 364, 3009–3021. https://doi.org/10.1098/rstb.2009.0185.

United Nations, Department of Economic and Social Affairs, P., 2019. World Population Ageing 2019: Highlights.

Van der Loos, H.F., 1995. Va/Stanford rehabilitation robotics research and development program: lessons learned in the application of robotics technology to the field of rehabilitation. IEEE Transactions on Rehabilitation Engineering 3, 46–55. https://doi.org/10.1109/86.372891.

Vilarinho, D., Theodosiou, A., Leitão, C., Leal-Junior, A., Domingues, M., Kalli, K., André, P., Antunes, P., Marques, C., 2017. Pofbg-embedded cork insole for plantar pressure monitoring. Sensors 17, 2924. https://doi.org/10.3390/s17122924.

Villa-Parra, A., Delisle-Rodriguez, D., Souza Lima, J., Frizera-Neto, A., Bastos, T., 2017. Knee impedance modulation to control an active orthosis using insole sensors. Sensors 17, 2751. https://doi.org/10.3390/s17122751.

Wallin, T.J., Pikul, J., Shepherd, R.F., 2018. 3d printing of soft robotic systems. Nature Reviews Materials 3, 84–100. https://doi.org/10.1038/s41578-018-0002-2.

Walsh, C., 2018. Human-in-the-loop development of soft wearable robots. https://doi.org/10.1038/s41578-018-0011-1.

Wong, C., Zhang, Z.Q., Lo, B., Yang, G.Z., 2015. Wearable sensing for solid biomechanics: a review. IEEE Sensors Journal 15, 2747–2760. https://doi.org/10.1109/JSEN.2015.2393883.

World Health Organization, 2011. World Health Statistics 2011.

World Health Organization, 2018. World Health Statistics 2018: monitoring health for the SDGs, sustainable development goals. Geneva.

Xu, J., Mitra, S., Van Hoof, C., Yazicioglu, R.F., Makinwa, K.A., 2017. Active electrodes for wearable eeg acquisition: review and electronics design methodology. IEEE Reviews in Biomedical Engineering 10, 187–198. https://doi.org/10.1109/RBME.2017.2656388.

Zhong, N., Liao, Q., Zhu, X., Zhao, M., Huang, Y., Chen, R., 2015. Temperature-independent polymer optical fiber evanescent wave sensor. Scientific Reports 5, 1–10. https://doi.org/10.1038/srep11508.

Zhong, N., Zhao, M., Zhong, L., Liao, Q., Zhu, X., Luo, B., Li, Y., 2016. A high-sensitivity fiber-optic evanescent wave sensor with a three-layer structure composed of Canada balsam doped with geo2. Biosensors and Bioelectronics 85, 876–882. https://doi.org/10.1016/j.bios.2016.06.002.

Chapter 2

Soft wearable robots

2.1 Soft robots: definitions and (bio)medical applications

Robotics offer advanced solutions in several areas, especially for industrial applications and manufacturing. The constant evolution of automation have led to the development of new applications as well as control paradigms in which the human intervention was no longer necessary or was drastically reduced Zhang et al. (2019). Since early 2000s, the fully-automated processes are a reality with applications in many manufacturing or process industries (Romeo et al., 2020). Early developments of automated systems mainly rely on robust robotic machines that perform predefined tasks without physical interaction with the users. In fact, the process automated industry in its early days can be considered as a dangerous environment for humans, where accidents have been reported when a human invades the robots workspace (Kirschgens et al., 2018), since these systems mainly relied on position control or even the force/torque control without additional sensors for physical interaction with humans or unexpected events on the environment. In cases of unexpected events, some robots have security or alarm systems, which are in constant evolution to provide safe operation in presence of humans or as a response to unexpected events (Kirschgens et al., 2018).

Nevertheless, it is possible to observe the appearance of robotic application in several environments, from cleaning robots at home to automatic management and sorting in warehouses. Humans and robots coexisting in the workplace provide an efficient and cohesive workforce where unique skills of both parties can be used to achieve effective and innovative goals (Laschi et al., 2016). In fact, the continuous demand of collaborative works between humans and robots as well as the novel designs in product manufacturing have led to the developments of control systems and robots' design in order to achieve higher dexterity to perform the activities (Phan et al., 2020).

The interaction between human and robotics systems can be monitored using computer vision systems (Jaimes and Sebe, 2007), physical sensors such as strain gauges (Villa-Parra et al., 2017) as well as acoustics and laser-based approaches (Jiménez et al., 2019) commonly used in obstacle avoidance for mobile robots. In such new control paradigm, the position or force closed-loop control that consider only the robot's states are improved by considering the interaction with the environment, resulting in the stiffness control of the robotics joints (Rocon et al., 2008). The additional parameter of the physical interaction between

https://doi.org/10.1016/B978-0-32-385952-3.00010-X

human and robot given rise to the popularization of impedance and admittance controllers (Keemink et al., 2018), which lead to an improved human-robot interaction as summarized in Huo et al. (2016).

In this sense, robots are expected to be seen in many tasks that involve direct contact and interaction with humans. Using soft matter to build such devices is the current trend and will allow pushing the use of robots to new limits, applications, and (natural) environments (Wang and Iida, 2015; Shen, 2016; Rus and Tolley, 2018).

Considering the assistive devices, it is possible to observe an important evolution from passive instruments (such as cranes, wheelchairs, and walkers) and passive structures for joint stabilization and support (such as the knee and elbow braces) to complex and automated solutions (including wearable robotic devices) for rehabilitation and functional compensation (Rocon and Pons, 2011).

Although rigid link and more traditional approaches to robotics can also be seen as robots inspired by animals with hard skeletons, they require meticulous programming and extensive feedback to avoid collisions and dangerous situations when dealing with humans, objects, and unpredictable situations. Bioinspired soft robots are built elastic/moldable materials that can adapt to their surroundings (Shen, 2016).

The requirements of more complex and closer human-robot interaction resulted on the development of novel compliant actuators to improve the commonly used rigid actuators. Thus technologies such as pneumatic muscles (Hamaya et al., 2019), series elastic actuators (Leal Junior et al., 2016), and cable-driven actuators (Bhattacherjee et al., 2018) were explored and proposed for the development of innovative applications for robotics, aiming at a safer interaction between human and robot, higher controllability and smaller impact energy in case of accidents and unintended contacts. It is also worth noting that continuous development of manufacturing techniques and materials processing allowed the development of soft structures for robotic systems, which in some cases, may substitute the rigid metallic structures commonly found in early reports of industrial robots (Ansari et al., 2017), enhancing compliance and safety in operation and interaction with human subjects.

Soft robotics is a growing research field that is born by the combination of robotics and soft materials and textiles (Walsh, 2018). Other authors define soft robots, in terms of the Young's modulus of the material (Rus and Tolley, 2018), as "systems that are capable of autonomous behavior, and that are primarily composed of materials with moduli in the range of that of soft biological materials." The RoboSoft[1] community defines soft robots as "devices which can actively interact with the environment and which can undergo 'large' deformations relying on inherent or structural compliance" (Cianchetti et al., 2015). Other definitions of soft robots, including previous uses of the term referring to robots with rigid links and mechanically compliant joints with variable stiffness

1 RoboSoft: A Coordination Action for Soft Robotics, http://www.robosoftca.eu/.

(Albu-Schaffer et al., 2004) can be found in Laschi et al. (2016), where a spectrum of soft robots approaches is presented: from devices with rigid links and joints with a few compliant structures to completely soft devices.

Regardless of the definition, soft robotic devices may offer several benefits, which include energy absorption for stability, physical robustness, and human safe operation. Some implementations of soft robots are also able to implement embodied intelligence principles (Pfeifer et al., 2007) with low cost fabrication techniques.

Such devices are particularly useful for manipulating delicate objects: Novel soft robotic grasping technologies are expected to support the sustainable intensification of agriculture and drive manufacturing productivity throughout the food chain (Duckett et al., 2018). This can be empowered by the current trend on interconnected robotic systems working seamlessly alongside their human coworkers in farms and food factories.

The use of soft materials in robotics has also enabled innovative robot capabilities and expanded the possibilities for medical and biomedical applications, including soft tools for surgery and drug delivery, wearable and assistive devices, prostheses, artificial organs, and active simulators for training and biomechanical studies (Cianchetti et al., 2018). Soft robotics are especially interesting for medical applications that require compliance, soft interaction with patients, and mechanical compatibility with the human body, considering the previously mentioned internal and external uses.

In this sense, biocompatibility of materials is a key aspect for the successful design of soft robots in close interaction with human subjects. From allergies and contact reactions caused by the external use of such soft materials to immediate immune response and the possibility of rejection from long-term implantation, it is important to guarantee body acceptability when designing soft robots (Cianchetti et al., 2018).

Considering the previously mentioned applications, the development of robotic devices for surgery is an important area for the application of soft devices. As minimally invasive surgery became the gold standard and natural orifice transluminal endoscopic surgery (NOTES) emerged as an interesting approach (Vitiello et al., 2013), the development of soft actuation technologies have been explored to implement flexible, accurate, and safe surgical tools (Cianchetti et al., 2018). In this sense, flexible fluidic actuators (FFAs), smart materials and cable-driven flexible mechanisms have been explored as actuators for soft surgical devices (Le et al., 2016). Such technologies along with others, such as shape memory materials (SMMs), electroactive polymers (EAPs), tendon driven actuators and electrorheological and magnetorheological materials (ERMs and MRMs) also allowed the development of important soft solutions for a range of interesting applications.

Another remarkably interesting field of application is the development of soft materials and robotic devices for drug delivery. Many soft materials used in such applications are based on biocompatible materials, which degrade over time releasing the medication inside the patient's body (Cianchetti et al., 2018). Microrobots (Nelson et al., 2010) can also be considered as a viable alternative for minimally invasive medicine. Such devices must be able to move and interact with their environment, reaching a targeted area, providing drug administration for a preestablished duration, and to degrade (or be removed) without causing adverse effects. Magnetically actuated microrobots for in vivo biomedical applications (Gultepe et al., 2013; Leong et al., 2009; Zhang et al., 2009) and three-dimensional magnetic manipulation systems (Schuerle et al., 2013; Kummer et al., 2010) have been proposed for biomedical applications. The fabrication of an untethered, self-folding, soft microrobotic platform to perform targeted, on-demand delivery of biological agents is presented in Fusco et al. (2014).

Soft robots can also be used for mimicking the human body. In this context, soft robots can improve the functionality and acceptability of limb protheses, replicate physiological functions in artificial organs and be used as body-part simulators, reducing animal and patient tests (Cianchetti et al., 2018).

Regarding the scope of this book, the authors would like to focus on implementations of soft robots for rehabilitation and assistance, especially considering the wearable devices. Therefore Section 2.2 will address such applications of soft robots with more detail and aim at illustrating the importance of such devices for expanding safe interaction and cooperation with patients with motor dysfunctions or other related conditions.

2.2 Soft robots for rehabilitation and functional compensation

Functional compensation and rehabilitation robots are an important alternative to meet the needs of an aging society. Rigid robotic devices are already successfully being used in many applications, where they have demonstrated important benefits for restoring motor functions (Tefertiller et al., 2011). The so-called exoskeletons have been widely used in a range of rehabilitation and functional compensation strategies, such as restoring ambulatory walking while delivering (partial) body weight support (Esquenazi et al., 2012), providing gait rehabilitation (Banala et al., 2007) and assistance (Dollar and Herr, 2007) for motor-impaired patients, and characterizing the impedance of the user's joints (Lee et al., 2014).

Going beyond load carriage, exoskeletons can provide consistent and intensive recovery therapies over longer periods (Huang and Krakauer, 2009), regardless of the therapist's fatigue level (Lo and Xie, 2012). Considering robotic devices for upper limb rehabilitation, end-effector rehabilitation robots such as the MIT-MANUS (Krebs et al., 2003) have undergone extensive clinical testing to evaluate and demonstrate their effectiveness as rehabilitative devices (Lum et al., 2002; Krebs et al., 1999). Such positive results drove the appearance of commercial solutions and shifted upper limb robotic therapy research toward exoskeleton robots, allowing the controlled application of torque to individual joints separately (Lo and Xie, 2012). Devices for upper limb rehabilitation, such as the ARMin III (Nef et al., 2009), and for providing function, such as tremor assessment and compensation (Rocon et al., 2007) have been extensively presented over the past two decades. Commercially available rehabilitation devices for the upper limb are also found and include the Armeo products (Hocoma AG, Switzerland), and the MyoPro Orthosis (Myomo Inc., USA). Full body solutions, such as HAL-5 (Cyberdyne Inc., Japan), also emerged as alternatives designed to assist the disabled and elderly in their daily tasks, and to support workers with physically demanding jobs. HAL for medical user and therapy medical services (limited to lower limb applications) has been used to provide treatments for functional improvement of patients with a diversity of conditions, including stroke (Fukuda et al., 2015).

Soft robots emerged as a natural and important evolution of exoskeletal robotic devices, expanding the possibilities of safe and effective human-robot interaction, especially considering that an exchange of mechanical power occurs during both rehabilitation therapies and motor assistance. There is a growing interest in materials science, robotics, and medical research communities on the development of soft wearable robots (Walsh, 2018).

Impedance and admittance control of robots based on rigid links demonstrated to be a viable alternative for achieving adaptability in rehabilitation and assistive robotic devices. Admittance control can be used for a broad range of human-robot physical interaction applications and strategies (Keemink et al., 2018). An admittance control strategy was presented by Wu et al. (2018) to induce the active participation of patients while maximizing the use of recovered motor functions during rehabilitation. Gait rehabilitation based on variable admittance control of an exoskeleton was presented by Taherifar et al. (2018). Biomimetic actuators designed for modulating stiffness and, therefore, achieving a more natural interaction and control of the user's limbs are also focus of extensive research.

Some important aspects should be considered when using soft actuator technologies (such as Mckibben actuators, FFAs, and SMAs) for controlling rigid devices. First, the mechanical design of the robot must avoid macro- and micro-misalignments between the human and robot joints. Macro-misalignments can occur if there is an oversimplification of the exoskeleton joints for interacting with specific human joints (Rocon et al., 2008). Generally, this could occur when an exoskeleton joint or joint group has less degrees of freedom than the human joint or joint group. The most important consequence of macro-misalignments in exoskeleton devices is a limitation on the available workspace for a specific joint or joint group.

Micro-misalignments, on the other hand, can occur even if there is a correct match between the number of degrees of freedom of the exoskeleton and the user. They occur because it is not possible to align the exoskeleton perfectly to the human joints (especially due to intersubject variability). This noncoincident joint rotation axes can generate unphysiological loads that can damage soft tissues (Cianchetti et al., 2018) or shear forces between the exoskeleton attachment point and the human limb (Rocon et al., 2008).

Another negative aspect of using rigid interfaces with soft actuators is the interference with the natural motion of the user, usually caused by using mechanical and mechatronic components with substantial inertia in the exoskeleton design. Adding mass to the to the legs, for instance, will increase the metabolic cost during acceleration and deceleration. This could, additionally, disrupt the natural biomechanics of walking or cause discomfort (Asbeck et al., 2014).

Reliable soft mechatronic devices are currently being used for gradually replacing (traditional) rigid interfaces connected by bioinspired actuators and active parts are being designed to be in close contact with the user (Cianchetti et al., 2018). This addresses both previously presented issues, as no rigid external mechanical structures are used. Compared to rigid devices, soft robots are compliant and present low weight, allowing safe interaction with the human for medical and wearable applications. Additionally, such devices can be designed with low profile, being able to be worn underneath clothing (Asbeck et al., 2014). Soft materials also minimizes restrictions to the wearer and eliminate the need of precisely aligning the robot and biological joints (Walsh, 2018).

In this sense, different actuator technologies, such as cable-driven mechanisms, pneumatic actuators (or artificial muscles), and Bowden systems, have been used successfully for designing full body soft exosuits (Wehner et al., 2013) for gait assistance and different soft devices for upper-limbs, including fingers (Maeder-York et al., 2014; Delph et al., 2013), wrist (Villoslada et al., 2015), elbow (Copaci et al., 2017), and shoulders (Galiana et al., 2012). Such actuators are used to generate forces across joints to assist the underlying biological muscles with exceptionally low inertia and low restriction to the natural movements of the wearer. The concept of *soft clothing-like exosuits* was introduced in the context of locomotion assistance in Wehner et al. (2013). Such devices use the bone structure to support compressive loads, generating forces in parallel with the muscles and use textiles to interface with the human body (Asbeck et al., 2014). Some examples of soft actuator technologies are shown in Fig. 2.1.

Here, it is important to note that the lack of external rigid structures will have impact on applications that require (full) body weight support for locomotion (Neuhaus et al., 2011) or load carriage (Walsh et al., 2007), where an additional parallel path to the ground is required to avoid overloading the wearer's joints. This could be considered a limitation of soft wearable robots or, at least, a reason for selecting different approaches (such as more traditional rigid devices) for specific applications.

The incorporation of soft sensors and the design and use of soft materials for wearable devices are essential for achieving functionality, performance, comfort, and usability. Additionally, it is important to mention that there are crucial

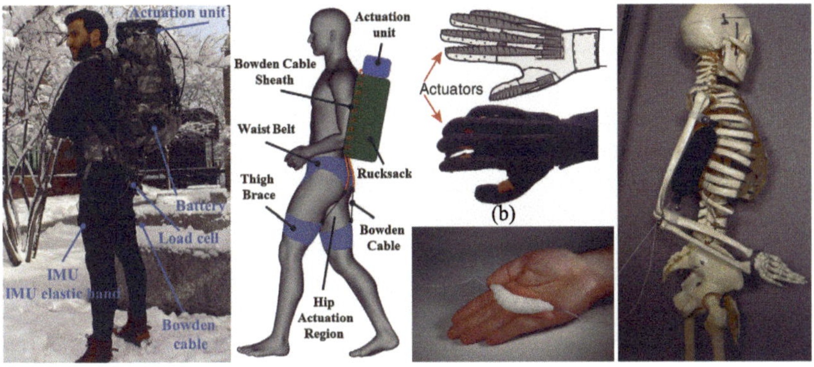

FIGURE 2.1 Examples of actuator technologies for soft robots. (a) Cable-driven exosuit (Panizzolo et al., 2019). (b) Soft robotic glove with fabric-based actuators (Cappello et al., 2018). (c) Soft artificial muscle, composed of ethanol distributed throughout the solid silicone elastomer matrix (Miriyev et al., 2017). (d) Soft artificial McKibben-type artificial muscle implemented as a biceps (Miriyev et al., 2017). All images reproduced under Creative Commons Attribution 4.0 International License.

challenges regarding actuation, sensing, and control that need to be overcome to achieve actual benefits to the users.

Not only the wearable devices benefit from this trend of soft technologies and soft robots for different assistive applications are found on the literature. Soft robotic manipulators exhibit compliance and dexterity ensuring safe human–robot interaction. This makes them suitable for supporting people with motor impairments in better performing activities of daily living (ADLs). Basic activities of daily living (also referred as BADLs) have a great impact on the quality of life and independence (Katz, 1983; Andersen et al., 2004), but can be put at risk due to motor impairments, especially in patients with neurological disorders (Yap et al., 2017). Considering that most of these activities have to be performed with great frequency (several times a day), in close physical interaction with the human subject and currently depend (mostly) on the support of a family member or healthcare professionals, the design of soft solutions for ADLs will provide safe interaction and great functionality for an important number of users.

An example of a soft shower arm to assist the elderly in the bathing task is presented by Ansari et al. (2017). A serial interconnection of three modules is presented, where the proximal segment is built of cable-based actuation to compensate for gravitational effects and the central and distal segments used hybrid actuation (based on McKibben and cable-based actuators) to autonomously reach delicate body parts. The customizability of soft structures also enables the development of insoles customized for each user (Leal-Junior et al., 2019b) as well as the customization of assistive devices (ten Kate et al., 2017). It is also worth noting that wearable systems can be used as a training tool of both human and machine aiming on rehabilitation exercises on the human side, whereas for the robotic system, the interaction with the operator can result in the task learning for the robot that would perform the same activity in autonomous mode after proper training (Teramae et al., 2018).

One popular wearable application of such soft robotic structure is on the development of gloves and grippers for assistance and rehabilitation (Rus and Tolley, 2015). Yap et al. (2017) presents a soft robotic glove to provide hand function assistance using fabric-reinforced soft pneumatic actuators. Haptics and silicone actuators based on air pressurization and compatible with human tissues are addressed to achieve an intrinsically soft and compliant device. In this sense, the authors present a soft robotic glove for grasping assistance during ADL aiming at assisting stroke survivors. In fact, the development of hand exoskeletons have been thoroughly studied due to the importance of reach and grasp activities as well as object manipulation, where such motor functions can be lost due to different clinical conditions. Thus hand exoskeletons of different actuation technologies, such as pneumatic, hydraulic, tendon-driven, and elastomer-based were proposed (Shahid et al., 2018), where the proposed systems can aid the user on grasp and manipulation activities by the increasing the control and the grasp capabilities of the users fingers.

2.3 Human-in-the-loop design of soft structures and healthcare systems

2.3.1 Human-in-the-loop systems

As mentioned in Chapter 1, the development of wearable robots have three main purposes: augmenting the human capacities, restoring a physical disability (related to loss of a limb or degenerative diseases), or assisting in the development of some tasks that were not efficiently performed due to weakness in the skeletal muscles generally related to diseases or the natural aging process. In wearable robotics, the user (human) is generally considered in the control loop, since position or force control in the robotics joints without taking into account the user's movement intention or interaction forces can result in injuries for the user (Forner-Cordero et al., 2008). Thus, the wearable robots control includes human-in-the-loop features when the physical interaction with the robot is considered on its control. In addition, the customizability offered by manufacturing techniques such as the additive layer manufacturing in which adaptable and/or customized robotic structures can be designed for each user in a manner that the human is also included in the robotic system design, resulting in novel optimized mechatronic solutions not only for healthcare but also in different fields.

The so-called human-in-the loop design refers to different fields of novel technologies for intelligent or user-friendly systems. Such technologies are related to the integration of humans in the design of robotic systems, including the optimization of the robotics structures based on the user's requirements (ten

Kate et al., 2017), development of machine learning approaches with human-aided training (Budd et al., 2019) and development of control strategies that consider the physical or cognitive interaction between human and machine (Hussain et al., 2016). In fact, many of the developments in human-in-the-loop systems are related to data analytics in processes well aligned to the trends in Industry 4.0, especially the cyber-physical systems (Sousa Nunes et al., 2015), which is motivated by the possibility of creating new scenarios on organizational relations when there is the analysis of not only the data related to the process, but also the feedback of humans involved in the process (Budd et al., 2019). This development is driven by the fact that humans have enhanced perception, preferences, and abstraction based on qualitative aspects, which aid in the development of machine learning. However, such approach has limitations and constraints related to the ethics concerns arisen from including human in some data analytics loop or processes as well as the technical limitations of the available platforms (Guo et al., 2017). Despite these limitations, human-in-the-loop design is in constant evolution, being one of the core aspects on the development of smart environments and cyber physical systems as thoroughly discussed in Sousa Nunes et al. (2015). A deep discussion on human-in-the-loop for data analytics is beyond the scope of this book, which is focused on the development of wearable systems considering the users in all stages of wearable technologies development, applications, and implementations. Thus the human-in-the-loop aspects that are further discussed in this section are related to some machine learning applications due to its applicability on novel control designs for robotics systems and the software design of such mechatronic solutions as well as control paradigms that included human physical and/or cognitive interaction. In addition, the aspects of robot hardware design considering its customization and adaptation to each user and conditions are also discussed.

In machine learning or deep learning applications, the use of human-in-the-loop approaches can be beneficial in medical diagnosis as well as image interpretation and similar classification algorithm, in which a feedback of a human operator is necessary for the achievement of optimum performance in interpretative cases (Budd et al., 2019). Despite the evolution and widespread of autonomous deep learning in many tasks in different fields as summarized in Makarenko (2016), humans still outperform machine learning algorithms in classification and interpretation applications, especially if an expert in a predefined application is compared with the algorithm (Holzinger, 2016), which is especially desired in medical images interpretation and diagnosis, where a classification error can lead to serious implications on the patient's treatment (Budd et al., 2019). The inclusion of the human factor in the machine learning algorithms also finds applications in arts and musical fields, where the human perception and psychometrics factors are beneficial. In robotics, human-in-the-loop approaches result in optimization of the robotics structures (Gul et al., 2018), control of collaborative robotic units (Fan et al., 2018) as well as optimization of metabolic cost on the human-robot interaction for wearable robots employed in gait assistance applications (Zhang et al., 2017).

Conventional position controllers are replaced to or combined with control strategies that include the human-robot interaction forces. To that extent, two commonly employed control strategies for human-robot interaction systems are the impedance and admittance controllers (Keemink et al., 2018). Actually, there are many control strategies for human-robot physical interaction, which include the feedback-based approaches such as hybrid force/position control as well as stiffness control, damping control (both related to impedance control) (Rocon et al., 2008). Fig. 2.2(a) shows the block diagram of position and impedance controllers, where in impedance controller, the force input is converted to position according to the system dynamics. Considering the interaction between the robot and human, there is an interaction loop as shown in Fig. 2.2(b), where sensors for the interaction between the robotic system and its user are positioned in the interface between them and such information is considered on the feedback in the impedance control (Rocon et al., 2008). It is worth to mention that the control systems presented in Fig. 2.2 are used in wearable robots since the early 2000s. Further developments on sensors systems, actuation, machine learning, and optimization algorithms resulted in novel human-in-the-loop systems are discussed in Section 2.3.2.

The challenges of human-in-the-loop approaches for control systems are related to the complexity of humans, where the physiological, psychological, and behavioral aspects are difficult to be anticipated and modeled (Munir et al., 2013). This feature indicates important challenges in the control systems implementations, where closed-loop systems generally place demands on the sensors systems that need to account for the physical or cognitive interaction between the human and robot in control systems as well as the identification of the user's parameters in the case of human-in-the-loop design of wearable robots. In order to further explore the human-in-the-loop approach, Section 2.3.2 presents an overview of application and current trends of such approach, which finds many applications in industrial manufacturing, collaborative robotics, cyber-physical system, soft robotics, and as the main topic in this book, the wearable robotics, where the soft robotics structures are commonly employed due to its customizability. Details on human-in-the-loop design and control systems for wearable robotics are discussed in Section 2.3.3, where the main concepts of wearable and soft robotics systems are further discussed in the scope of the human-in-the-loop design. Finally, the technologies the enable developing human-in-the-loop wearable robotic systems are discussed in Section 2.3.2, which included the technologies for material processing and fabrication with high customization, enabling the adaption of such novel robotics structures for different users as well as the integration of sensors and actuators in these customized structures.

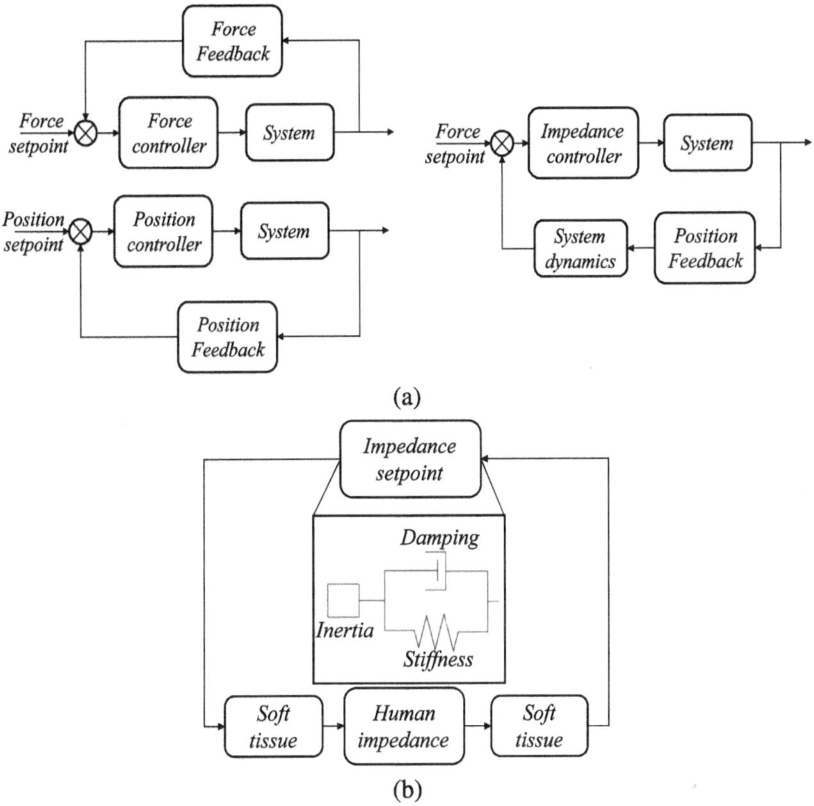

FIGURE 2.2 (a) Block diagram of closed-loop force, position, and impedance controllers. (b) Closed-loop block diagram for human-in-the-loop impedance control.

2.3.2 Human-in-the-loop applications and current trends

The wide range of applications and developments using human-in-the-loop approaches lead to a myriad of reports from image classification (Wang et al., 2019) to wearable soft robots design (Walsh, 2018). As one of the current trends in machine learning algorithms, the use of experienced professionals can aid on the algorithm training and optimization of the algorithm's interpretation, especially in medical diagnosis as presented in Budd et al. (2019) using human-in-the-loop in deep learning approaches for medical images analysis. In fact, including humans as feedback or in training stages help to improve current artificial intelligence systems, where humans can also be considered as sensors in some modeling and simulation processes of social interaction (Bosse and Engel, 2019) as well as the digitalization of working scenarios (Thiele et al., 2016) with the possibility of multiple reconfigurations of the artificial intelligence algorithms based on the human feedback (Grønsund and Aanestad, 2020).

One of the most prominent applications of human-in-the-loop approaches is on the cyber-physical systems, which are a current trend in Industry 4.0 with many developments reported in the last few years (Lee, 2015). Such developments are also closely related to the internet of things technologies in collaborative networks using the integration of human-in-the-loop approaches (Garrido-Hidalgo et al., 2018). The integration of cyber-physical systems and the decision making capabilities of the human-integrated interfaces resulted in significant advances in smart industrial environments by means of collaborative assembly (Ruiz Garcia et al., 2019). In this case, the human cognitive capabilities enhance the system's performances due to human traits such as memory reflection learning, problem solving, and training; reasoning and belief; interaction and communication, among others naturally achieved capacities of humans that can be integrated in the algorithms (Emmanouilidis et al., 2019).

The human-in-the-loop applications in this context are divided into human control and human monitoring applications, where the aforementioned use of humans in feedback for machine learning algorithms is regarded as the human monitoring application. On the other hand, the use of humans on supervisory or direct control of interfaces and devices is regarded as the human control application (Sousa Nunes et al., 2015). In a supervisory control loop, the human operator control or supports the decisions of the machine in a collaborative approach, where neither the operator has the overload of all decisions nor the machine has the unsupervised decisions (Gross et al., 2017). In a monitoring application, the human agent roles in a smart manufacturing environment is the data acquisition from human perception directly input in the algorithm (i.e., using humans as sensors), state estimation input in the algorithm by the operator through memory, calculation and reasoning (Cimini et al., 2020).

The developments in the collaborative control include the cyber-physical robotic systems, where modular robots cells can be programmed, configured, or automated wit high flexibility and customizability Michniewicz and Reinhart (2014). Thus human-machine interfaces can be used on the control of multiple robotics cells using a configuration in which the human acts as a supervisor in the control loop (Orsag et al., 2017). These robotic cyber-physical systems have been used in different applications, including the monitoring of transmission lines with multiple robots cells (Fan et al., 2018) as well as collaborative approaches for surgical interventions, where these applications are well aligned with the requirements of the internet of robotic things commonly used in smart environments as summarized in Romeo et al. (2020). In addition, virtual reality also plays an important role in the development of immersive cyber-physical systems in which the human can interact with the robot in virtual environments (Rahman, 2018). Such approach enables the development of natural interfaces for remote human-robot collaboration, which is especially desired in hazardous environments (Liu and Wang, 2020). The control of joint orientations can also be achieved with admittance controllers in human-in-the-loop approaches using the operator as a feedback for the robots orientation (Perrusquía and Yu, 2020);

moreover, control paradigms such as the adaptive stiffness with the interaction of human operator in order to provide to the robot the capability of performing dexterous tasks.

In general, the wearable robots can be regarded as human-in-the-loop systems, since the device is worn by the operator, where the physical interaction between the human and the robot is considered in the robot's control (Huo et al., 2016). In addition, in some control strategies the cognitive interaction between the user and the device is considered (Bueno et al., 2008) as well as the design and dimensions of the wearable robot is customized for each user (Walsh, 2018). In fact, wearable robots are mainly related to advanced healthcare solutions, where such evolution in the healthcare is motivated by the advances in medicine and quality of life as well as the demands placed by the current worldwide scenario of population aging, as depicted in Chapter 1. To that extent and also due to the advances in Industry 4.0, the so-called Healthcare 4.0 has experienced a widespread in the areas presented in Fig. 2.3. Although the Healthcare 4.0 is related to many aspects such as self-management wellness, smart pharmaceuticals, and medical intake monitoring (as shown in Fig. 2.3, the aspects related to the wearable robotics are mainly the assisted living and rehabilitation. It is worth noting that if wearable sensors systems are also considered, the wearable technologies cover broader region of the Healthcare 4.0 approaches indicated in Fig. 2.3, since wearable sensors can be used on the monitoring of physiological (and pathological) signals (Koyama et al., 2018), provide tools for disease monitoring and telemedicine (Rangasamy et al., 2011) as well as the healthcare personalization (Ahad et al., 2019). In addition, the connectivity of such devices enables their use as cloud-based health information systems. Thus, wearable technologies cover almost all aspects involved in Healthcare 4.0, as indicated in Fig. 2.3.

The wearable devices are human-in-the-loop solutions with many applications in healthcare, as shown in Fig. 2.3, where current developments in soft robotics further enhance the applicability of such wearable systems and provide a deeper connection between the user and the wearable devices, since these new developments in soft robotics technology enable the rapid prototyping of customized structures (Wallin et al., 2018). The application of human-in-the-loop design approaches in wearable devices is further explored in Section 2.3.3

2.3.3 Human-in-the-loop design in soft wearable robots

The development of customized structures for locomotion assistance has been proposed throughout the years, where the system design has direct impact on the user's locomotion or any other activity that the wearable robot was designed to assist (Asbeck et al., 2014). In conventional rigid robots, the transparent interaction between human is achieved through compliant actuators (dos Santos et al., 2015) or using control strategies that consider the human-robot interaction forces in the device impedance/admittance control in approaches that involve

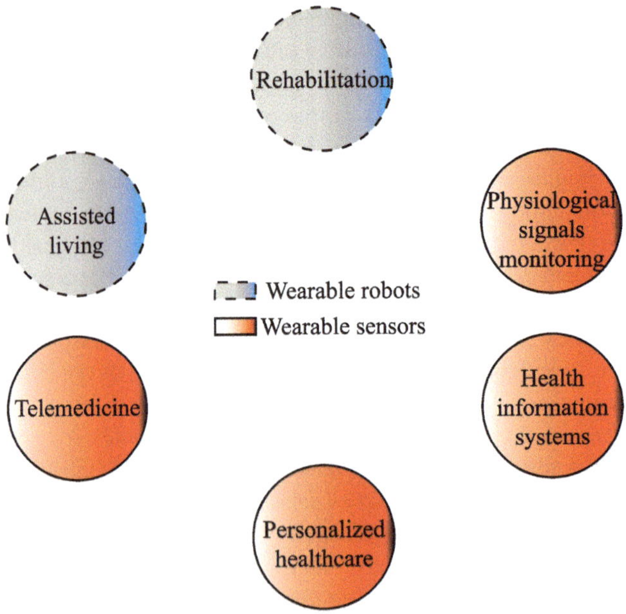

FIGURE 2.3 The impact of wearable devices in Healthcare 4.0 applications.

the stiffness or damping control (Rocon et al., 2008). The control of mechanical impedance can also be achieved using systems with dynamic impedance, which is achieved using soft structures and actuators that allow a myriad of shapes, geometries, stiffness control, and the possibility of controlling or customizing these parameters for each user, application, or operation condition (Laschi et al., 2016).

The use of human-in-the-loop approaches on the exoskeleton devices can lead to a customizable and optimized solution for gait assistance (Walsh, 2018). To that extent, a transparent exoskeleton can be obtained, where there is an optimum assistance during the gait (Zhang et al., 2017). The optimum assistance is obtained considering the minimal metabolic cost of the user when his/her movements are assisted by the exoskeleton (Ding et al., 2018). In this case, there is the inclusion of user's metabolic data as well as kinematic and kinetic parameters on the control system to obtain a control approach in which the desired trajectories of the limbs are obtained with the minimal metabolic costs (Fang and Yuan, 2020).

As one of the core technologies for soft robotics and human-in-the-loop design, the additive layer manufacturing offers multiple advantages in the rapid prototyping of customized structures. Such advantages also include the reduced environmental impact with the possibility of using biodegradable materials on the printing and the simplification of supply chain due to the use of fewer components and the possibility of distributed manufacturing, i.e., the products are manufactured near the customers (Huang et al., 2013). Also, due to these advantages, the additive layer manufacturing is one important technology on Industry 4.0. The additive layer manufacturing is also based on the integration of the manufacturing process with the computer-aided design (CAD), where the CAD 3D solid model is converted to a file format suitable with the manufacturing machine. Then the file is uploaded in the machine, where basic manipulation can be performed, such as scaling, object orientation changes, and infill configuration as well as printing orientation. Finally, the part is fabricated using one of the many possible technologies of additive layer manufacturing (Horn and Harrysson, 2012).

There are some additive layer manufacturing (or 3D printing) technologies; among them, the fused deposition modeling (FDM) is a popular approach that experienced a major widespread in the last few years (Huang et al., 2013). In this case, a thermoplastic polymer is heated at the printer nozzle in a molten state that allows it to be deposited layer-by-layer, resulting in the desired structure (Ngo et al., 2018). The simplicity, low cost, and the fast manufacturing make FDM an attractive technology for the development of multifarious structures. However, drawbacks such as low mechanical resistance and limited number of materials motivate the development and widespread of different 3D printing technologies. Recently, the developments in photopolymerizable resins and liquid crystal displays (LCD) and projectors make the stereolithography (SLA) approach a popular method, despite being one of the earliest methods proposed for additive layer manufacturing (Ngo et al., 2018). In this method, UV-curing monomers are placed in a container, which is subjected to UV source with the geometry desired for each layer using a LCD projector. This technique allows the development of complex parts with high resolution (around 10 μm) and quality (Ngo et al., 2018). Although it is a slow process, there is also a possibility of using biocompatible materials for biomedical bioprinting. It is also worth noting that there are other additive layer manufacturing techniques such as the direct energy deposition used on the metal and alloys printing as well as the power bed fusion that allows high resolution, but at a high cost (Ngo et al., 2018).

The evolution of 3D printing is one of the core enablers of the soft robotics technologies, where the 3D printing methods (mostly FDM and SLA) have been applied on the development of different wearable and nonwearable soft robots (Gul et al., 2018). In addition, replica moulding technique is still applicable for customized structures, including the ones in soft robotics, where there is also the possibility of using the aforementioned additive layer manufacturing techniques in conjunction with different materials in order to achieve a system with the desired mechanical properties (Wallin et al., 2018). As the robotic system also needs an actuator for its activation, different actuation technologies have been employed in soft robotics applications. These actuators include shape memory alloys, where there is a change of the material geometrical parameters and material features in response to temperature variations, which resulted in origami robots (Rus and Tolley, 2018) and in support or small scale structures with limited stress transmission (Cianchetti et al., 2018). In a similar class of actuators, shape memory polymers also present geometric changes under temperature stimuli with the same drawbacks of small output force/strain, despite their lower cost (Rus and Tolley, 2018). Another polymer actuation approach is the electroactive polymers, including the dielectric elastomers, where the material deforms when subjected to electric fields, which is a principle similar to the ones of electro or magnetorheological fluids that can be used on the development of fluidic actuators (De Vicente et al., 2011). In order to achieve higher power densities, actuators technologies such as hydraulic and pneumatic are employed, where artificial muscles can be developed using inflatable structures controlled by fluids (Cianchetti et al., 2018). As an alternative for soft and flexible actuators with high controllability, the use of tendon-driven actuators in soft robotics was proposed (Kastor et al., 2020). In this case, the tendons are remotely pulled by a motor, providing the desired movement to the flexible structure. The main drawback in this approach is the tendon dynamic responses, which generally is not considered in the control or is included in the control loop as a black box model.

The development of actuators in soft robotics is followed by the evolution in sensors technologies, as the robotic systems operate in closed-loop control; flexible sensors are used to provide a feedback on the actuators states. To that extent, developments in stretchable electronics have enable the use of resistive and capacitive sensors in soft structures (Wallin et al., 2018), where different substrates can be used to achieve higher flexibility of the sensors (Costa et al., 2019). Furthermore, flexible piezoresistive materials can be used on pressure and force sensing as well as magnetic sensors for kinematics assessment in soft robotics (Rus and Tolley, 2018). In the last few years, optoelectronic sensors also emerged as an alternative for soft robotics sensing using low cost approaches such as the light intensity variation with the possibility of applying such sensors in stretchable materials as well as in different geometries (especially slab and cylindrical waveguides) (Wallin et al., 2018).

The current advances in 3D printing in conjunction with the developments in soft actuators and flexible sensors resulted in the integration of sensors and actuators in soft structures (Zolfagharian et al., 2020). Thus, it is possible to embed the sensors structures in the 3D printed material, which provides additional functionalities for the structure. This approach results in 3D printed sensors for the assessment of strain, pressure, magnetic field, flow, temperature, and humidity as summarized in Zolfagharian et al. (2020). These integrated solutions can also incorporate the soft robotic actuator, resulting in embedded and customized solutions that can be adapted for each user or application. Therefore in addition to the three-dimensional properties and geometrics of the device, such approach also enable programmable transformations in the printed structure, which is well aligned with the definition of 4D printing (Zolfagharian et al., 2020). Furthermore, it is also possible to add multiple materials in the printing process to obtain actuators and smart structures with controlled properties

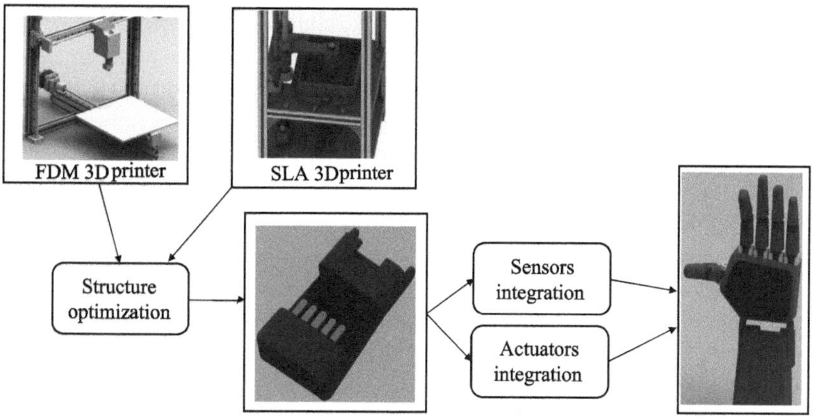

FIGURE 2.4 Schematic representation of the wearable robot fabrication using 3D printers.

(Ge et al., 2016). Fig. 2.4 shows the schematic representation of soft structures fabrication via additive layer manufacturing, actuator technologies, and sensors approaches for soft robotics and human-in-the-loop design. The integration of these three key components, i.e., structure, actuators, and sensors, is also presented.

The use of optical fiber sensors technologies can further enhance the performance of the integrated structures due to the intrinsic advantages of optical fiber sensing technologies, including electromagnetic field immunity, galvanic isolation, multiplexing capabilities, and small dimensions (Othonos and Kalli, 1999). These features enable the embedment of optical fibers in the 3D printed structures, where integrated solutions such as the one presented in Fig. 2.4 can be obtained (Zolfagharian et al., 2020). It is also worth noting that optical fibers can be fabricated directly from the 3D printers, which enable a further step on their integration in soft structures, where not only the properties of the structure can be tailored based on its user, but also the optical fiber dimensions and properties can be optimized for each application or for each user. In addition, there are many optical fiber sensing approaches (including the low cost optoelectronic systems aforementioned) that results in higher resolution and accuracy when compared with conventional electronic sensors (Othonos and Kalli, 1999). All these aspects will be further explored and discussed in the next chapters of this book, where the fundamentals of optical fibers and their sensing approaches will be depicted. Moreover, the many developments of optical fiber sensors in wearable applications will presented and discussed in detail.

2.4 Current trends and future approaches in wearable soft robots

There are different challenges and an encouraging range of future possibilities in biomedical soft robots. Biocompatible materials and bioinspired applications are essential for the development of flexible solutions for surgery, endoscopy, and drug delivery systems. The interaction of different areas, such as engineering, materials science, biology and medicine, mathematics, and modeling will drive the creation of new soft robotics applications and are imperative for promoting technological progress (Laschi et al., 2016).

Although this chapter aimed at presenting the main advantages and a range of possibilities of soft robotic devices, being completely soft may introduce challenging disadvantages. By looking at nature, it is possible to observe that soft animals are usually small or are found in a medium that support their bodies, such as in water or underground (Laschi et al., 2016). Larger animals usually need a skeleton for bodyweight support (Kim et al., 2013). This could point toward the necessity of merging rigid structures to soft robotics for assistive or rehabilitation devices that are designed for certain disabilities.

Realizing effective soft robots require the integration of materials and functionalities including sensing, actuation, powering, control, and processing. The combination of innovative fabrication techniques, such as micro/nano-molding (Heckele and Schomburg, 2004; Saha et al., 2016), soft lithography (Qin et al., 2010), and 3D printing (Wu et al., 2011), Wallin et al. (2018), poses as promising solutions for soft robotics (Laschi et al., 2016). Current developments of soft materials and additive manufacturing technologies, such as 3D printing, fused deposition modeling and direct ink writing, have enabled new capabilities and the performance of complex 3D movements in soft robots (Wallin et al., 2018). Advances in manufacturing technologies allied with material science will also allow robotic devices that appear and behave like biological systems.

Improved sensing and actuation technologies will also be necessary to achieve functional, portable, and biocompatible soft robotic solutions. Soft robots require stretchable sensors that traditionally rely on the electrical properties of materials and composites for measuring signals. Resistive and capacitive sensors, as an example, are the most common choices for measuring strain in soft robots. Nevertheless, such sensors may present important drawbacks regarding hysteresis, fabrication complexity, and incompatibilities with soft actuators. The use of optical sensors based on stretchable waveguides and built with advanced manufacturing techniques may present an alternative to such problems, while exhibiting high precision and low hysteresis (Wallin et al., 2018). Strain sensing in a soft prosthetic hand based on silicone-polyurethane waveguides was presented in Zhao et al. (2016). Stereolithography has also been applied to develop miniature pH sensors (Yin et al., 2016).

Polymer optical fiber (POF) sensors are an interesting alternative to electrical/electronic sensor, while providing crucial advantages related to high flexibility, lower Young's modulus, high elastic limits, and impact resistance. A broad range of sensing solutions for biomedical and healthcare applications was proposed by the authors in Leal-Junior et al. (2019a) and include movement analysis, monitoring physiological parameters, and the instrumentation of (rigid)

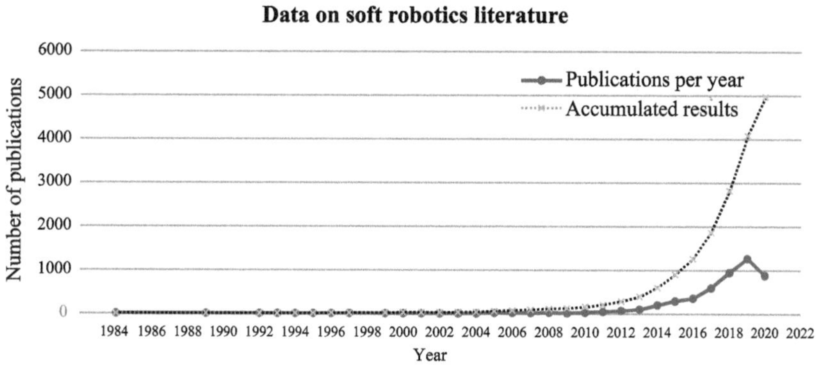

FIGURE 2.5 Data on soft robotics literature obtained from Scopus in August 2020. Search query: (TITLE-ABS-KEY("soft robot*")OR TITLE-ABS-KEY("soft bodied robot*")OR (TITLE-ABS-KEY("soft material*")AND TITLE-ABS-KEY(robot*))).

robotic devices. It is possible to foresee a widespread use of POF sensors in soft robotics, as we intend to demonstrate through this book.

Most of the current actuation technologies are not suitable for the next generation of soft wearable assistive devices. Common actuator technologies for soft robotics are flexible fluidic actuators, shape memory materials, electroactive polymers, electrorheological and magnetorheological materials, and tendon driven actuators (Cianchetti et al., 2018). Such actuator approaches are applied in different applications and present important drawbacks for practical use. In this sense, actuator technologies are a very active field of research for soft robotic applications.

Simulation and control of soft robots are also important aspects to assist on the design and adoption of such devices in real contexts. This requires a better understanding of soft materials and their interaction with the environment to produce desired robotic behaviors. The morphology of a robot alters its control requirements, from determining behaviors that can be performed to regulating the amount of control required for such behaviors (Paul, 2006). Lifelike abilities to change shape, stiffen, row, self-heal, develop, and evolve are expected to allow soft robots to adapt their behavior and morphology to specific tasks (Laschi et al., 2016).

Regarding power source technologies, there is also a need for achieving solutions designed to specifically work with soft robots. Futuristic approaches show the possibility of using the human energy source to sustain the functionality of wearable devices or prosthesis (Cianchetti et al., 2018), where a chemical power source could be provided by the wearer (Roseman et al., 2015). In this sense, the interaction of different areas, such as engineering and materials science, biology, and chemistry will be essential to achieve alternative power sources.

Soft robotics is a highly active field of research that has been growing fast in the last decade. Fig. 2.5 presents the evolution of publications in the topic of soft robots since 1984 up to August 2020 (data retrieved from Scopus database). From a value close to zero in 2004, the accumulated results in 2020 is remarkably high, reaching almost 5000 publications with more than 1000 in the last year (2019).

References

Ahad, A., Tahir, M., Yau, K.L.A., 2019. 5g-based smart healthcare network: architecture, taxonomy, challenges and future research directions. IEEE Access 7, 100747–100762. https://doi.org/10.1109/ACCESS.2019.2930628.

Albu-Schaffer, A., Fischer, M., Schreiber, G., Schoeppe, F., Hirzinger, G., 2004. Soft robotics: what Cartesian stiffness can obtain with passively compliant, uncoupled joints? In: 2004 IEEE/RSJ International Conference on Intelligent Robots and Systems (IROS) (IEEE Cat. No. 04CH37566). IEEE, Sendai, Japan, pp. 3295–3301.

Andersen, C.K., Wittrup-Jensen, K.U., Lolk, A., Andersen, K., Kragh-Sørensen, P., 2004. Ability to perform activities of daily living is the main factor affecting quality of life in patients with dementia. Health and Quality of Life Outcomes 2, 52. https://doi.org/10.1186/1477-7525-2-52.

Ansari, Y., Manti, M., Falotico, E., Mollard, Y., Cianchetti, M., Laschi, C., 2017. Towards the development of a soft manipulator as an assistive robot for personal care of elderly people. International Journal of Advanced Robotic Systems 14, 172988141668713. https://doi.org/10.1177/1729881416687132.

Asbeck, A.T., De Rossi, S.M., Galiana, I., Ding, Y., Walsh, C.J., 2014. Stronger, smarter, softer: next-generation wearable robots. IEEE Robotics & Automation Magazine 21, 22–33. https://doi.org/10.1109/MRA.2014.2360283.

Banala, S.K., Agrawal, S.K., Scholz, J.P., 2007. Active leg exoskeleton (ALEX) for gait rehabilitation of motor-impaired patients. In: IEEE 10th International Conference on Rehabilitation Robotics, ICORR'07, pp. 401–407.

Bhattacherjee, S., Chattopadhayay, S., Rao, V., Seth, S., Mukherjee, S., Sengupta, A., Bhaumik, S., 2018. Kinematics and teleoperation of tendon driven continuum robot. Procedia Computer Science 133, 879–886. https://doi.org/10.1016/j.procs.2018.07.106.

Bosse, S., Engel, U., 2019. Real-time human-in-the-loop simulation with mobile agents, chat bots, and crowd sensing for smart cities. Sensors (Switzerland) 19. https://doi.org/10.3390/s19204356.

Budd, S., Robinson, E.C., Kainz, B., 2019. A survey on active learning and human-in-the-loop deep learning for medical image analysis.

Bueno, L., Brunetti, F., Frizera, A., Pons, J.L., 2008. Human–robot cognitive interaction, 1998.

Cappello, L., Meyer, J.T., Galloway, K.C., Peisner, J.D., Granberry, R., Wagner, D.A., Engelhardt, S., Paganoni, S., Walsh, C.J., 2018. Assisting hand function after spinal cord injury with a fabric-based soft robotic glove. Journal of NeuroEngineering and Rehabilitation 15, 59. https://doi.org/10.1186/s12984-018-0391-x.

Cianchetti, M., Calisti, M., Margheri, L., Kuba, M., Laschi, C., 2015. Bioinspired locomotion and grasping in water: the soft eight-arm OCTOPUS robot. Bioinspiration & Biomimetics 10, 035003. https://doi.org/10.1088/1748-3190/10/3/035003.

Cianchetti, M., Laschi, C., Menciassi, A., Dario, P., 2018. Biomedical applications of soft robotics. Nature Reviews Materials 3, 143–153. https://doi.org/10.1038/s41578-018-0022-y.

Cimini, C., Pirola, F., Pinto, R., Cavalieri, S., 2020. A human-in-the-loop manufacturing control architecture for the next generation of production systems. Journal of Manufacturing Systems 54, 258–271. https://doi.org/10.1016/j.jmsy.2020.01.002.

Copaci, D., Cano, E., Moreno, L., Blanco, D., 2017. New design of a soft robotics wearable elbow exoskeleton based on shape memory alloy wire actuators. Applied Bionics and Biomechanics 2017, 1–11. https://doi.org/10.1155/2017/1605101.

Costa, J.C., Spina, F., Lugoda, P., Garcia-Garcia, L., Roggen, D., Münzenrieder, N., 2019. Flexible sensors—from materials to applications. Technologies 7, 35. https://doi.org/10.3390/technologies7020035.

De Vicente, J., Klingenberg, D.J., Hidalgo-Alvarez, R., 2011. Magnetorheological fluids: a review. Soft Matter 7, 3701–3710. https://doi.org/10.1039/c0sm01221a.

Delph, M.A., Fischer, S.A., Gauthier, P.W., Luna, C.H.M., Clancy, E.A., Fischer, G.S., 2013. A soft robotic exomusculature glove with integrated sEMG sensing for hand rehabilitation. In: 2013 IEEE 13th International Conference on Rehabilitation Robotics (ICORR). IEEE, Seattle, WA, pp. 1–7.

Ding, Y., Kim, M., Kuindersma, S., Walsh, C.J., 2018. Human-in-the-loop optimization of hip assistance with a soft exosuit during walking. Science Robotics 3, 1–9. https://doi.org/10.1126/scirobotics.aar5438.

Dollar, A.M., Herr, H., 2007. Active orthoses for the lower-limbs: challeges and state of the art. In: Proceedings of the 2007 IEEE 10th International Conference on Rehabilitation Robotics.

Duckett, T., Pearson, S., Blackmore, S., Grieve, B., Chen, W.H., Cielniak, G., Cleaversmith, J., Dai, J., Davis, S., Fox, C., From, P., Georgilas, I., Gill, R., Gould, I., Hanheide, M., Hunter, A., Iida, F., Mihalyova, L., Nefti-Meziani, S., Neumann, G., Paoletti, P., Pridmore, T., Ross, D., Smith, M., Stoelen, M., Swainson, M., Wane, S., Wilson, P., Wright, I., Yang, G.Z., 2018. Agricultural robotics: the future of robotic agriculture. arXiv:1806.06762.

Emmanouilidis, C., Pistofidis, P., Bertoncelj, L., Katsouros, V., Fournaris, A., Koulamas, C., Ruiz-Carcel, C., 2019. Enabling the human in the loop: linked data and knowledge in industrial cyber-physical systems. Annual Reviews in Control 47, 249–265. https://doi.org/10.1016/j.arcontrol.2019.03.004.

Esquenazi, A., Talaty, M., Packel, A., Saulino, M., 2012. The ReWalk powered exoskeleton to restore ambulatory function to individuals with thoracic-level motor-complete spinal cord injury. American Journal of Physical Medicine & Rehabilitation 91, 911–921. https://doi.org/10.1097/PHM.0b013e318269d9a3.

Fan, F., Wu, G., Wang, M., Cao, Q., Yang, S., 2018. Multi-robot cyber physical system for sensing environmental variables of transmission line. Sensors (Switzerland) 18. https://doi.org/10.3390/s18093146.

Fang, J., Yuan, Y., 2020. Human-in-the-loop optimization of wearable robots to reduce the human metabolic energy cost in physical movements. Robotics and Autonomous Systems 127, 103495. https://doi.org/10.1016/j.robot.2020.103495.

Forner-Cordero, A., Pons, J.L., Wisse, M., Ceres, R., Calderón, L., 2008. Basis for bioinspiration and biomimetism in wearable robots. Wearable Robots: Biomechatronic Exoskeletons, 17–45. https://doi.org/10.1002/9780470987667.ch2.

Fukuda, H., Samura, K., Hamada, O., Saita, K., Ogata, T., Shiota, E., Sankai, Y., Inoue, T., 2015. Effectiveness of acute phase hybrid assistive limb rehabilitation in stroke patients classified by paralysis severity. Neurologia Medico-Chirurgicală 55, 487–492. https://doi.org/10.2176/nmc.oa.2014-0431.

Fusco, S., Sakar, M.S., Kennedy, S., Peters, C., Bottani, R., Starsich, F., Mao, A., Sotiriou, G.A., Pané, S., Pratsinis, S.E., Mooney, D., Nelson, B.J., 2014. An integrated microrobotic platform for on-demand, targeted therapeutic interventions. Advanced Materials 26, 952–957. https://doi.org/10.1002/adma.201304098.

Galiana, I., Hammond, F.L., Howe, R.D., Popovic, M.B., 2012. Wearable soft robotic device for post-stroke shoulder rehabilitation: identifying misalignments. In: 2012 IEEE/RSJ International Conference on Intelligent Robots and Systems. IEEE, Vilamoura-Algarve, Portugal, pp. 317–322.

Garrido-Hidalgo, C., Hortelano, D., Roda-Sanchez, L., Olivares, T., Ruiz, M.C., Lopez, V., 2018. Iot heterogeneous mesh network deployment for human-in-the-loop challenges towards a social and sustainable industry 4.0. IEEE Access 6, 28417–28437. https://doi.org/10.1109/ACCESS.2018.2836677.

Ge, Q., Sakhaei, A.H., Lee, H., Dunn, C.K., Fang, N.X., Dunn, M.L., 2016. Multimaterial 4d printing with tailorable shape memory polymers. Scientific Reports 6, 1–11. https://doi.org/10.1038/srep31110.

Gross, K.C., Baclawski, K., Chan, E.S., Gawlick, D., Ghoneimy, A., Liu, Z.H., 2017. A supervisory control loop with prognostics for human-in-the-loop decision support and control applications. In: 2017 IEEE Conference on Cognitive and Computational Aspects of Situation Management, CogSIMA 2017.

Grønsund, T., Aanestad, M., 2020. Augmenting the algorithm: emerging human-in-the-loop work configurations. Journal of Strategic Information Systems 29, 101614. https://doi.org/10.1016/j.jsis.2020.101614.

Gul, J.Z., Sajid, M., Rehman, M.M., Siddiqui, G.U., Shah, I., Kim, K.H., Lee, J.W., Choi, K.H., 2018. 3d printing for soft robotics – a review. Science and Technology of Advanced Materials 19, 243–262. https://doi.org/10.1080/14686996.2018.1431862.

Gultepe, E., Randhawa, J.S., Kadam, S., Yamanaka, S., Selaru, F.M., Shin, E.J., Kalloo, A.N., Gracias, D.H., 2013. Biopsy with thermally-responsive untethered microtools. Advanced Materials 25, 514–519. https://doi.org/10.1002/adma.201203348.

Guo, Y., Hu, X., Hu, B., Cheng, J., Zhou, M., Kwok, R.Y., 2017. Mobile cyber physical systems: current challenges and future networking applications. IEEE Access 6, 12360–12368. https://doi.org/10.1109/ACCESS.2017.2782881.

Hamaya, M., Matsubara, T., Teramae, T., Noda, T., Morimoto, J., 2019. Design of physical user–robot interactions for model identification of soft actuators on exoskeleton robots. International Journal of Robotics Research, 1–14. https://doi.org/10.1177/0278364919853618.

Heckele, M., Schomburg, W.K., 2004. Review on micro molding of thermoplastic polymers. Journal of Micromechanics and Microengineering 14, R1–R14. https://doi.org/10.1088/0960-1317/14/3/R01.

Holzinger, A., 2016. Interactive machine learning for health informatics: when do we need the human-in-the-loop? Brain Informatics 3, 119–131. https://doi.org/10.1007/s40708-016-0042-6.

Horn, T.J., Harrysson, O.L., 2012. Overview of current additive manufacturing technologies and selected applications. Science Progress 95, 255–282. https://doi.org/10.3184/003685012X13420984463047.

Huang, S.H., Liu, P., Mokasdar, A., Hou, L., 2013. Additive manufacturing and its societal impact: a literature review. International Journal of Advanced Manufacturing Technology 67, 1191–1203. https://doi.org/10.1007/s00170-012-4558-5.

Huang, V.S., Krakauer, J.W., 2009. Robotic neurorehabilitation: a computational motor learning perspective. Journal of NeuroEngineering and Rehabilitation 6, 5. https://doi.org/10.1186/1743-0003-6-5.

Huo, W., Mohammed, S., Moreno, J.C., Amirat, Y., 2016. Lower limb wearable robots for assistance and rehabilitation: a state of the art. IEEE Systems Journal 10, 1068–1081. https://doi.org/10.1109/JSYST.2014.2351491.

Hussain, I., Spagnoletti, G., Salvietti, G., Prattichizzo, D., 2016. An emg interface for the control of motion and compliance of a supernumerary robotic finger. Frontiers in Neurorobotics 10, 1–13. https://doi.org/10.3389/fnbot.2016.00018.

Jaimes, A., Sebe, N., 2007. Multimodal human-computer interaction: a survey. Computer Vision and Image Understanding 108, 116–134. https://doi.org/10.1016/j.cviu.2006.10.019.

Jiménez, M.F., Monllor, M., Frizera, A., Bastos, T., Roberti, F., Carelli, R., 2019. Admittance controller with spatial modulation for assisted locomotion using a smart walker. Journal of Intelligent & Robotic Systems 94, 621–637. https://doi.org/10.1007/s10846-018-0854-0.

Kastor, N., Mukherjee, R., Cohen, E., Vikas, V., Trimmer, B.A., White, R.D., 2020. Design and manufacturing of tendon-driven soft foam robots. Robotica 38, 88–105. https://doi.org/10.1017/S0263574719000481.

ten Kate, J., Smit, G., Breedveld, P., 2017. 3d-printed upper limb prostheses: a review. Disability and Rehabilitation: Assistive Technology 12, 300–314. https://doi.org/10.1080/17483107.2016.1253117.

Katz, S., 1983. Assessing self-maintenance: activities of daily living, mobility, and instrumental activities of daily living. Journal of the American Geriatrics Society 31, 721–727. https://doi.org/10.1111/j.1532-5415.1983.tb03391.x.
Keemink, A.Q., van der Kooij, H., Stienen, A.H., 2018. Admittance control for physical human–robot interaction. International Journal of Robotics Research 37, 1421–1444. https://doi.org/10.1177/0278364918768950.
Kim, S., Laschi, C., Trimmer, B., 2013. Soft robotics: a bioinspired evolution in robotics. Trends in Biotechnology 31, 287–294. https://doi.org/10.1016/j.tibtech.2013.03.002.
Kirschgens, I.A., Ugarte, I.Z., Uriarte, E.G., Rosas, A.M., Vilches, V.M., 2018. Robot hazards: from safety to security.
Koyama, Y., Nishiyama, M., Watanabe, K., 2018. Smart textile using hetero-core optical fiber for heartbeat and respiration monitoring. IEEE Sensors Journal 18, 6175–6180. https://doi.org/10.1109/JSEN.2018.2847333.
Krebs, H., Hogan, N., Volpe, B., Aisen, M., Edelstein, L., Diels, C., 1999. Overview of clinical trials with MIT-MANUS: a robot-aided neuro-rehabilitation facility. Technology and Health Care 7, 419–423. https://doi.org/10.3233/THC-1999-7606.
Krebs, H.I., Palazzolo, J.J., Dipietro, L., Ferraro, M., Krol, J., Rannekleiv, K., Volpe, B.T., Hogan, N., 2003. Rehabilitation robotics: performance-based progressive robot-assisted therapy. Autonomous Robots 15, 7–20.
Kummer, M.P., Abbott, J.J., Kratochvil, B.E., Borer, R., Sengul, A., Nelson, B.J., 2010. OctoMag: an electromagnetic system for 5-DOF wireless micromanipulation. IEEE Transactions on Robotics 26, 1006–1017. https://doi.org/10.1109/TRO.2010.2073030.
Laschi, C., Mazzolai, B., Cianchetti, M., 2016. Soft robotics: technologies and systems pushing the boundaries of robot abilities. Science Robotics 1, eaah3690. https://doi.org/10.1126/scirobotics.aah3690.
Le, H.M., Do, T.N., Phee, S.J., 2016. A survey on actuators-driven surgical robots. Sensors and Actuators. A, Physical 247, 323–354. https://doi.org/10.1016/j.sna.2016.06.010.
Leal Junior, A.G., de Andrade, R.M., Filho, A.B., 2016. In: Series Elastic Actuator: Design, Analysis and Comparison. InTech.
Leal-Junior, A.G., Diaz, C.A., Avellar, L.M., Pontes, M.J., Marques, C., Frizera, A., 2019a. Polymer optical fiber sensors in healthcare applications: a comprehensive review. Sensors 19, 3156. https://doi.org/10.3390/s19143156.
Leal-Junior, A.G., Díaz, C.R., Marques, C., Pontes, M.J., Frizera, A., 2019b. 3d-printed pof insole: development and applications of a low-cost, highly customizable device for plantar pressure and ground reaction forces monitoring. Optics and Laser Technology 116, 256–264. https://doi.org/10.1016/j.optlastec.2019.03.035.
Lee, E.A., 2015. The past, present and future of cyber-physical systems: a focus on models. Sensors (Switzerland) 15, 4837–4869. https://doi.org/10.3390/s150304837.
Lee, H., Krebs, H.I., Hogan, N., 2014. Multivariable dynamic ankle mechanical impedance with active muscles. IEEE Transactions on Neural Systems and Rehabilitation Engineering 22, 971–981. https://doi.org/10.1109/TNSRE.2014.2328235.
Leong, T.G., Randall, C.L., Benson, B.R., Bassik, N., Stern, G.M., Gracias, D.H., 2009. Tetherless thermobiochemically actuated microgrippers. Proceedings of the National Academy of Sciences 106, 703–708. https://doi.org/10.1073/pnas.0807698106.
Liu, H., Wang, L., 2020. Remote human–robot collaboration: a cyber-physical system application for hazard manufacturing environment. Journal of Manufacturing Systems 54, 24–34. https://doi.org/10.1016/j.jmsy.2019.11.001.
Lo, H.S., Xie, S.Q., 2012. Exoskeleton robots for upper-limb rehabilitation: state of the art and future prospects. Medical Engineering & Physics 34, 261–268. https://doi.org/10.1016/j.medengphy.2011.10.004.
Lum, P.S., Burgar, C.G., Shor, P.C., Majmundar, M., Van der Loos, M., 2002. Robot-assisted movement training compared with conventional therapy techniques for the rehabilitation of upper-limb motor function after stroke. Archives of Physical Medicine and Rehabilitation 83, 952–959. https://doi.org/10.1053/apmr.2001.33101.

Maeder-York, P., Clites, T., Boggs, E., Neff, R., Polygerinos, P., Holland, D., Stirling, L., Galloway, K., Wee, C., Walsh, C., 2014. Biologically inspired soft robot for thumb rehabilitation. Journal of Medical Devices 8, 020933. https://doi.org/10.1115/1.4027031.

Makarenko, V.A., 2016. Deep learning algorithms for signal recognition in long perimeter monitoring distributed fiber optic sensors. IEEE International Workshop on Machine Learning for Signal Processing, MLSP 2016-Novem, 1–6. https://doi.org/10.1109/MLSP.2016.7738863.

Michniewicz, J., Reinhart, G., 2014. Cyber-physical robotics – automated analysis, programming and configuration of robot cells based on cyber-physical-systems. Procedia Technology 15, 566–575. https://doi.org/10.1016/j.protcy.2014.09.017.

Miriyev, A., Stack, K., Lipson, H., 2017. Soft material for soft actuators. Nature Communications 8, 596. https://doi.org/10.1038/s41467-017-00685-3.

Munir, S., Stankovic, J.a., Liang, C.J.M., Lin, S., 2013. Cyber physical system challenges for human-in-the-loop control. In: The 8th International Workshop on Feedback Computing.

Nef, T., Guidali, M., Riener, R., 2009. ARMin III – arm therapy exoskeleton with an ergonomic shoulder actuation. Applied Bionics and Biomechanics 6, 127–142. https://doi.org/10.1080/11762320902840179.

Nelson, B.J., Kaliakatsos, I.K., Abbott, J.J., 2010. Microrobots for minimally invasive medicine. Annual Review of Biomedical Engineering 12, 55–85. https://doi.org/10.1146/annurev-bioeng-010510-103409.

Neuhaus, P.D., Noorden, J.H., Craig, T.J., Torres, T., Kirschbaum, J., Pratt, J.E., 2011. Design and evaluation of Mina: a robotic orthosis for paraplegics. In: 2011 IEEE International Conference on Rehabilitation Robotics. IEEE, Zurich, pp. 1–8.

Ngo, T.D., Kashani, A., Imbalzano, G., Nguyen, K.T., Hui, D., 2018. Additive manufacturing (3d printing): a review of materials, methods, applications and challenges. Composites. Part B, Engineering 143, 172–196. https://doi.org/10.1016/j.compositesb.2018.02.012.

Orsag, M., Haus, T., Tolić, D., Ivanović, A., Car, M., Palunko, I., Bogdan, S., 2017. Human-in-the-loop control of multi-agent aerial systems. In: 2016 European Control Conference, ECC 2016, pp. 2139–2145.

Othonos, A., Kalli, K., 1999. Fiber Bragg gratings: fundamentals and applications in telecommunications and sensing. In: Fiber Bragg Gratings: Fundamentals and Applications in Telecommunications and Sensing. Artech House Inc., Norwood.

Panizzolo, F.A., Freisinger, G.M., Karavas, N., Eckert-Erdheim, A.M., Siviy, C., Long, A., Zifchock, R.A., LaFiandra, M.E., Walsh, C.J., 2019. Metabolic cost adaptations during training with a soft exosuit assisting the hip joint. Scientific Reports 9, 9779. https://doi.org/10.1038/s41598-019-45914-5.

Paul, C., 2006. Morphological computation: a basis for the analysis of morphology and control requirements. Robotics and Autonomous Systems 54, 619–630. https://doi.org/10.1016/j.robot.2006.03.003.

Perrusquía, A., Yu, W., 2020. Human-in-the-loop control using Euler angles. Journal of Intelligent and Robotic Systems: Theory and Applications 97, 271–285. https://doi.org/10.1007/s10846-019-01058-2.

Pfeifer, R., Bongard, J., Grand, S., 2007. How the Body Shapes the Way We Think: A New View of Intelligence. MIT Press, Cambridge, Mass. ocm65644796.

Phan, G.h., Hansen, C., Tommasino, P., Budhota, A., Mohan, D.M., Hussain, A., Burdet, E., Campolo, D., 2020. Estimating human wrist stiffness during a tooling task. Sensors 20, 3260. https://doi.org/10.3390/s20113260.

Qin, D., Xia, Y., Whitesides, G.M., 2010. Soft lithography for micro- and nanoscale patterning. Nature Protocols 5, 491–502. https://doi.org/10.1038/nprot.2009.234.

Rahman, S.M., 2018. Cyber-physical-social system between a humanoid robot and a virtual human through a shared platform for adaptive agent ecology. IEEE/CAA Journal of Automatica Sinica 5, 190–203. https://doi.org/10.1109/JAS.2017.7510760.

Rangasamy, M., Balasubramaniam, A., Krishnarajan, D., Kante, N.R., Kumar, N.S., 2011. Role of telemedicine in health care system: a review. Pharmaceutical Research, 1–12.

Rocon, E., Belda-Lois, J., Ruiz, A., Manto, M., Moreno, J., Pons, J., 2007. Design and validation of a rehabilitation robotic exoskeleton for tremor assessment and suppression. IEEE Transactions on Neural Systems and Rehabilitation Engineering 15, 367–378. https://doi.org/10.1109/TNSRE.2007.903917.

Rocon, E., Pons, J.L., 2011. Exoskeletons in Rehabilitation Robotics. Springer Tracts in Advanced Robotics, vol. 69. Springer, Berlin, Heidelberg.

Rocon, E., Ruiz, A.F., Raya, R., Schiele, A., Pons, J.L., Belda-Lois, J.M., Poveda, R., Vivas, M.J., Moreno, J.C., 2008. Human–robot physical interaction. In: Pons, J.L. (Ed.), Wearable Robots. John Wiley & Sons, Ltd, Chichester, UK, pp. 127–163.

Romeo, L., Petitti, A., Marani, R., Milella, A., 2020. Internet of robotic things in smart domains: applications and challenges. Sensors (Switzerland) 20, 1–23. https://doi.org/10.3390/s20123355.

Roseman, J.M., Lin, J., Ramakrishnan, S., Rosenstein, J.K., Shepard, K.L., 2015. Hybrid integrated biological–solid-state system powered with adenosine triphosphate. Nature Communications 6, 10070. https://doi.org/10.1038/ncomms10070.

Ruiz Garcia, M.A., Rojas, R., Gualtieri, L., Rauch, E., Matt, D., 2019. A human-in-the-loop cyber-physical system for collaborative assembly in smart manufacturing. Procedia CIRP 81, 600–605. https://doi.org/10.1016/j.procir.2019.03.162.

Rus, D., Tolley, M.T., 2015. Design, fabrication and control of soft robots. Nature 521, 467–475. https://doi.org/10.1038/nature14543.

Rus, D., Tolley, M.T., 2018. Design, fabrication and control of origami robots. Nature Reviews Materials 3, 101–112. https://doi.org/10.1038/s41578-018-0009-8.

Saha, B., Toh, W.Q., Liu, E., Tor, S.B., Hardt, D.E., Lee, J., 2016. A review on the importance of surface coating of micro/nano-mold in micro/nano-molding processes. Journal of Micromechanics and Microengineering 26, 013002. https://doi.org/10.1088/0960-1317/26/1/013002.

dos Santos, W.M., Caurin, G.A.P., Siqueira, A.A.G., 2015. Design and control of an active knee orthosis driven by a rotary series elastic actuator. Control Engineering Practice, 1–12. https://doi.org/10.1016/j.conengprac.2015.09.008.

Schuerle, S., Erni, S., Flink, M., Kratochvil, B.E., Nelson, B.J., 2013. Three-dimensional magnetic manipulation of micro- and nanostructures for applications in life sciences. IEEE Transactions on Magnetics 49, 321–330. https://doi.org/10.1109/TMAG.2012.2224693.

Shahid, T., Gouwanda, D., Nurzaman, S.G., Gopalai, A.A., 2018. Moving toward soft robotics: a decade review of the design of hand exoskeletons. Biomimetics 3. https://doi.org/10.3390/biomimetics3030017.

Shen, H., 2016. Meet the soft, cuddly robots of the future. Nature 530, 24–26.

Sousa Nunes, D.S., Zhang, P., Sa Silva, J., 2015. A survey on human-in-the-loop applications towards an Internet of all. IEEE Communications Surveys and Tutorials 17, 944–965. https://doi.org/10.1109/COMST.2015.2398816.

Taherifar, A., Vossoughi, G., Ghafari, A.S., 2018. Variable admittance control of the exoskeleton for gait rehabilitation based on a novel strength metric. Robotica 36, 427–447. https://doi.org/10.1017/S0263574717000480.

Tefertiller, C., Pharo, B., Evans, N., Winchester, P., 2011. Efficacy of rehabilitation robotics for walking training in neurological disorders: a review. The Journal of Rehabilitation Research and Development 48, 387. https://doi.org/10.1682/JRRD.2010.04.0055.

Teramae, T., Ishihara, K., Babič, J., Morimoto, J., Oztop, E., 2018. Human-in-the-loop control and task learning for pneumatically actuated muscle based robots. Frontiers in Neurorobotics 12, 1–10. https://doi.org/10.3389/fnbot.2018.00071.

Thiele, T., Sommer, T., Schröder, S., Richert, A., Jeschke, S., 2016. Human-in-the-loop processes as enabler for data analytics in digitalized organizations. Human-in-the-Loop Processes as Enabler for Data Analytics in Digitalized Organizations, 1–11. https://doi.org/10.18420/muc2016-ws11-0004.

Villa-Parra, A., Delisle-Rodriguez, D., Souza Lima, J., Frizera-Neto, A., Bastos, T., 2017. Knee impedance modulation to control an active orthosis using insole sensors. Sensors 17, 2751. https://doi.org/10.3390/s17122751.

Villoslada, A., Flores, A., Copaci, D., Blanco, D., Moreno, L., 2015. High-displacement flexible shape memory alloy actuator for soft wearable robots. Robotics and Autonomous Systems 73, 91–101. https://doi.org/10.1016/j.robot.2014.09.026.

Vitiello, V., Lee, Su-Lin, Cundy, T.P., Yang, Guang-Zhong, 2013. Emerging robotic platforms for minimally invasive surgery. IEEE Reviews in Biomedical Engineering 6, 111–126. https://doi.org/10.1109/RBME.2012.2236311.

Wallin, T.J., Pikul, J., Shepherd, R.F., 2018. 3d printing of soft robotic systems. Nature Reviews Materials 3, 84–100. https://doi.org/10.1038/s41578-018-0002-2.

Walsh, C., 2018. Human-in-the-loop development of soft wearable robots. Nature Reviews Materials 3, 78–80. https://doi.org/10.1038/s41578-018-0011-1.

Walsh, C.J., Endo, K., Herr, H., 2007. A quasi-passive leg exoskeleton for load-carrying augmentation. International Journal of Humanoid Robotics 04, 487–506. https://doi.org/10.1142/S0219843607001126.

Wang, L., Iida, F., 2015. Deformation in soft-matter robotics: a categorization and quantitative characterization. IEEE Robotics & Automation Magazine 22, 125–139. https://doi.org/10.1109/MRA.2015.2448277.

Wang, P., Peng, D., Li, L., Chen, L., Wu, C., Wang, X., Childs, P., Guo, Y., 2019. Human-in-the-loop design with machine learning. In: Proceedings of the International Conference on Engineering Design, ICED 2019-Augus, pp. 2577–2586.

Wehner, M., Quinlivan, B., Aubin, P.M., Martinez-Villalpando, E., Baumann, M., Stirling, L., Holt, K., Wood, R., Walsh, C., 2013. A lightweight soft exosuit for gait assistance. In: 2013 IEEE International Conference on Robotics and Automation. IEEE, Karlsruhe, Germany, pp. 3362–3369.

Wu, Q., Wang, X., Chen, B., Wu, H., 2018. Development of a minimal-intervention-based admittance control strategy for upper extremity rehabilitation exoskeleton. IEEE Transactions on Systems, Man, and Cybernetics: Systems 48, 1005–1016. https://doi.org/10.1109/TSMC.2017.2771227.

Wu, W., DeConinck, A., Lewis, J.A., 2011. Omnidirectional printing of 3D microvascular networks. Advanced Materials 23, H178–H183. https://doi.org/10.1002/adma.201004625.

Yap, H.K., Lim, J.H., Nasrallah, F., Yeow, C.H., 2017. Design and preliminary feasibility study of a soft robotic glove for hand function assistance in stroke survivors. Frontiers in Neuroscience 11, 547. https://doi.org/10.3389/fnins.2017.00547.

Yin, M.J., Yao, M., Gao, S., Zhang, A.P., Tam, H.Y., Wai, P.K.A., 2016. Rapid 3D patterning of poly(acrylic acid) ionic hydrogel for miniature pH. Sensors. Advanced Materials 28, 1394–1399. https://doi.org/10.1002/adma.201504021.

Zhang, J., Fiers, P., Witte, K.A., Jackson, R.W., Poggensee, K.L., Atkeson, C.G., Collins, S.H., 2017. Human-in-the-loop optimization of exoskeleton assistance during walking. Science 356, 1280–1284. https://doi.org/10.1126/science.aal5054.

Zhang, L., Abbott, J.J., Dong, L., Kratochvil, B.E., Bell, D., Nelson, B.J., 2009. Artificial bacterial flagella: fabrication and magnetic control. Applied Physics Letters 94, 064107. https://doi.org/10.1063/1.3079655.

Zhang, Z., Wang, X., Liu, J., Dai, C., Sun, Y., 2019. Robotic micromanipulation: fundamentals and applications. Annual Review of Control, Robotics, and Autonomous Systems 2, 181–203. https://doi.org/10.1146/annurev-control-053018-023755.

Zhao, H., O'Brien, K., Li, S., Shepherd, R.F., 2016. Optoelectronically innervated soft prosthetic hand via stretchable optical waveguides. Science Robotics 1. https://doi.org/10.1126/scirobotics.aai7529.

Zolfagharian, A., Kaynak, A., Kouzani, A., 2020. Closed-loop 4d-printed soft robots. Materials and Design 188, 108411. https://doi.org/10.1016/j.matdes.2019.108411.

Chapter 3

Gait analysis: overview, trends, and challenges☆

3.1 Human gait

Human locomotion is, in simple terms, the displacement of the body from one place to other. Independent and functional locomotion consists of oscillatory movements of the body segments to guarantee stability and coordination (Cappozzo, 1984).

According to Kirtley (2006), gait is defined as "any method of locomotion characterized by periods of loading and unloading of the limbs." This includes, among other modalities, running, hopping, and skipping, but walking is the most frequently used gait. It is the most convenient means of traveling short distances (Perry, 1992) and is also associated with independence and essential for performing several of the activities of daily living and social activities.

Furthermore, gait is the body's natural means of locomotion that involves a change in place, position, and posture relative to some point in the environment using mainly the trunk and the lower limbs (Hamill and Knutzen, 2006). This results from a process that involves the central nervous system, peripheral nerves, muscles, bones, and joints (Whittle, 2014). Other over-ground locomotion modalities include climbing stairs and running.

Although walking is the most habitual and essential activity for daily life and social participation, this activity is a complex dynamic task. Its learning takes place in successive stages; consequently, each person shows particular gait characteristics (Montero-Odasso et al., 2009). The capacity to performance activities of daily living, such as walking, determines a person's functional ability (Senden et al., 2012) and a functional performance of gait patterns reflects directly on the ability to develop independently in the community.

In this sense, it is important to note that about 15% of the world's population live with some disability condition, of which 2%–4% suffer significant functional problems.[1] Different conditions may alter biomechanical characteristics that define a healthy gait pattern. Frequently, these motor impairments significantly impact a person's ability to function independently in daily life.

For instance, stroke is the third leading cause of disability in the world (Johnson et al., 2016). Stroke is a brain dysfunction that occurs due to a disturbance in

☆ This chapter is carried out with the participation of Laura Susana Vargas Valencia.

[1] World report on disability, 2011. https://www.who.int/disabilities/world_report/2011/report/en/.

https://doi.org/10.1016/B978-0-32-385952-3.00011-1

the blood supply flowing to the brain that may cause irreversible damages (Sims and Muyderman, 2010). According to the World Health Organization (WHO), 15 million people experience stroke annually worldwide, and of these, 5 million are permanently incapacitated. Stroke consequences depend on which part of the brain is damaged and how seriously it is affected.[2] Some post-stroke motor-related symptoms includes gait and balance disturbance, coordination problems, functional limb weakness, muscle spasticity, and total or partial inability of one side of the body (Gargano et al., 2011).

Moreover, road traffic crashes, sport accidents, falls, infections, cancer, or tumors can cause complete or incomplete damage to the spinal cord, disturbing the normal autonomic and sensory-motor function. Every year, between 250 and 500 thousand people suffer a spinal cord injury (SCI) worldwide.[3,4] Paralysis, muscle function deficit, weakness, numbness, and sensation loss below the level of the injury are some of the symptoms that these people may experience (Thomas and Zijdewind, 2006).

Furthermore, mobility impairments can often be associated with the elderly population. Until 2050, the projection of people aged over 60 years is expected to grow to 2.1 billion (United Nations and Affairs, 2017). Diseases are a frequent cause for motor disability in people aged between 65 and 84 years old, such as Alzheimer's dementia and Parkinson's disease. Also stroke, joint injury, musculoskeletal deformations, and impairments after an orthopedic surgery frequently result in motor impairments in elderly people (Alexander and Goldberg, 2005). Moreover, age is one of the main risk factors for falls, which is a major public health problem (W.H. Organization, 2008). In fact, an estimation of 646 thousand individuals' death due to falls occurs worldwide each year, of which over 80% are in low- and middle-income countries.

Although bones, connective tissues and joints may begin to deteriorate in population over the age of 60, young individuals are also susceptible to conditions such as osteoarthritis and ligament injuries (Amoako and Pujalte, 2014). Overuse of the joints during very active jobs or high-impact/contact sports, and obesity are also risk factors (Foundation, 2018). Osteoarthritis (OA) is one of the most frequent joint diseases, which ranks fifth between all disability conditions worldwide (Foundation, 2018). Joint cartilage wear-and-tear (with eventual loss), surrounding tissue degeneration and bones rubbing together characterize osteoarthritis. Its symptoms include pain, tenderness, loss of flexibility, and stiffness, which decrease range of motion (Santos et al., 2011; Nakamura et al., 2016). Regarding ligament injuries, anterior cruciate ligament (ACL) injury is the most common trauma to the knee, affecting elderly population and

2 World Health Organization, Cardiovascular Diseases. https://www.who.int/health-topics/cardiovascular-diseases/.

3 World Health Organization, Spinal cord injury. https://www.who.int/en/news-room/fact-sheets/detail/spinal-cord-injury.

4 Spinal cord injury: as many as 500 000 people suffer each year. https://www.who.int/mediacentre/news/releases/2013/spinal-cord-injury-20131202/en/.

especially young athletes. Complete or near complete tearing of the ACL may occur with or without contact, producing deep pain, loss of range of motion with difficulty to straight the affected knee, and instability.

Additionally, motor dysfunction can arise during pregnancy, at birth or after birth up to about age of three. Cerebral palsy (CP) is the most common motor disability in childhood, which affects about 1 in 500 neonates, with an estimated prevalence of 17 million people worldwide (Graham et al., 2016). CP may affect the normal development of infant brain and cause physical disability mainly in the areas motion-related. The most common symptoms of this disorder are involuntary movements, spasticity, balance, problem, and unsteady gait (Rosenbaum et al., 2007).

To assist people with gait impairments, clinicians and researchers employ treatments using physical therapy and often assistive devices are prescribed. Such devices can be passive or active, depending on their interaction with the user. Among them include walking canes, crutches, manual and autonomous wheelchairs, passive and smart walkers, prostheses and orthoses, and more recently, robotic exoskeletons and devices based on soft robotics (Martins et al., 2012).

In this sense, several research groups have been dedicated to develop robotic exoskeletons to accelerate the functional recovery of the gait or to provide new levels of recovery through alternative approaches. As a result, the concept of wearable robots (Pons, 2008) came up, understood as mechatronic devices that act in parallel to the user's limbs, in order to supply energy to them, assisting or replacing their movement. This assistive technology aims to complement the functional motor ability of body segments and restore impaired function (Pons, 2008). This subject was broadly addressed in Chapter 1 along with the more recent advances in assistive wearable soft robots, presented in Chapter 2.

According to Pons in (Pons, 2008), the concept of "wearable" does not certainly mean that the robot is ambulatory or portable, as in many cases, the lack of light and small technologies, especially actuators and energy sources, can be a limitation. In this context, a more sophisticated sensory network is also needed. This network should include wearable motion capture systems to improve the interaction between the user and the exoskeleton.

In addition, when devices such as exoskeletons are used for rehabilitation, continuous monitoring of the device is needed to ensure its correct working. This task has to be performed even in environments where users carry out their daily activities. Therefore, the measurement of kinematic and kinetic parameters is a fundamental part of the development of exoskeletons (Tawakal Hasnain Baluch et al., 2012) and soft robots, since these parameters are used for the study of biomechanics, to assess progress during recovery and performance on functional compensation devices. These parameters are also part of the control strategies of robotic systems (Accoto et al., 2013). Thus the development of sensory systems, embedded or in parallel to exoskeletons and soft wearable devices, to estimate kinematic and dynamic parameters during assistance is of fundamental importance to achieve functional and versatile assistance devices.

3.2 Gait cycle: definitions and phases

A complete description of human gait includes kinematic and kinetic data. Gait could also be understood as a chain of successive events that means a cyclic pattern of movement repeated over time (Vaughan et al., 1999). Therefore the gait cycle is the basic unit to characterize the way of walking, assuming that successive cycles will be reasonably similar if not the same. This cycle can be subdivided in two primary phases (Fig. 3.1). The *stance phase* (approximately 60% of the gait cycle) is defined when the foot is in contact to the ground and the *swing phase* (approximately 40% of the gait cycle), period of time that the foot is not in contact with the ground and the leg is moving forward and preparing for the next contact.

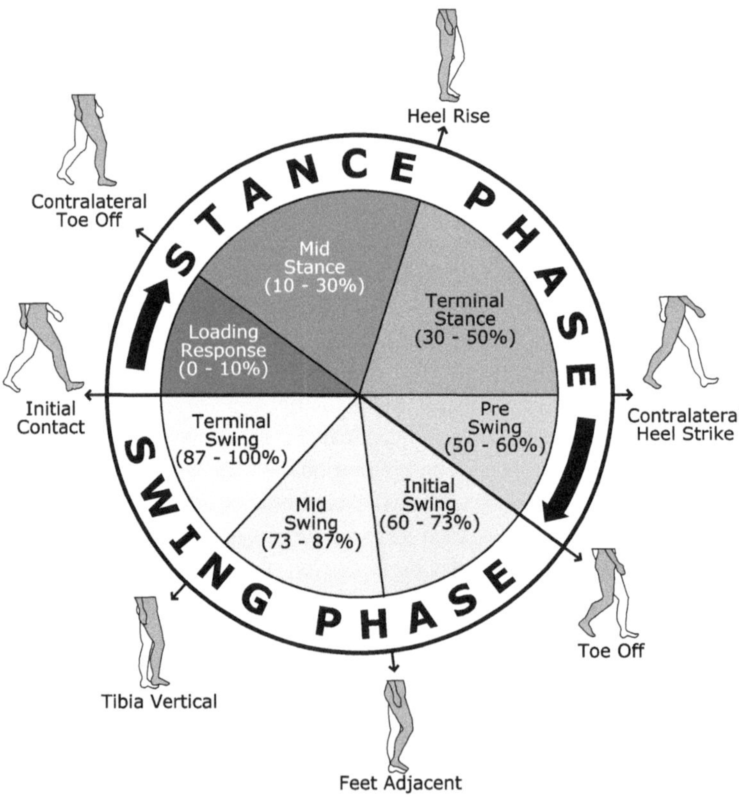

FIGURE 3.1 Schematic representation of the gait cycle and its subdivisions.

Each phase is subdivided in periods or subphases. The stance phase or "support phase" comprises the following subphases (Whittle, 2014; Tao et al., 2012): loading response, mid-stance, terminal stance, and preswing. The swing phase is subdivided into initial swing, mid-swing, and terminal swing.

Moreover, seven events subdivide the gait cycle in subphases (Whittle, 2014):

1. initial contact or heel strike, when the heel contacts with the ground;
2. contralateral toe off, this is toe off on the other foot;
3. heel rise, also called "heel off," when the heel begins to lift from the ground;
4. contralateral heel strike, this is a heel strike on the other foot;
5. toe off, when the foot leaves the floor;
6. foot adjacent, this is the time at which the swinging limb passes the stance limb and
7. tibia vertical, when the tibia of swinging limb corresponds with the vertical axis.

From this general description, the gait study can be usually approached from two perspectives: the study of kinematics and/or kinetics. Kinematics is understood how the study of the motion of bodies without consideration of the causes that produce it. On the other hand, kinetics is the study of the relationship between the movement of bodies and its causes, namely forces and torques.

3.2.1 Kinematics and dynamics of human gait

A variety of parameters could be expressed in terms of percentage of the gait cycle. Kinematic parameters of human gait consist of those related to displacements, velocities, and accelerations, specifically the lower limb joint angles. In addition, spatiotemporal parameters such as gait speed, step length, stride length, stance time, swing time, and cadence are commonly analyzed.

Several spatiotemporal parameters are important descriptors of human gait and are constantly analyzed to assess function and mobility in clinical practice. *Step* and *stride* lengths are linear distances between both feet when are in contact with the floor and between two successive placements of the same foot (two-step lengths), respectively (see Fig. 3.2). The *stance time* is defined as the duration of the time between heel strike and toe off of the same foot. It comprises single support and double support. In the same manner, the *swing time* is measured as the duration of the time between toe off and next heel strike of the same foot. *Cadence* is the number of steps per time unit, usually expressed in steps per minute (steps/min). *Gait average speed* is the distance traveled by the full body in a time period, generally expressed in meters per second (m/s) and is the result of the product of cadence and step length.

When studying gait, angular displacements, velocities, and accelerations of the hip, knee, and ankle joints are ones of the kinematic parameters of most interest. These angular displacements are experimented on three reference planes (Whittle, 2014). The sagittal plane divides symmetrically the body

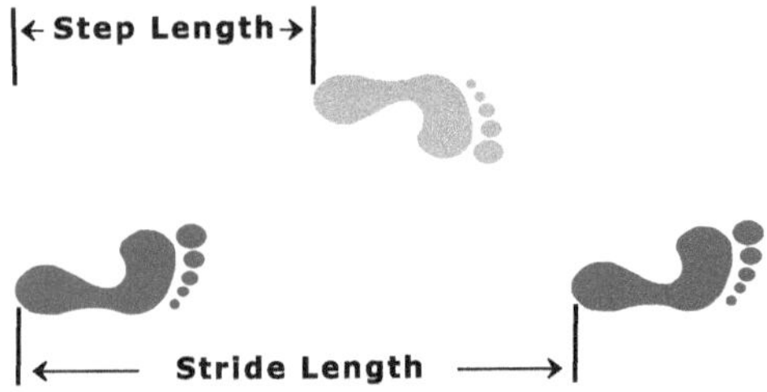

FIGURE 3.2 Schematic representation of step and stride length definitions.

through the vertical, into right and left sides. The frontal plane (or coronal plane) divides the body into front (anterior) and back (posterior) portions. Finally, the transverse plane (or horizontal plane) divides the body into superior (cranial) and inferior (caudal) portions.

Joint movement are described according to the plane where they occur. For instance, in the sagittal plane, hip extension, and knee flexion are movements in the anterior direction and hip flexion and knee extension occur in the posterior direction. Moreover, ankle dorsiflexion (flexion) refers to the movement of the foot in the upward direction when standing, and ankle plantarflexion is the opposite direction.

In order to study the dynamics of human gait, it is imperative to know all internal and external forces and torques acting on the body. This task has some high grade of complexity due to the measurement of kinetics of joints is not carried out directly with current technologies.

Therefore these parameters could be estimated by using the kinematic data along with the position and orientation of the body segments and measuring the ground reaction force (GRF), through force platforms, and the point of application of this force (Winter, 2009). Using anthropometry, the body segment lengths, centers of mass (COM) positions and its mass can be determined. Anthropometric measurements, kinematics, and external forces could be inputs of the link-segment model (Winter, 2009) and, using an inverse solution, the joint reaction forces and muscle moments could be calculated.

In the next section, a summary of kinematic and dynamic measurement systems used for the analysis if human gait are presented.

3.3 Gait analysis systems: fixed systems and wearable sensors

Gait analysis is the application of anatomical and biomechanical principles to understand and characterize systematically the human locomotion (Winter, 2009). Researchers use a variety of techniques to determine kinematic, kinetic and metabolic parameters, muscles mechanics, and electromyography according to the applicability.

In clinical scenarios, gait analysis is traditionally based on subjective approaches mainly through observation. Consequently, the assessment results and treatment decision depends on the expertise of the evaluator. Since this assessment may present significant inter and intraobserver variability, in addition to gait impairment characteristics that cannot be identified by the human eye, systems that allow analyzing gait in a more reliable and repeatable way are needed.

Research laboratories and clinics may include sophisticated gait analysis devices, known as gold standard systems, based on optical processing and floor sensors. Such systems normally consist of fixed instrumentation and work in controlled, structured, and delimited spaces (Muro-De-La-Herran et al., 2014).

The optic sensors used for kinematic analysis of human gait include analog, digital and time-of-flight (ToF) cameras (Manca et al., 2010; Bovi et al., 2011; Derawi et al., 2011; Nguyen and Meunier, 2014), laser range scanners (Clark et al., 2013; Gipsman et al., 2014) and infrared sensors (Xue et al., 2010). The type of technology and the methods applied define whether body markers are required or not (Ceseracciu et al., 2014).

For the purpose of reconstruction and analysis of human body kinematics using multicamera systems, stereophotogrammetry is perhaps the most sophisticated technique. Applying this technique, bone positions and orientations and relative movements between adjacent bones (joint kinematics) can be estimated. Basically, this technique consists of that infrared cameras detect a set of markers placed on the body according to gait analysis protocols. Within those protocols are some variations of the Conventional gait model (Baker et al., 2003; Baker, 2006), indicated as Newington–Gage–Davis model (Davis et al., 1991), the Helen Hayes model (Kadaba et al., 1990) and the VCM (Vicon Clinical Manager) model. Other available protocols include LAMB (Rabuffetti and Crenna, 2004), CAST (Cappozzo et al., 1995) and the Foot model.

Most of the models consider body segment as a rigid body, leaving aside the soft tissue problem. Thus, it can be assumed that all markers have a position and orientation fixed relative to associated bony segment. In such a way, technical and anatomical coordinate systems can be defined using mathematical models in order to estimate joint kinematics.

Within the floor sensors, there are the force/torque platforms and pressure measurement systems, which are equipped with pressure and ground reaction force (GRF) sensors to measure the force applied by the subject while walking (Hunt et al., 2006; Roerdink et al., 2008). Combining such measurements with anthropometric measurements and kinematics, it is possible to obtain a complete dynamic parameters, including joint torques and power estimations, that are necessary for a complete biomechanical analysis of the human gait.

The most common commercial optical motion capture systems are Vicon System (Oxford metrics, UK) and BTS devices (BTS Bioengineering company, Italy). Among others are also Qualisys (Qualisys AB, Sweden), OptiTrack (NaturalPoint Inc., USA), and Clinical 3DMA (STT Systems, Spain). Integrated solutions to clinical gait analysis generally consist of a minimum of 6 high precision infrared cameras (8 cameras is more recommended), and some of them may include modular sensory floor composed by 2 or more force plates, video system and electromyography (EMG) recording.

Although optical instruments are widely used and reliable, these systems present some disadvantages, such as complex set-ups, volume-limited, and high-cost. In addition to these disadvantages, these optical systems perhaps are not suitable to real-time ambulatory applications, making them impractical for telerehabilitation (Kong et al., 2013). Furthermore, with the rising healthcare revolution, in-home monitoring (Onder et al., 2012), telemedicine (W.H. Organization, 2010), and telerehabilitation (Hailey et al., 2011) are already a reality. Thus as traditional 3D camera systems are expensive and need time consuming procedures for public and even private health systems.

Currently, technologies used to analyze lower limb motion are migrating from bulky-fixed, dedicated, and high cost devices to light, compact, wearable, and relatively low-cost sensors. Several research groups currently invest efforts in the development and validation of wearable motion capture systems to provide reliable tools for therapists and clinicians to conduct gait assessment either during diagnosis and/or rehabilitation phases including external environments. These systems are referred to as "wearable" due to their condition of portability and can be placed on the body in a simpler way without interfering with the user's movements.

Wearable systems include electrogoniometers (Kumar et al., 2009; Sato et al., 2010), extensometers, and electromyography (EMG) (Freed et al., 2011). Also, these systems involve pressure and force sensors, such as instrumented shoes (Bae and Tomizuka, 2013) or insoles (De Rossi et al., 2011). Other systems consist of accelerometers (Yang and Hsu, 2010; Yang et al., 2011), gyroscopes, and magnetometers or their combination creating inertial sensors (Luinge and Veltink, 2005; Rodríguez-Martín et al., 2013; Tadano et al., 2013; Alonge et al., 2014). Inertial sensor or IMUs (Inertial Measurement Units) are an in-rising alternative as wearable systems for motion tracking. For gait analysis, each body segment intended to be tracked should have at least one IMU placed on it. Each IMU should provide measurements, usually in three dimensions, of the angular velocity, acceleration, and magnetic field vector in its local sensor frame.

Using inertial sensors, different approaches to estimate body segment orientation and, therefore, estimating joint angles are presented in the literature. By using only gyroscopes, an accumulative error may be generated due to the integration of the angular velocity (Luinge and Veltink, 2005; Tong and Granat, 1999). To increase accuracy, it is suggested to add accelerometers to determine the direction of the local vertical and use fusion algorithms such a Kalman filter. Meanwhile, magnetometers may contribute with stability in the horizontal plane avoiding heading drifts, but magnetic disturbances affect highly their performance (Roetenberg et al., 2007). Additionally, several researches have presented different alternatives (Zimmermann et al., 2018; Narváez et al., 2018; Vargas-Valencia et al., 2016; Palermo et al., 2014; Seel et al., 2014; Cutti et al.,

2010; Favre et al., 2009; Picerno et al., 2008) to reduce the misalignment errors, which refer to the alignment between the IMU coordinate system (or axes) and the anatomically-defined coordinate system (or joint motion axes). These approaches aim to provide more reliable and repeatable procedures when using IMUs for gait analysis.

Alternatively, the use of soft and flexible wearable devices, as previously discussed in Chapter 2, are emerging as interesting approaches for a more complete and continuous analysis of human gait in a more diverse set of scenarios. In this senses, polymer optical fiber sensors present some advantages that make them attractive to develop wearable systems. They are lightweight, flexible, compact, and, unlike some microelectromechanical systems, immune to an electromagnetic field. For instance, POF-based systems have demonstrated its adaptability and accuracy to measure joint angle in exoskeleton applications (Leal-Junior et al., 2018b). In addition, this technology presents multiplexing features, which allow to measure different parameters with a single sensor (Peters, 2010), for example angle and temperature (Leal-Junior et al., 2018a). The aforementioned advantages make optical fiber sensors an interesting option to be embedded in textiles for sensor applications (Krehel et al., 2014), enabling the fabrication of monitoring systems with better appearance and nonintrusive for users movements. Furthermore, these promising technologies, when are made of malleable materials, such polymeric optical fibers, are successfully applied in soft robotics (Li et al., 2019).

References

Accoto, D., Carpino, G., Sergi, F., Tagliamonte, N.L., Zollo, L., Guglielmelli, E., 2013. Design and characterization of a novel high-power series elastic actuator for a lower limb robotic orthosis. International Journal of Advanced Robotic Systems 10, 359.

Alexander, N.B., Goldberg, A., 2005. Gait disorders: search for multiple causes. Cleveland Clinic Journal of Medicine 72, 586.

Alonge, F., Cucco, E., D'Ippolito, F., Pulizzotto, A., 2014. The use of accelerometers and gyroscopes to estimate hip and knee angles on gait analysis. Sensors 14, 8430–8446.

Amoako, A.O., Pujalte, G.G.A., 2014. Osteoarthritis in young, active, and athletic individuals. Clinical Medicine Insights: Arthritis and Musculoskeletal Disorders 7. CMAMD–S14386.

Bae, J., Tomizuka, M., 2013. A tele-monitoring system for gait rehabilitation with an inertial measurement unit and a shoe-type ground reaction force sensor. Mechatronics 23, 646–651.

Baker, R., 2006. Gait analysis methods in rehabilitation. Journal of NeuroEngineering and Rehabilitation 3, 4.

Baker, R., Rodda, J., et al., 2003. All you ever wanted to know about the conventional gait model but were afraid to ask. Melbourne: Women and Children's Health.

Bovi, G., Rabuffetti, M., Mazzoleni, P., Ferrarin, M., 2011. A multiple-task gait analysis approach: kinematic, kinetic and emg reference data for healthy young and adult subjects. Gait & Posture 33, 6–13.

Cappozzo, A., 1984. Gait analysis methodology. Human Movement Science 3, 27–50.

Cappozzo, A., Catani, F., Della Croce, U., Leardini, A., 1995. Position and orientation in space of bones during movement: anatomical frame definition and determination. Clinical Biomechanics 10, 171–178.

Ceseracciu, E., Sawacha, Z., Cobelli, C., 2014. Comparison of markerless and marker-based motion capture technologies through simultaneous data collection during gait: proof of concept. PLoS ONE 9, e87640.

Clark, R.A., Pua, Y.H., Bryant, A.L., Hunt, M.A., 2013. Validity of the Microsoft kinect for providing lateral trunk lean feedback during gait retraining. Gait & Posture 38, 1064–1066.

Cutti, A.G., Ferrari, A., Garofalo, P., Raggi, M., Cappello, A., Ferrari, A., 2010. 'outwalk': a protocol for clinical gait analysis based on inertial and magnetic sensors. Medical & Biological Engineering & Computing 48, 17.

Davis III, R.B., Ounpuu, S., Tyburski, D., Gage, J.R., 1991. A gait analysis data collection and reduction technique. Human Movement Science 10, 575–587.

De Rossi, S.M.M., Vitiello, N., Lenzi, T., Ronsse, R., Koopman, B., Persichetti, A., Vecchi, F., Ijspeert, A.J., Van der Kooij, H., Carrozza, M.C., 2011. Sensing pressure distribution on a lower-limb exoskeleton physical human-machine interface. Sensors 11, 207–227.

Derawi, M.O., Ali, H., Cheikh, F.A., 2011. Gait recognition using time-of-flight sensor. In: BIOSIG 2011–Proceedings of the Biometrics Special Interest Group.

Favre, J., Aissaoui, R., Jolles, B.M., de Guise, J.A., Aminian, K., 2009. Functional calibration procedure for 3d knee joint angle description using inertial sensors. Journal of Biomechanics 42, 2330–2335.

Foundation, A., 2018. Arthritis by the numbers – book of trusted facts & figures. https://www.arthritis.org/Documents/Sections/About-Arthritis/arthritis-facts-stats-figures.pdf.

Freed, A., Chan, A.D., Lemaire, E.D., Parush, A., 2011. Wearable emg analysis for rehabilitation (wear)-surface electromyography in clinical gait analysis. In: 2011 IEEE International Symposium on Medical Measurements and Applications. IEEE, pp. 601–604.

Gargano, J.W., Wehner, S., Reeves, M.J., 2011. Presenting symptoms and onset-to-arrival time in patients with acute stroke and transient ischemic attack. Journal of Stroke and Cerebrovascular Diseases 20, 494–502.

Gipsman, A., Rauschert, L., Daneshvar, M., Knott, P., 2014. Evaluating the reproducibility of motion analysis scanning of the spine during walking. Advances in Medicine 2014.

Graham, K., Rosenbaum, P., Paneth, N., Dan, B., Lin, J.P., Damiano, L.D., Becher, J., Gaebler, D., Colver, A., Reddihough, D., Crompton, K., Lieber, R., 2016. Cerebral palsy. Nature Reviews Disease Primers 2, 15082. https://doi.org/10.1038/nrdp.2015.82.

Hailey, D., Roine, R., Ohinmaa, A., Dennett, L., 2011. Evidence of benefit from telerehabilitation in routine care: a systematic review. Journal of Telemedicine and Telecare 17, 281–287.

Hamill, J., Knutzen, K.M., 2006. Biomechanical Basis of Human Movement. Lippincott Williams & Wilkins.

Hunt, M.A., Birmingham, T.B., Giffin, J.R., Jenkyn, T.R., 2006. Associations among knee adduction moment, frontal plane ground reaction force, and lever arm during walking in patients with knee osteoarthritis. Journal of Biomechanics 39, 2213–2220.

Johnson, W., Onuma, O., Owolabi, M., Sachdev, S., 2016. Stroke: a global response is needed. Bulletin of the World Health Organization 94, 634.

Kadaba, M.P., Ramakrishnan, H., Wootten, M., 1990. Measurement of lower extremity kinematics during level walking. Journal of Orthopaedic Research 8, 383–392.

Kirtley, C., 2006. Clinical Gait Analysis. Elsevier. https://linkinghub.elsevier.com/retrieve/pii/B9780443100093X50012.

Kong, W., Sessa, S., Cosentino, S., Zecca, M., Saito, K., Wang, C., Imtiaz, U., Lin, Z., Bartolomeo, L., Ishii, H., et al., 2013. Development of a real-time imu-based motion capture system for gait rehabilitation. In: 2013 IEEE International Conference on Robotics and Biomimetics (ROBIO). IEEE, pp. 2100–2105.

Krehel, M., Wolf, M., Boesel, L.F., Rossi, R.M., Bona, G.L., Scherer, L.J., 2014. Development of a luminous textile for reflective pulse oximetry measurements. Biomedical Optics Express 5, 2537–2547.

Kumar, N., Pankaj, D., Mahajan, A., Kumar, A., Sohi, B., 2009. Evaluation of normal gait using electro-goniometer. Journal of Scientific & Industrial Research 68, 696–698.

Leal-Junior, A., Frizera, A., Marques, C., Pontes, M.J., 2018a. Polymer-optical-fiber-based sensor system for simultaneous measurement of angle and temperature. Applied Optics 57, 1717–1723.
Leal-Junior, A.G., Frizera, A., Vargas-Valencia, L., dos Santos, W.M., Bó, A.P., Siqueira, A.A., Pontes, M.J., 2018b. Polymer optical fiber sensors in wearable devices: toward novel instrumentation approaches for gait assistance devices. IEEE Sensors Journal 18, 7085–7092.
Li, H., Li, H., Lou, X., Meng, F., Zhu, L., 2019. Soft optical fiber curvature sensor for finger joint angle proprioception. Optik 179, 298–304.
Luinge, H.J., Veltink, P.H., 2005. Measuring orientation of human body segments using miniature gyroscopes and accelerometers. Medical & Biological Engineering & Computing 43, 273–282.
Manca, M., Leardini, A., Cavazza, S., Ferraresi, G., Marchi, P., Zanaga, E., Benedetti, M.G., 2010. Repeatability of a new protocol for gait analysis in adult subjects. Gait & Posture 32, 282–284.
Martins, M.M., Santos, C.P., Frizera-Neto, A., Ceres, R., 2012. Assistive mobility devices focusing on smart walkers: classification and review. Robotics and Autonomous Systems 60, 548–562.
Montero-Odasso, M., Casas, A., Hansen, K.T., Bilski, P., Gutmanis, I., Wells, J.L., Borrie, M.J., 2009. Quantitative gait analysis under dual-task in older people with mild cognitive impairment: a reliability study. Journal of NeuroEngineering and Rehabilitation 6, 35.
Muro-De-La-Herran, A., Garcia-Zapirain, B., Mendez-Zorrilla, A., 2014. Gait analysis methods: an overview of wearable and non-wearable systems, highlighting clinical applications. Sensors 14, 3362–3394.
Nakamura, Y., Uchiyama, S., Kamimura, M., Komatsu, M., Ikegami, S., Kato, H., 2016. Bone alterations are associated with ankle osteoarthritis joint pain. Scientific Reports 6, 18717.
Narváez, F., Árbito, F., Proaño, R., 2018. A quaternion-based method to imu-to-body alignment for gait analysis. In: International Conference on Digital Human Modeling and Applications in Health, Safety, Ergonomics and Risk Management. Springer, pp. 217–231.
Nguyen, H.A., Meunier, J., 2014. Gait analysis from video: camcorders vs. kinect. In: International Conference Image Analysis and Recognition. Springer, pp. 66–73.
Onder, G., Carpenter, I., Finne-Soveri, H., Gindin, J., Frijters, D., Henrard, J.C., Nikolaus, T., Topinkova, E., Tosato, M., Liperoti, R., et al., 2012. Assessment of nursing home residents in Europe: the services and health for elderly in long term care (shelter) study. BMC Health Services Research 12, 5.
Organization, W.H., Ageing, W.H.O., Unit, L.C., 2008. WHO global report on falls prevention in older age. World Health Organization.
Organization, W.H., et al., 2010. Telemedicine: opportunities and developments in member states. Report on the second global survey on eHealth. World Health Organization.
Palermo, E., Rossi, S., Marini, F., Patanè, F., Cappa, P., 2014. Experimental evaluation of accuracy and repeatability of a novel body-to-sensor calibration procedure for inertial sensor-based gait analysis. Measurement 52, 145–155.
Perry, J., 1992. Gait Analysis: Normal and Pathological Function. Slack Incorporated.
Peters, K., 2010. Polymer optical fiber sensors—a review. Smart Materials and Structures 20, 013002.
Picerno, P., Cereatti, A., Cappozzo, A., 2008. Joint kinematics estimate using wearable inertial and magnetic sensing modules. Gait & Posture 28, 588–595.
Pons, J.L., 2008. Wearable Robots: Biomechatronic Exoskeletons. John Wiley & Sons.
Rabuffetti, M., Crenna, P., 2004. A modular protocol for the analysis of movement in children. Gait & Posture 20, S77–S78.
Rodríguez-Martín, D., Pérez-López, C., Samà, A., Cabestany, J., Català, A., 2013. A wearable inertial measurement unit for long-term monitoring in the dependency care area. Sensors 13, 14079–14104.
Roerdink, M., Lamoth, C.J., Beek, P.J., et al., 2008. Online gait event detection using a large force platform embedded in a treadmill. Journal of Biomechanics 41, 2628–2632.
Roetenberg, D., Baten, C.T., Veltink, P.H., 2007. Estimating body segment orientation by applying inertial and magnetic sensing near ferromagnetic materials. IEEE Transactions on Neural Systems and Rehabilitation Engineering 15, 469–471.

Rosenbaum, P., Paneth, N., Leviton, A., Goldstein, M., Bax, M., Damiano, D., Dan, B., Jacobsson, B., 2007. A report: the definition and classification of cerebral palsy April 2006. Developmental Medicine and Child Neurology. Supplement 109, 8–14.

Santos, F.C., Souza, P.M.R.d., Toniolo Neto, J., Atallah, Á.N., 2011. Treatment of pain associated to knee osteoarthritis in the elderly: a randomized double-blind clinical trial with lysine clonixinate. Revista Dor 12, 6–14.

Sato, T.d.O., Hansson, G.Å, Coury, H.J.C.G., 2010. Goniometer crosstalk compensation for knee joint applications. Sensors 10, 9994–10005.

Seel, T., Raisch, J., Schauer, T., 2014. Imu-based joint angle measurement for gait analysis. Sensors 14, 6891–6909.

Senden, R., Savelberg, H., Grimm, B., Heyligers, I., Meijer, K., 2012. Accelerometry-based gait analysis, an additional objective approach to screen subjects at risk for falling. Gait & Posture 36, 296–300.

Sims, N.R., Muyderman, H., 2010. Mitochondria, oxidative metabolism and cell death in stroke. Biochimica et Biophysica Acta. Molecular Basis of Disease 1802, 80–91.

Tadano, S., Takeda, R., Miyagawa, H., 2013. Three dimensional gait analysis using wearable acceleration and gyro sensors based on quaternion calculations. Sensors 13, 9321–9343.

Tao, W., Liu, T., Zheng, R., Feng, H., 2012. Gait analysis using wearable sensors. Sensors 12, 2255–2283.

Tawakal Hasnain Baluch, A.M., Iqbal, J., Izhar, U., Khan, U.S., 2012. Kinematic and dynamic analysis of a lower limb exoskeleton. World Academy of Science, Engineering and Technology 6, 812–816.

Thomas, C.K., Zijdewind, I., 2006. Fatigue of muscles weakened by death of motoneurons. Muscle & Nerve: Official Journal of the American Association of Electrodiagnostic Medicine 33, 21–41.

Tong, K., Granat, M.H., 1999. A practical gait analysis system using gyroscopes. Medical Engineering & Physics 21, 87–94.

United Nations, D.o.E., Affairs, S., 2017. World population ageing 2017: highlights.

Vargas-Valencia, L.S., Elias, A., Rocon, E., Bastos-Filho, T., Frizera, A., 2016. An imu-to-body alignment method applied to human gait analysis. Sensors 16, 2090.

Vaughan, C.L., Davis, B.L., O'Connor, J.C., 1999. Dynamics of Human Gait, vol. 731, 2 ed. Kiboho.

Whittle, M.W., 2014. Gait Analysis: An Introduction. Butterworth-Heinemann.

Winter, D.A., 2009. Biomechanics and Motor Control of Human Movement. John Wiley & Sons.

Xue, Z., Ming, D., Song, W., Wan, B., Jin, S., 2010. Infrared gait recognition based on wavelet transform and support vector machine. Pattern Recognition 43, 2904–2910.

Yang, C.C., Hsu, Y.L., 2010. A review of accelerometry-based wearable motion detectors for physical activity monitoring. Sensors 10, 7772–7788.

Yang, C.C., Hsu, Y.L., Shih, K.S., Lu, J.M., 2011. Real-time gait cycle parameter recognition using a wearable accelerometry system. Sensors 11, 7314–7326.

Zimmermann, T., Taetz, B., Bleser, G., 2018. Imu-to-segment assignment and orientation alignment for the lower body using deep learning. Sensors 18, 302.

Part II

Introduction to optical fiber sensing

Chapter 4

Optical fiber fundaments and overview

4.1 Historical perspective

One of the first practical developments of light in telecommunications dated back in the 18th century when there was the demonstration of a semaphore from Claude Chappe (Agrawal, 2016). However, this device needs a visible line of sight between emitter and receiver to perform the communication, which is an important drawback that hinders the development of such technology. Then the investigation of reflection and refraction in different materials were performed, where the total internal reflection phenomenon was demonstrated in a plexiglass material (Tricker, 2002). Due to the total internal reflection, the light is confined in a transparent material surrounded by a lower refractive index medium (Tricker, 2002). This effect was employed in the beginning of the 20th century for microscope illumination using quartz rods under bending (Ziemann et al., 2008).

These first developments that form the foundation for current optical fiber technology have the drawback of high optical losses. In order to tackle this limitation, a material layer surrounding the region at which the light is confined was created, the so-called cladding layer. In this development, the concept of optical fiber core and cladding was created, where the cladding has lower refractive index when compared to the optical fiber core (Large et al., 2008). In this case, the fiber bundles fabricated have lower crosstalk between fibers, since each fiber is protected by the cladding layer. In addition, this approach resulted in optical fibers with optical attenuation of around 1 dB/m, which makes them suitable for medical imaging (and other applications with signal transmission to short lengths) (Agrawal, 2016). However, this optical loss is too high for communication purposes. Aiming at further reducing the optical attenuation in the first optical fibers, Charles Kao investigated the light transmission properties in different materials and the performed investigation indicated that the optical losses are closely related to materials' impurities (Tricker, 2002). The impurities reduction resulted in optical losses some orders of magnitude smaller than the ones previously proposed, 20 dB/km. The optical fiber developments are also closely related to the laser invention, where it was possible to use a light source with narrow wavelength that allows the use in different wavelengths in room temperatures, which enable the widespread of optical fiber technology.

Optical Fiber Sensors for the Next Generation of Rehabilitation Robotics
https://doi.org/10.1016/B978-0-32-385952-3.00013-5

Continuous improvements in optical fiber materials and fabrication methods with reduction of impurities, thermal gradients on the fabrication and transparent materials development resulted in the single mode fibers (SMFs) with optical attenuation as low as 0.15 dB/km considering the 1550 nm wavelength window, commonly used in telecommunication and sensing applications (Ziemann et al., 2008). As such low optical attenuation was achieved with silica glass fibers, concurrent developments in polymer optical materials have been proposed to obtain systems with low optical attenuation. Although the demonstrations of total internal reflection were made in a polymer waveguide, polymer optical fibers (POFs) suffer from a high optical attenuation, which hinder their development for many years. The first developments in POFs dated back in 1950 with cladding made of liquid beeswax, which was then substituted by cured polymer resins (Ziemann et al., 2008).

The investigation of light attenuation mechanisms in polymers have been conducted to enable practical application of POFs, motivated by its mechanical features, such as high flexibility (even with high core diameters), impact toughness and nonbrittle nature (in contrast with silica optical fibers) (Ziemann et al., 2008). Polymer materials are intrinsically less efficient for light transmission than glass waveguides due to higher absorption losses, which includes the C-H bonds vibration that becomes significant in infrared wavelength region (Koike and Asai, 2009). As one approach to mitigate this issue, the hydrogen is substituted by fluorine bonds, which resulted in lower transmission losses as well as the possibility of using the fibers in the infrared region, where most of optical components are developed, since it is the region with lowest optical losses in silica fibers and also motivated by the development of Erbium-doped fiber amplifiers (EDFAs). However, practical limitations in the fluorinated POF development are related to the difficulty and expensive nature of the process (Makino et al., 2012). Despite this limitations, the fluorinated POFs are widely commercially available and can be used for short communications with optical fiber cables with lengths in the order of hundreds meters (Ziemann et al., 2008). Nevertheless, poly methyl methacrylate (PMMA) is the most commonly used material for POFs and is used in the visible wavelength range (especially on 650 nm), where their advantages of material features and lower temperatures on optical fiber fabrication have enabled the widespread of POFs in illumination, short range communications and sensing; the latter will be explored in the next chapters.

The timeline of optical fiber development includes the first report on total internal reflection as well as modern single mode optical fibers and graded-index fibers, made of silica and POF. The timeline indicates the decades of advances in optical fiber technology that indicates continuous advances, where modern systems also include optical fiber doping with optically active materials, such as Erbium (Liu et al., 2010), and nanoparticles of different types (Veber et al., 2019) to provide additional applications and functionalities for the optical waveguides. The light propagation principles in waveguides are depicted in

Section 4.2, whereas the different types and geometries of optical fibers as well as their properties are depicted in Section 4.3. It is important to note that the optical fiber developments are closely related to the developments of active and passive optical components (such as lasers, attenuators, filter, and optical couplers), which are described in Section 4.4. The fiber fabrication methods are described and compared in Section 4.5.

4.2 Light propagation in optical waveguides

The properties of light transmission, which include interferences, polarization, refraction and transmission modes are explained in wave model, analyzed in different wavelength ranges, short infrared for silica optical fibers and visible for POFs. Fig. 4.1 shows the electromagnetic spectrum regions used in optical fibers applications. Although it is not represented in Fig. 4.1, mid-infrared can be also used for medical images as gas sensing applications. However, the most commonly used region is on the short-infrared, specifically in C-band, comprising of the region between 1530 nm and 1565 nm, region at which the erbium-doped fiber amplifiers (EDFAs) operate. As POFs have an even higher optical attenuation in the short-infrared region, mostly of POFs (exception to perfluorinated POFs) operate at near-infrared, specifically on the 850 nm band or in the visible wavelength region. The ultraviolet (UV) range is employed by pulsed or continuous wave (CW) lasers for micromachining, microstructures, or gratings inscription in optical fibers.

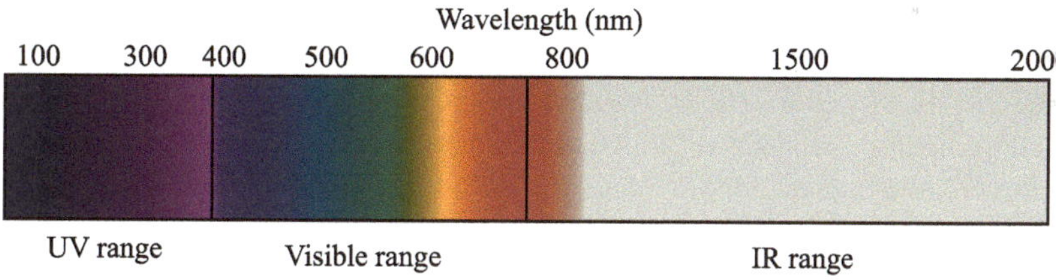

FIGURE 4.1 Electromagnetic spectrum regions for optical fiber applications.

The light propagates with total internal reflection in an optical fiber (or optical waveguide) and it follows the well-known Snell law presented in Eq. (4.1):

$$\frac{\sin(\theta_i)}{\sin(\theta_t)} = \frac{n_2}{n_1}, \tag{4.1}$$

where n_2 is the refractive index of the more refractive medium, whereas n_1 is the refractive index of the less refractive medium. Furthermore, θ_i is the incidence angle and θ_t is the transmitted angle. The Snell law can be applied to estimate the critical angle of the light propagation inside the fiber, which is the incident angle when the transmitted angle is 90°.

Moreover, in an optical waveguide, only the light rays with angles lower than the acceptance angle are propagated inside the fiber. The acceptance angle (θ_{max}) is a function of the core refractive index (n_{co}), external refractive index (or cladding refractive index) (n_{ext}) and the critical angle (θ_c) as shown in Eq. (4.2):

$$\frac{sin(\theta_{max})}{sin(\theta_c)} = \frac{n_{co}}{n_{ext}}. \tag{4.2}$$

Following the Snell law and the requirements for total internal reflection, the optical comprises of transparent materials for core and cladding layers, where the core material has higher refractive index than the cladding material. In this case, in angles within the acceptance angle ranges (see Eq. (4.2)), there is a reflection on the interface between core and cladding. If the light rays reach the core/cladding interface with angles higher than the acceptance angle, there is no total reflection and part of the signal is refracted to the cladding region (Ziemann et al., 2008). It is also worth to mention that the propagation properties of waveguides are directly related to its shape. Although there is a widespread of planar waveguides, the analysis is focused on cylindrical waveguides, commonly used on optical fibers.

As an important parameter on light propagation analysis, a mode is defined as an electromagnetic field pattern that propagates in a waveguide (Large et al., 2008), regardless to its intensity distribution, since optical losses lead to reduction in the magnitude, without changing the distribution pattern. The propagation modes have their own phase velocity, related to the speed of light in vacuum and effective refractive index, where the latter is obtained from the propagation constant of each mode, which is related to the wavelength and angular frequency (Large et al., 2008). Another important characteristic of propagation modes is their classification, divided into radiation, bound, evanescent and leaky modes (Ziemann et al., 2008). As modes with higher significance for the transmission analysis in solid core optical fibers, bound and radiation modes have real numbers for the propagation constant. Furthermore, bound and radiation modes are localized and delocalized, respectively, and propagate indefinitely without loss (Large et al., 2008). In microstructured optical fibers (MOFs), as discussed in Section 4.3, leaky modes are significant on light propagation in MOFs; they have complex propagation constants, where their propagation show an exponential decay as the distance (or propagation length) increases. In contrast evanescent modes have only the imaginary part of the propagation constant and show low propagation with exponential decay in short distance (in a range of a few wavelengths). The evanescent modes do not play an important role in communications applications, but are important for some sensing approaches, especially on biosensing applications (Mohammed et al., 2019).

Additionally, the acceptance angle is important to determine the optical fiber numerical aperture (NA), which is the sine of the acceptance angle and it represents the amount of light that an optical system can receive through the acceptance cone (Ziemann et al., 2008). POFs have higher NA than silica fibers. POFs have higher signal attenuation and higher modal dispersion, which leads to smaller bandwidth, since there is a time difference between the light propagation with angles close to 0° and propagation with angles close to the critical angle (Bilro et al., 2012).

Optical fibers are also classified in single mode and multimode waveguides. This classification relates the number of propagation modes inside the fiber. An optical fiber is multimode if its normalized frequency is higher than 2.405; otherwise, it is single mode (Ziemann et al., 2008). Eq. (4.3) presents the normalized frequency (V) as a function of the core radius (a), wavelength (λ), and numerical aperture (NA):

$$V = \frac{2\pi a}{\lambda} NA. \tag{4.3}$$

The number of modes (N) on a step index fiber is presented in Eq. (4.4). For graded-index fiber, the number of modes is smaller, as the number of modes is divided by a higher factor as shown in Eq. (4.5):

$$N_m = \frac{V^2}{2}, \tag{4.4}$$

$$N_m = \frac{V^2}{4}. \tag{4.5}$$

For example, typical value of the number of modes for wavelength of 650 nm is 2.8 million in a conventional step index POF with PMMA core with 0.98 mm of diameter (Ziemann et al., 2008). Each mode represents one particular solution of the Maxwell equation. Since it has so many modes and POF sensor generally do not have long lengths of fiber, the geometrical optics approach can be employed on the sensor analysis (Bilro et al., 2012). It is also worth noting that the number of modes is related to the normalized frequency and, as shown in Eq. (4.3), this parameter depends on the wavelength. Thus, the optical fibers can be single mode only for a wavelength region, where the wavelength in the interface between single and multimode operation is known as the cutoff-wavelength, as shown in Fig. 4.2 for a silica optical fiber with 125 μm of core diameter and NA of 0.14. Considering Eq. (4.3), it is also possible to infer that the optical fiber dimensions also has direct influence on the number of modes, since higher fiber core radius results in higher values of normalized frequency. Similarly, the numerical aperture also plays an important role on the number of modes estimation. As conventionally employed PMMA POFs have 1 mm core diameter and NA of around 0.45, such fibers present large multimode operation. In addition, numerical aperture can be calculated from the squared differences between core and cladding refractive indices. For this reason, the refractive index difference between core and cladding should be as small as possible to achieve a single mode operation.

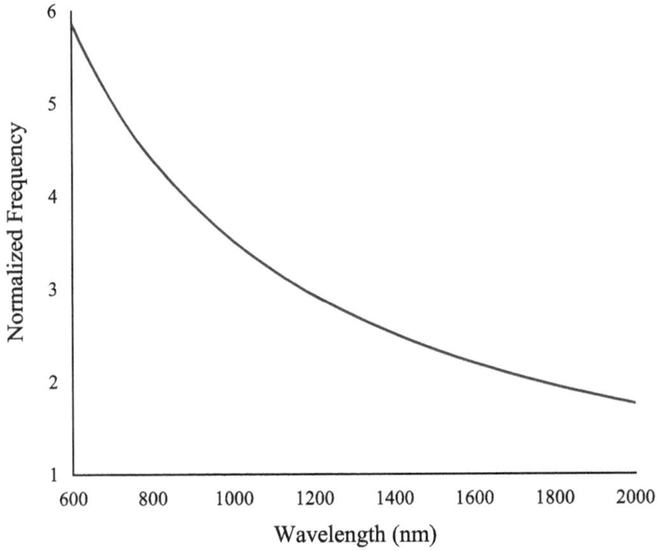

FIGURE 4.2 Analytical analysis of the number of modes as a function of the parameters involved on the normalized frequency estimation.

4.3 Optical fiber properties and types

The light propagation properties and fundaments previously discussed are directly applied on optical fiber technology. In a single mode optical fiber, there is two propagation states, related to two orthogonal polarization directions that propagate with the same velocity if the fiber is symmetrical and with homogeneous material (Ziemann et al., 2008). Perturbations on the symmetry and homogeneity conditions lead to different effective indices for the propagation states, leading to different polarization states, which is a basic premise of birefringent optical fibers (Large et al., 2008). Thus, the birefringence leads to different propagation velocities for the modes, a fast and a slow axis propagation. It is important to mention that in practical or experimental applications even the conventional single mode optical fibers do not have perfect symmetry or material homogeneity, which leads to minor birefringence conditions, resulting in polarization mode dispersion for applications that involve long lengths of optical fibers (tens of kilometers) (Agrawal, 2016). In optical fibers, there are small differences between the cladding and core (to obtain a single or few mode operation as discussed in Section 4.3). This feature enables the approximation of weak guidance that results in a simplification on the mathematical modeling and optical fiber experimental analysis, as the analysis can be reduced to a single scalar wave equation (Large et al., 2008).

In multimode fibers, the propagation conditions of the different modes are considered, since these conditions can influence the fiber propagation parameters. In this case, there is the light propagation in different modes, where some of these modes are not launched with a uniform power distribution and these modes can exchange energies between them, resulting in a harder determination and modeling of the light propagation in such fibers (Large et al., 2008). For the analysis of the propagation in such optical fibers, the uniform and equilibrium mode distribution are used. In addition, an important parameter for all optical fibers is their attenuation, which is conventionally measured using the cut-back technique or other techniques summarized and discussed in Ziemann et al. (2008). In the optical power and attenuation analysis, there is an exponential decay of the input optical power as it travels in a predefined length of the optical fiber, where the exponential decay depends on the fiber length and the attenuation coefficient related to each optical fiber. It is also worth to mention that the multimode optical fibers have additional parameters for the attenuation analysis. In conventional multimode POFs or high-NA optical fibers, there are path differences in propagation directions, which results in higher attenuation on modes with higher propagation angles. In addition, the mode-dependent attenuation can also be related to the attenuation in the core-cladding interface, since the propagation modes that reflect after longer lengths of propagation in the cladding can present higher attenuation due to the cladding material attenuation coefficient (Ziemann et al., 2008). Furthermore, the mode coupling can occur in multimode optical fibers in which there is an energy transference from one propagation direction to other ones. This effect is related to nonuniformities that can create scattering regions in the optical fiber. The mode coupling also can occur in the core-cladding interface and depend on the angle of propagation.

The high attenuation is the main drawback of the PMMA POFs compared with silica fibers, as mentioned earlier in this document, since higher numerical aperture leads to a higher attenuation. There are two mechanisms for the light attenuation in an optical fiber: extrinsic and intrinsic processes. Intrinsic losses are inherent of the fiber composition. The two major intrinsic losses are radiation absorption due to electronic transitions on the atomic bonds and Rayleigh scattering, which is determined by fluctuations on the material composition, density, and orientation, which happen due to the material anisotropy (Bilro et al., 2012). Extrinsic losses are caused by factors that can be controlled (but not always eliminated) in the manufacture process, installation, or by environmental control. The most common extrinsic losses are caused by impurities absorption, microfractures, structural imperfections, and radiation losses due micro and macrobends (Ziemann et al., 2008).

As another important parameter in optical fiber analysis, especially for long range communications, the dispersion is related to differences in travel times of the different modes within an optical fiber. As each mode has its own propagation condition (related to the wavelength, polarization, and propagation path), the differences between each mode can result in differential delays on the light propagation. Due to this behavior, there is a reduction of the modulation amplitude at higher frequencies, which leads a low-pass filter behavior of optical fibers. Thus the bandwidth of an optical system is the frequency at which the amplitude of the period optical signal is half of its input amplitude (Ziemann et al., 2008). Fig. 4.3 shows the representation of an optical pulse before and after traveling the optical fiber (input and output), where it can be seen a broadening in the signal due to the dispersion, which also leads to an amplitude reduction (in addition to the amplitude reduction due to optical attenuation in the optical fiber).

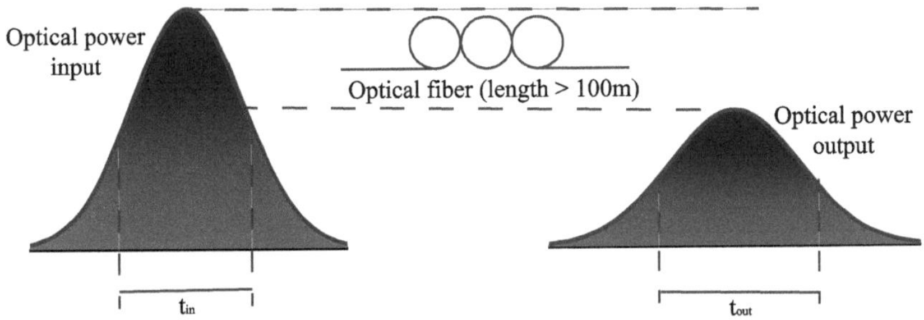

FIGURE 4.3 Optical signal with Gaussian shape input and output in an optical fiber.

It is also worth noting that in multimode fibers, the multiple propagation paths lead to another dispersion source, the mode dispersion, which is related to the differences in the propagation paths that also result in slightly different propagation times. As the pulses arrive at the fiber end at different times, there is a pulse broadening as well as amplitude reduction. Similarly, there is also the chromatic dispersion, which is related to the light rays that penetrate the cladding region when there is no total internal reflection, leading to different propagation velocities for these rays that, in consequence, result in pulse broadening. As it is directly related to the material dispersion properties, the longer wavelengths generally propagate with a higher velocity, which also indicates the pulse broadening (shape variation) as a function of the wavelength (Agrawal, 2016).

Solid core optical fibers have a circular cross-section with core and cladding layers. In addition, commercial solid core fibers also have a jacket (or overcladding). Most of the optical signal is propagated through the core section, which has a higher refractive index than the cladding. Light intensity losses happen due to absorption and frustrated total internal reflection when in contact with a surface (Bilro et al., 2012). For this reason, the core is surrounded by a cladding layer with a refractive index lower than the one found at core. The jacket provides mechanical protection for the POF and it is usually made of polyethylene (Zubia and Arrue, 2001). Fig. 4.4 shows the cross sectional area of a solid core fiber with a single core.

The core has two major types of refractive index profiles: step index profile (SI) and graded index profile (GI). In a step index profile, the core is homogeneous and presents a uniform refraction index. On the other hand, graded index

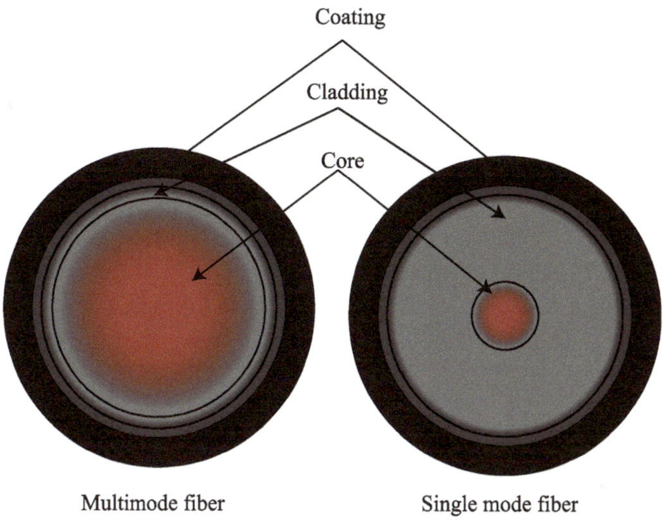

FIGURE 4.4 Cross-sectional view of optical fibers (figure out of scale).

profile fibers have a distribution gradient of the core refractive index, which increases with the distance between the fiber axis and its extremity. Moreover, due to this refractive index variation in a GI profile, the ray propagation does not occur in a straight line (Bilro et al., 2012), leading to lower attenuation of perfluorinated POFs with GI profiles than the one of the SI. However, the fabrication of such GI profile is more costly. There is also a multistep profile created to present an alternative to the GI profile, which is difficult to manufacture. This fiber presents a multilayer profile and every layer has its own refraction index, resulting in a nonstraight ray propagation (Ziemann et al., 2008).

In addition to the conventional single core optical fibers, there is also the multicore fibers, where the cladding layer incorporates multiple cores, each one supporting a predefined number of modes (depending on the fiber dimensions, refractive indices and wavelength). In this case, the multicore optical fiber can extend the bandwidth capabilities of the waveguide, considering the radius of each core as well as their distances between them (Ziemann et al., 2008). Multicore optical fiber have been recently employed in many sensors applications, where it is possible to inscribe gratings or interferometric structures within each core, which results in the possibility of sensing multiple parameters in a small region of the fiber. Furthermore, this approach is commonly used on shape sensing applications, since the multiple sensors in different cores can provide a 3D shape reconstruction of the fiber as well as can indicate the vector bending or torsion (Barrera et al., 2018). Fig. 4.5 shows the cross-sectional view of a multicore optical fiber with 7 cores.

Contrasting with the aforementioned solid core optical fibers, microstructured optical fibers (MOFs) present a pattern of holes throughout the fiber

(Markos et al., 2017; Large et al., 2009). Such pattern has a defined geometry and pitch between the holes, where the ratio between the hole diameter and pitch defines its modal operation. If the ratio between the hole diameter and pitch is lower than 0.43, the MOF is endlessly single-mode (Markos et al., 2013), which means that the optical fiber is single mode at all frequencies (Yuan et al., 2011). Another advantage of MOFs is the possibility of holding gas or a biological sample in the holes for evanescent-wave sensing (Jensen et al., 2005). For these reasons, MOFs are extensively employed in different sensing applications, especially those involving FBGs with microstructured polymer optical fibers due to their endlessly single-mode operation (Oliveira et al., 2015; Webb, 2015), which is especially important for POFs, since most of these fibers are multimode. A typical mPOF made of PMMA with 3 ring hexagonal structure is shown in Fig. 4.5. In addition, the first single mode POF was proposed in 1991 by Kuzyk et al. (1991). Then, after some years of research, summarized in the review work by Zubia and Arrue (2001), a single mode step index POF was reported in Woyessa et al. (2016), which had a core made of TOPAS and a cladding of Zeonex for humidity insensitive and high temperatures operation when compared to other POFs, where the operating temperature is lower than 110 °C (Woyessa et al., 2016).

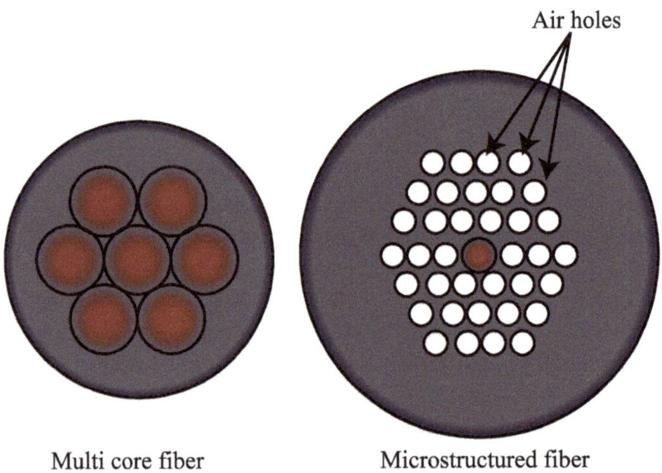

FIGURE 4.5 Cross-sectional view of a multicore optical fiber and a MOF.

4.4 Passive and active components in optical fiber systems

The telecommunications and sensing systems with optical fibers employ different active and passive components depending on the application. In general, the optical fiber system needs an input and components for signal detection, which are the light sources and photodetectors (or optical spectrum analyzers), respectively. Furthermore, there are other components such as optical filters, couplers, modulators, attenuators, and amplifiers that are commonly used in many optical fiber systems. Some of these components of optical fiber sensors systems are discussed in this section, namely light sources, photodetectors, optical couplers, optical circulators, and optical spectrum analyzers (OSAs). The discussion is focused on the operation principle description of components commonly used for sensors applications, especially for the applications in wearable robotics and devices. Interested readers should refer to (Chuang, 2009) for a detailed description of active and passive photonic devices.

4.4.1 Light sources

Light sources are core devices in optical fiber system regardless to the intended application, since they are the input of the system. Despite the many advances in lasers developments, including the vertical cavity surface emitting laser diodes, resonant cavity light emitting diodes (LEDs), distributed Bragg reflectors (DBR) and even on optical fiber lasers, this section is focused on the developments of the most basic and widely used light sources for the applications discussed in the next chapters. Thus, this section presents the luminescence emitting diodes, Fabry–Perot laser diodes and superluminescence light emitting diodes (SLEDs). Interested readers on other types of light sources should refer to specialized literature such as (Chuang, 2009).

A conventional structure of a LED comprises of two layers of a predefined semiconductor, where these layers result in a p-n junction, which result in the light emission with efficiency higher than 50%. However, the LED structure, presented in Fig. 4.6, lead to the light emission in all directions (Smith et al., 2001). Thus a LED is not a spatially coherent light source. As the semiconductors have large refractive indices (as high as 3), the light rays need smaller angles considering the normal axis (almost vertical) of the outer surface (see Fig. 4.6) to reach out the surface. In order to extent the light source capabilities, the laser diodes were proposed with similar structure of the LEDs with a p-n junction combined with a double hetero structure. There is resonator mirrors to create a resonant cavity in the case of the Fabry–Perot laser diode, as shown in Fig. 4.6. The laser diode has higher efficiency than LEDs with smaller spectral width, which is important for communications systems as well as some sensors applications, especially on the spectral interrogation of optical fiber sensors (as discussed in Chapter 6). Finally, a SLED also has similar constructive parameters as the aforementioned light sources. The difference in this case is the presence of an antireflection coating in one side of the structure to inhibit the creation of a cavity and enables lateral light emission, as shown in Fig. 4.6.

4.4.2 Photodetectors

Similar to the light sources, the photodetectors are also widely used in optical fiber system, as they are responsible for the signal acquisition at one end of the

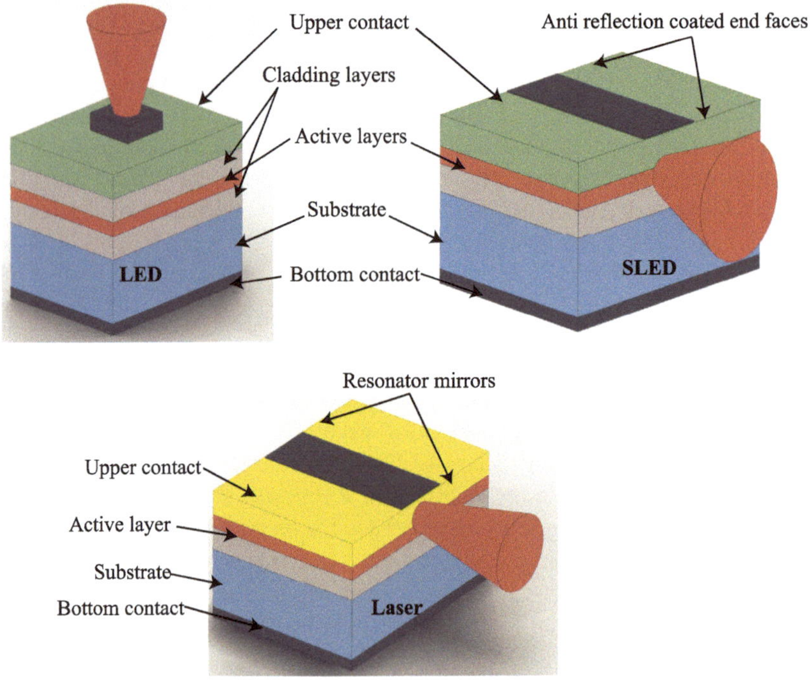

FIGURE 4.6 Structures of LEDs, SLEDs, and semiconductor lasers.

optical fiber under test. Despite many construction alternatives for photodiodes (such as the avalanche and metal-semiconductor-metal photodiodes) and different types of photodetectors in general, this section is focused on pin-photodiodes and phototransistors. Interested readers are suggested to check (Smith et al., 2001) for detailed description of different photodetectors.

In general, these components absorbs light quanta and converts them into electron-hole pairs, where all photons energies above the band gap are acquired. Thus the photodiodes produce a current flow when there is a light absorption with two main types: (i) solar cell in which the current is produced when there is a light incidence on the sensitive region. (ii) Photoconductor in which the light incidence results in a reduction on the reverse bias current (Ziemann et al., 2008). As the photodiodes essentially have restrictions on the current flow as a function of the optical intensity, these components commonly use transimpedance amplifiers (TIA) in order to increase the signal amplitude as well as convert the current variations into voltage. In contrast, phototransistors have their base terminal exposed and is activated when there is the incidence of photons on the device. The phototransistor can be represented as a photodiode with its photocurrent into the base of a transistor, which has different possibilities of connection. The three terminals can be connected on the common emitter or common collector configurations, as shown in Fig. 4.7. The differences between both configurations is the state transitions, which changes from high to low state in common emitter configuration, whereas the opposite occurs in the common collector configuration (Lynch et al., 2016). Fig. 4.7 shows the schematic representation of photodiodes and phototransistors, which indicate the similarity of both components. However, their fabrication as well as the performance and characteristics are different. Photodiodes have faster response in a wider frequency range than the phototransistors. In contrast, phototransistors have intrinsic amplification, which excludes the necessity of TIA or other external circuits for amplification, resulting in simpler circuits for signal acquisition (only the phototransistor and a resistor). The higher temperature stability is also an advantage of photodiodes. Although the fast response and wider frequency is an important advantage of photodiodes for telecommunication applications, such features are not crucial for sensors applications when there is short lengths of optical fibers (in the order of a few meters), especially when the sensing parameters have low-frequency dynamics (below the kHz range). For this reason, the applications presented throughout this book employs both components for light detection (photodiodes and phototransistors).

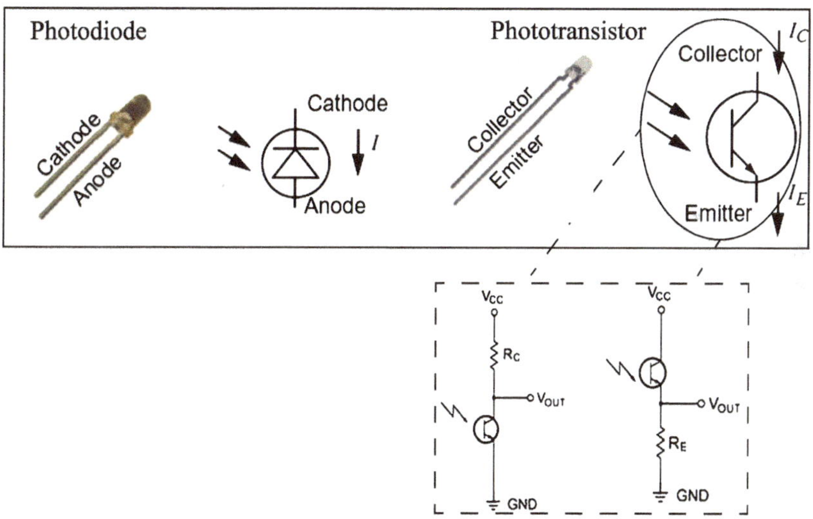

FIGURE 4.7 Schematic representation of photodiodes and phototransistors. Figure inset shows the configurations of phototransistors.

4.4.3 Optical couplers

Optical couplers (or splitters) are photonic devices enable of dividing an optical signal from one port to other ports, as shown in Fig. 4.8. A commonly used configuration has one input and two outputs (1x2), i.e., the optical signal is divided into two paths (or two optical fiber cables), where such division occurs with a fixed proportion (Ma et al., 2002). It is important to mention that different configurations are employed, e.g., two inputs/two outputs (2x2), one input/four outputs (1x4), and two inputs/three outputs (2x3). The coupling ratio (or splitting proportions) depends on the coupler configuration, which is the ratio that the input optical signals are divided between the outputs, i.e., a 50:50 coupling ratio in a 1x2 coupler indicates that half of the input optical power is coupled to each port. Other commonly employed coupling ratios are 90:10, 80:20, and 70:30. In addition to the coupling ratio, the insertion losses, directivity (or optical return loss), and excess loss are analyzed. There is also the possibility of analyzing the polarization dependent loss if the proposed application involves differences in the polarization states. The insertion losses are the ratio between the input and output optical powers at one port of the device, whereas the directivity defines the ratio of the input signal that is lost internally on the passive fiber. Furthermore, the excess loss is the ratio between the input and output power (considering all output ports) (Smith et al., 2001).

There are different technologies for optical couplers, which include the construction of special waveguides with multiple input and output paths, light coupling principle between fiber bundles and couplers based on side polished fibers, their operation principles are summarized in Ziemann et al. (2008). A widely used approach for optical couplers fabrication is based on the coupling between optical fibers. The operation principle of the light coupler employed on the compensation technique is shown in Fig. 4.8. The fibers are twisted and the core of the fibers are very close to each other. The fiber in which the light enters is referred from now on as the active fiber, whereas the fiber in which the light is coupled is the passive fiber. The laser source provides the light for the active fiber, which suffers the macrobending radiation losses due to its low curvature radius. This also results in the polarization of the evanescent wave and the proximity between the cores of the active and passive fibers enable the light energy transference from the active to the passive fiber. Nevertheless, it has a strict control of the coupling length and optical power source wavelength due to the necessity of obtain an exact split ratio.

4.4.4 Optical circulators

Optical circulators have applications and operation with some similarities when compared with the optical couplers (discussed in Section 4.4.3). The fundamental operation difference between both devices is the nonreciprocal polarization of the optical signals that occurs in optical circulators due to the Faraday effect (Hui, 2020). Thus, in optical circulators, the light is transmitted between the ports in a predefined sequence, as shown in Fig. 4.9, where a 3-port configuration is presented. In this case, the Port 1 is the input that enters Port 2 and the signal reflected in this port is redirected to Port 3. Thus, there is no back reflection between ports, i.e., the optical signal from Port 2 does not return to Port 1. In sensors applications, this device is widely employed in sensors that operate

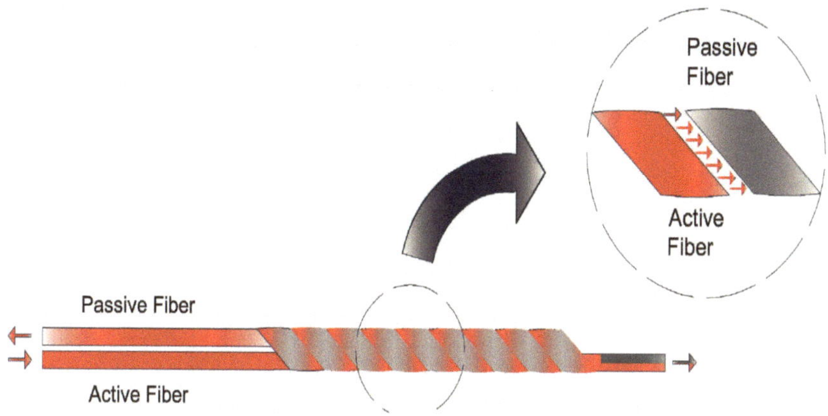

FIGURE 4.8 Operation principle of an optical coupler. The light enters on the active fiber and is coupled with the passive fiber on the twisted region.

in reflection mode, such as the fiber Bragg gratings (Diaz et al., 2019) discussed in Chapter 6.

The operation principle of an optical circulator is based on the use of magneto-optic materials used as Faraday rotators that changes the polarization of the input signal at each port of the device (Hui, 2020). In the propagation, the optical signal in Port 1 is split into two beams to be transmitted to two Faraday rotators, which rotates the beams to +45° and −45°. Thereafter, the beams pass through another Faraday rotator for additional +45° and −45° rotation, leading to orthogonal polarization states. Finally, both beams enter another beam displacement that performs the combination of the two beams that result in the reconstruction of the input signal with a 90° polarization rotation. The optical signal of the Port 1 output enters the Port 2, where the reflected signal passes through the opposite path as the ones that enter Port 1. However, the reflected signal beams have a 90° polarization rotation after passing through the rotators and displacers, which are recombined in Port 3, in a different spatial position than Port 1. Fig. 4.9 illustrates the operation principle of the optical circulator.

4.4.5 Spectrometers and optical spectrum analyzers

In many applications, there is the necessity of evaluating the optical power at each wavelength region, some sensors applications, such as interferometers, grating-based sensors, and some surface Plasmon resonance approaches; the operation principle of the sensor is based on the central or predefined wavelength variation as a function of a measurand (Guo et al., 2017b; Broadway et al., 2019). To that extent, optical spectrum analyzers (OSAs) and spectrometers are employed, since they are devices that provide the optical power at each wavelength component in a predefined range. Thus these devices are widely employed in sensors systems and the requirements for a wide wavelength range

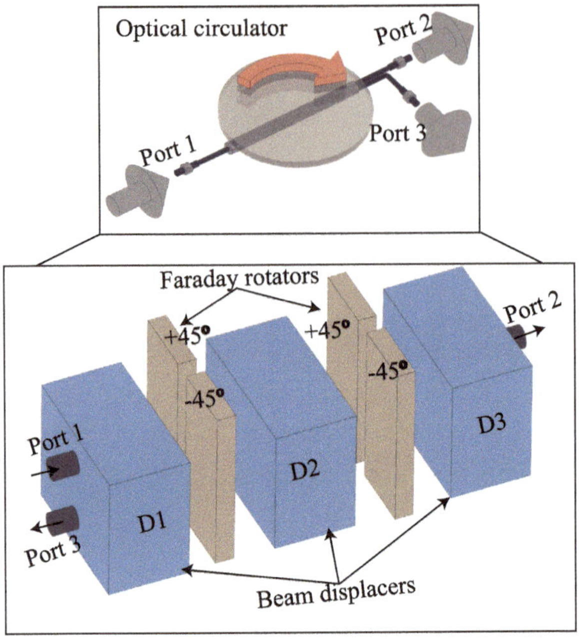

FIGURE 4.9 Operation principle of an optical circulator.

and accuracy as well as the resolution are continuously increasing, where some devices have picometer or even subpicometer wavelength resolution.

In spectrometers, the optical resolution and wavelength range are mainly defined by the diffraction grating used (see Fig. 4.10). The incident light beam is divided into different components corresponding to the diffraction grating properties, where such components are detected using a photodetector array or a charge-coupled device (CCD) camera. Thus, the diffraction grating is a key parameter that can be optimized for a given requirement of resolution and range. In this case, the blaze angle and the groove frequency are the main parameters in this component (Smith et al., 2001). The groove frequency is defined as the number of grooves per millimeter of the grating, whereas the blaze angle is the groove facet angle (as shown in Fig. 4.10), which has direct influence on the diffraction curve (Hui and O'Sullivan, 2009). For this reason, the blaze angle is adjusted to result in the highest peak efficiency for the desired wavelength range.

One of the possible configurations of OSAs is based on Fabry–Perot interferometer (FPI) scanning. Such configuration results in high spectral resolution (orders of magnitude higher than the ones of spectrometers or grating-based OSAs) (Hui and O'Sullivan, 2009). However, as a common tradeoff in these devices, the higher resolution generally results in lower wavelength range. The operation principle is based on the wavelength scanning by changing the FPI cavity length, which leads to a wavelength shift in the FPI output. The cavity length is mechanically changed using a piezoelectric actuator with submillimeter resolution. The FPI is connected to a photodetector that acquires the optical signal at each wavelength region during the FPI scanning, which is connected to an oscilloscope. A sawtooth waveform is used to synchronize the scanning and the oscilloscope. Fig. 4.10 also presents the schematic representation of the operation principles of an OSA based on the FPI scanning principle.

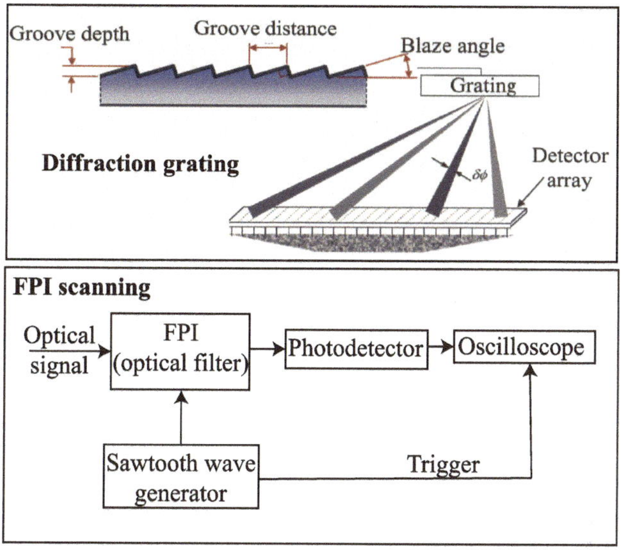

FIGURE 4.10 Operation principles of spectrometer and OSA.

4.5 Optical fiber fabrication and connection methods

A general description of optical fiber fabrication methods is presented, where the fabrication methods are described for silica and polymer optical fibers, since there are some differences in the fabrication, especially on the temperatures used on the material processing. In addition, due to the possibility of using different fabrication methods in POF fabrication due to their higher flexibility and lower processing temperatures. However, the POF fabrication methods discussed in this Section are related to step-index optical fibers. For a details on graded-index fiber fabrication, interested readers could refer to (Ziemann et al., 2008). Similarly, the connection methods or splicing between optical fibers are also discussed for both cases, i.e., silica and polymer materials. In this case, there are major differences in the connection methods especially due to the high difference in processing temperatures for these materials.

4.5.1 Fabrication methods

In general, the optical fibers are fabricated from a preform, for silica optical fiber, lengths of more than 1000 km are produced for optical fiber with 125 μm diameter. The preform can be fabricated through different methods in which a cylinder with higher diameter than the optical fiber is produced with the core/cladding proportions and refractive index profile of the optical fiber. The preform diameter is reduced to the one of the optical fiber using a draw tower to perform the automatic extrusion of the preform with controlled pulling pressure (or force) and temperature. Eq. (4.6) presents the relation between the optical fiber length (L_{fiber}), preform length ($L_{preform}$) and diameters of optical fiber and preform, defined as d_{fiber} and $d_{preform}$, respectively,

$$L_{fiber} = L_{preform} * \left(\frac{d_{preform}}{d_{fiber}}\right)^2. \tag{4.6}$$

The extrusion approach is also used on the fabrication of microstructured optical fibers. In the case of microstructured polymer optical fibers (mPOFs), the fabrication is generally performed with the drill and draw technique, which comprises of the preform fabrication in two steps. In the first step, commercial materials are casted into rods, where the desired air-hole pattern (with predefined hole diameter and pitch) is drilled in the rod. This process was reported for PMMA (Bundalo et al., 2014), polycarbonate (PC) (Woyessa et al., 2017), and cyclic olefin copolymers (COC) (Markos et al., 2013). Thereafter, the rod is sleeved into a tube of the same material as the rod, e.g., if a PMMA rod is drilled, it is inserted into a PMMA tube. An extrusion is performed in order to obtain a stable joint between the rod and tube. The fabricated preform is positioned in the draw tower, where the extrusion is performed with constant pressure (or force) and temperature, according to the extruded material. Thus in this case, the inner part of the fiber is extruded two times, whereas the outer part is extruded only one time, which does not lead to major differences in the optical fiber properties, since there is the same material in the inner and outer regions of the microstructured optical fiber. After the optical fiber fabrication, an annealing treatment is generally performed in order to provide a stress relaxation, reducing internal (and residual) stress created in the extrusion process. The annealing treatment can also increase the thermal stability of the POF (Woyessa et al., 2017), which can provide a higher repeatability on the properties after the fabrication and increase the material drawability. The effects of the annealing process in the optical fiber material proprieties are discussed in Chapter 5.

It is also worth noting that polymer fibers (and soft glass optical fibers) can be fabricated with different methods. For example, instead of drawing core and cladding from a single preform, it is possible to draw only the core from a preform (or a cylinder with higher diameter than the optical fiber) and applying the cladding by additional extrusion or enameling with the advantage of higher controllability of the polymerization process. This approach can also be used on the development of silica core optical fibers with polymer cladding. A similar approach is based on the two-head extrusion of core and cladding, where a coaxial extruder is used for the continuous intake of the core and cladding materials at different nozzles.

Although optical fibers are generally fabricated using heated extrusion or drawing process, such approach can be a time consuming and a rather inflexible manufacturing technique. As the preform is extruded in temperatures higher than 200 °C (Prado et al., 2017), it thereby exceeds the material glass transition temperature. The preform is heated and pulled under controlled tension, resulting in a stretched fiber yarn. However, in these processes, there is an asymmetric temperature profile along the fiber length, which can lead to a nonuniform diameter and asymmetric fiber properties, which are undesirable. To that extent, different approaches for POF fabrication includes the Light Polymerization Spinning (LPS) process, where polymer preforms are not used. In this case, the combination of monomers and other additives, make the POF manufacturing process highly scalable, repeatable, and customizable (Leal-Junior et al., 2018). In addition, highly flexible optical fiber can be obtained in this process with strain limits of up to 800% (Leal-Junior et al., 2018). The process is comprised of preparing a liquid mixture of monomers and additives through a dosing system. Then the liquid mixture passes through a mold with desired cylindrical shape, where the mixture with circular cross-section is polymerized by an ultraviolet (UV) lamp. Thereafter, the fiber is elongated until reaches its desired diameter and another UV curing is performed. Such process can reduce the residual stress on the fiber, which is stored as a spool after passing through a winder.

In a similar approach, the POF is fabricated without the UV curing step. In this case, thermosetting resins are used for an even smaller Young's modulus, in the order of a hundred kPa. In this case, flexible resins such as Polydimethylsiloxane precursors composed of monomer and curing agent are positioned inside a mold with the desired shape and diameter for the final fiber (Guo et al., 2017a) after the resin cure under a predefined temperature and time, depending on the resins properties. The demolding process is performed through the application of pressure (using pressurized water or mechanical tools) resulting in a highly flexible fiber core. The cladding can be fabricated using another resin (with different refractive index) or by solution doping. This process also enables the use of dye molecules and dopants to provide additional functionalities for the optical fiber, such as fluorescence in a predefined wavelength region (Leal-Junior et al., 2021).

As an alternative to traditional optical fiber manufacturing methods, the group of techniques commonly known as 3D printing or additive layer manufacturing (ALM) has revolutionized the manufacturing field in recent years (Horn and Harrysson, 2012). Its advantages are associated with the velocity in the manufacture of parts, in addition to the quality and possibility of customization. Additionally, 3D printing presents low cost and the possibility of recycling wasted material, which makes this technology considered an environmental friendly solution (Berman, 2012). Fused deposition modeling (FDM) is the most popular technique for 3D printing (Gul et al., 2018). In this technique, thermoplastic filaments are heated above their glass transition temperatures and are deposited at the desired location for construction of the desired geometry; these materials are deposited layer by layer (Gul et al., 2018). Theoretically, FDM is a suitable technique to manufacture preforms for POF production since the basic requirement for both POF preforms and 3D printer filament is the thermoplasticity of the material. The advantage of FDM in POFs manufacturing is the ease of reproducing complex shapes using this method. In addition, there is the additional advantage of being able to manufacture POFs with a wider range of materials, since the materials used in 3D printer filaments can be applied in the manufacture of optical fibers. Moreover, there is the possibility of mixing the polymers used in 3D printing for the manufacture of a preform, in which it may be possible to obtain control of some properties of the resulting fiber optic material such as refractive index, transparency, chemical composition, biodegradability, flexibility among others.

Over the years, the production of POF using 3D printing techniques has been proposed using different approaches and materials. A step index fiber was manufactured from two polymers using a 3D printer with 2 extrusion heads in Cook et al. (2016), where the fiber core material is the styrene-butadiene copolymer (SBP) and the shell is made of polyethylene terephthalate glycol-modified (PETG). Additionally, fibers with polycarbonate cores with different shapes and shells made of acrylonitrile butadiene styrene (ABS) were proposed in Zhao et al. (2017), where each part of the fiber was printed separately and then joined for its subsequent extrusion, which demonstrates the ability of 3D printing methods to obtain different forms for the optical fiber, and in this way, to customize properties. In addition, a microstructured fiber manufactured from preforms printed in 3D is demonstrated in Cook et al. (2015), which leads to the development of hollow core fibers from 3D printers, where a hollow core microstructured fiber for medium infrared, made from PETG preforms printed on a 3D printer was developed in Talataisong et al. (2018). Canning et al. (2016) presented a coreless fiber manufactured directly on the 3D printer, without the preform stage.

With the growing demand for environmentally friendly materials, components and equipment, allied with the possibility of controlling the properties of different materials through different manufacturing techniques (such as the 3D printing mentioned above). Recently, the development of optical fibers from biodegradable materials has been proposed not only as a sustainable correct alternative, but also for medical applications in vivo (Fu et al., 2018). A step-based optical fiber based on citrates was proposed in Shan et al. (2017), based on a two-stage process, where the cladding is produced from a mold and the core is injected in the second stage. The optical fiber produced was applied in fluorescence-based detection in vivo applications. In Fu et al. (2018), an optical fiber of poly-L-lactic acid (PLLA) was also proposed for in vivo applications, where fluorescence and fiber degradation were characterized.

4.5.2 Optical fiber connectorization approaches

The widespread of optical fiber technologies also result in a large number of connectors for optical fiber-based systems with different standards, depending on the applications. In sensors application, the connection between the optical fiber, generally modified with a microstructure or macrostructure for sensing of a desired parameter, with equipment for optical signal acquisition such as optical interrogators, photodetectors, and optical spectrum analyzers are performed through a standard connector. Commonly used connectors for sensors applications include the ones with physical contact (PC), which also include the ones with beveled end faces, such as the angled physical contact (APC) to reduce the reflection losses. Connectors generally have optical losses in the range of 0.1 dB and 0.5 dB, depending on the quality on the fabrication with the possibility of smaller losses as the technology evolves. Actually, POFs can use the same connectors as glass optical fibers. However, in large diameter POFs (the ones with about 1 mm diameter), connectorless approaches are also used due to the high numerical aperture of these optical fibers, which results in a lower cost system. In glass optical fiber cables, there are also additional Kevlar strands for strain relief and mechanical protection, which are not vital components for POF cables due to their inherent flexibility and fracture toughness. Connector losses due to Fresnel reflections, deviations on the positioning, diameter offset, and surface roughness of the optical fiber end face.

The process of splicing two fibers can be performed using different methods, depending on the optical fiber materials. As a common approach for both silica and polymer optical fibers, the connectorization between fibers (and the optical fiber directly connected to a connector) occurs with three primary steps: (i) optical fiber cleaving, (ii) splicing, and (iii) splice protection. In silica fibers, there are specialized optical fiber cleavers that use a diamond blade disc to cleave the optical fiber, positioned on the cleaver structure, with a predefined angle or perpendicular to the optical fiber, as generally occurs. It is worth to mention that mechanical splicing is also used on the silica fibers splicing, but generally the electric arc is used. In addition, the silica optical fiber splicing generally occurs in splicing machines that use electric arc (or filament) on the splicing due to the high processing temperatures of silica fibers (above 800 °C). In this case, both cleaved end facets of the optical fibers are placed inside the splicing machine that align the fibers using an actuation system. After the alignment, an electric arc welding is performed to splice both optical fibers at high temperatures. Then the fibers are removed, as the splicing region is more fragile, there are some approaches, generally based on mechanical protection, to enhance the robustness of the spliced region. As a common approach a plastic tube with metal protection is placed on the spliced region.

The aforementioned splicing and cleaving process are only applicable to silica optical fibers since there are major differences in the material properties when compared with POFs. In the latter, the correct cleaving of POFs is an important process prior to its connectorization since the cleaving parameters can influence the quality of the POF end facet (Stefani et al., 2012). Although it also occurs in silica fibers the control of temperature, speed, and angle have higher influence on the fiber end facet due to the POF material nonbrittle and soft nature. In this case, razor blades in temperature controlled systems are also employed to ensure that the cleaving is performed on a plate with suitable temperatures reported for each POF material (Bundalo et al., 2014; Woyessa et al., 2017; Markos et al., 2013). It is also worth to mention that systems with controlled cleaving speed are also desirable due to its influence on the fiber end facet and cleaving quality (Stefani et al., 2012).

Similarly, the POF splicing cannot occur in the commercial splicing machines, since their operation temperatures are much higher than the temperatures for POF processing. Thus mechanical splicing between fibers occurs when there is the necessity of connecting two POFs. However, in many cases, the POF is spliced to a silica optical fibers since the majority of cables and optical pigtails/connectors are manufactured using silica optical fibers. In this scenario a widely adopted technique for splicing both fibers is the butt-coupling method in which both fibers are cleaved (using the suitable method for each fiber). Then the cores of the optical fibers are aligned manually or automatically (using an alignment system). Thereafter, an optical adhesive (or UV-curing resin) is applied in-between the optical fibers. Finally, an UV lamp is used to perform the polymerization or UV curing of the resins, which result in the connection between both fibers. This method generally results in higher optical losses than the fusion splicing method. However, it still is the suitable method for the connection when POF sensors are applied, especially when multimode POFs are using. In this case, there is also the possibility of splicing the multimode POF to a multimode silica optical fiber in order to provide a smoother diameter reduction (between silica and polymer optical fiber), since the multimode silica fiber has higher core diameter than the single mode one, which can reduce the optical losses in the splice. In the butt-coupling method, a mechanical protection can be applied on the spliced region to reduce the risk of breakage and misalignments in the spliced region.

Another recent (and important) development on POF splicing technology is the direct connection between the POF and a connector (without using the butt-coupling method). This method can potentially reduce the optical losses in POF connection, which is a fundamental issue on the practical applications of POF sensors systems (Broadway et al., 2019). In this case, the POF is connected directly on the connector with the lateral and axial alignment of the POF inside the connector, which enable plug and play operation of POF sensors with reduced risk of breakage and physical supports for splicing protection, commonly used on the butt-coupling method since the breakage generally occurs in the splicing region.

References

Agrawal, G.P., 2016. Optical communication: its history and recent progress. https://doi.org/10.1007/978-3-319-31903-2.

Barrera, D., Madrigal, J., Sales, S., 2018. Long period gratings in multicore optical fibers for directional curvature sensor implementation. Journal of Lightwave Technology 36, 1063–1068. https://doi.org/10.1109/JLT.2017.2764951.

Berman, B., 2012. 3-D printing: the new industrial revolution. Business Horizons 55, 155–162. https://doi.org/10.1016/j.bushor.2011.11.003.

Bilro, L., Alberto, N., Pinto, J.L., Nogueira, R., 2012. Optical sensors based on plastic fibers. Sensors (Switzerland) 12, 12184–12207. https://doi.org/10.3390/s120912184.

Broadway, C., Min, R., Leal-Junior, A., Marques, C., Caucheteur, C., 2019. Toward commercial polymer fiber Bragg grating sensors: review and applications. Journal of Lightwave Technology 37. https://doi.org/10.1109/JLT.2018.2885957.

Bundalo, I.L., Nielsen, K., Markos, C., Bang, O., 2014. Bragg grating writing in pmma microstructured polymer optical fibers in less than 7 minutes. Optics Express 22, 5270. https://doi.org/10.1364/OE.22.005270.

Canning, J., Hossain, M.A., Han, C., Chartier, L., Cook, K., Athanaze, T., 2016. Drawing optical fibers from three-dimensional printers. Optics Letters 41, 5551. https://doi.org/10.1364/OL.41.005551.

Chuang, S.L., 2009. Physics of Photonic Devices, 2 ed. Wiley.

Cook, K., Balle, G., Canning, J., Chartier, L., Athanaze, T., Hossain, M.A., Han, C., Comatti, J.E., Luo, Y., Peng, G.D., 2016. Step-index optical fiber drawn from 3D printed preforms. Optics Letters 41, 4554. https://doi.org/10.1364/OL.41.004554.

Cook, K., Leon-Saval, S., Canning, J., Reid, Z., Hossain, M.A., Peng, G.D., 2015. Air-structured optical fibre drawn from a 3D-printed preform. Optics Letters 40, 3966–3969. https://doi.org/10.1117/12.2195466.

Diaz, C.A.R., Leal-Junior, A.G., Avellar, L.M., Antunes, P.F.C., Pontes, M.J., Marques, C.A., Frizera, A., Ribeiro, M.R.N., 2019. Perrogator: a portable energy-efficient interrogator for dynamic monitoring of wavelength-based sensors in wearable applications. Sensors 19 (13), 2962. https://doi.org/10.3390/s19132962.

Fu, R., Luo, W., Nazempour, R., Tan, D., Ding, H., Zhang, K., Yin, L., Guan, J., Sheng, X., 2018. Implantable and biodegradable poly(l-lactic acid) fibers for optical neural interfaces. Advanced Optical Materials 6, 1–8. https://doi.org/10.1002/adom.201700941.

Gul, J.Z., Sajid, M., Rehman, M.M., Siddiqui, G.U., Shah, I., Kim, K.H., Lee, J.W., Choi, K.H., 2018. 3D printing for soft robotics – a review. Science and Technology of Advanced Materials 19, 243–262. https://doi.org/10.1080/14686996.2018.1431862. https://www.tandfonline.com/doi/full/10.1080/14686996.2018.1431862.

Guo, J., Niu, M., Yang, C., 2017a. Highly flexible and stretchable optical strain sensing for human motion detection. Optica 4, 1285. https://doi.org/10.1364/optica.4.001285.

Guo, T., Gonzalez-Vila, A., Loyez, M., Caucheteur, C., 2017b. Plasmonic optical fiber-grating immunosensing: a review. Sensors (Switzerland) 17, 1–20. https://doi.org/10.3390/s17122732.

Horn, T.J., Harrysson, O.L., 2012. Overview of current additive manufacturing technologies and selected applications. Science Progress 95, 255–282. https://doi.org/10.3184/003685012X13420984463047.

Hui, R., 2020. Passive optical components. In: Introduction to Fiber-Optic Communications. Elsevier, pp. 209–297.

Hui, R., O'Sullivan, M., 2009. Basic instrumentation for optical measurement. Fiber Optic Measurement Techniques, 129–258. https://doi.org/10.1016/b978-0-12-373865-3.00002-1.

Jensen, J.B., Hoiby, P.E., Emiliyanov, G., Bang, O., Pedersen, L.H., Bjarklev, A., 2005. Selective detection of antibodies in microstructured polymer optical fibers. Optics Express 13, 5883. https://doi.org/10.1364/OPEX.13.005883.

Koike, Y., Asai, M., 2009. The future of plastic optical fiber. NPG Asia Materials 1, 22–28. https://doi.org/10.1038/asiamat.2009.2.

Kuzyk, M.G., Paek, U.C., Dirk, C.W., 1991. Guest-host fibers for nonlinear optics. Applied Physics Letters 59, 902.

Large, M.C.J., Moran, J., Ye, L., 2009. The role of viscoelastic properties in strain testing using microstructured polymer optical fibres (mpof). Measurement Science and Technology 20, 034014. https://doi.org/10.1088/0957-0233/20/3/034014.

Large, M.C.J., Poladian, L., Barton, G.W., van Eijkelenborg, M.A., 2008. Microstructured Polymer Optical Fibres. Springer US, Boston, MA.

Leal-Junior, A., Guo, J., Min, R., Fernandes, A.J., Frizera, A., Marques, C., 2021. Photonic smart bandage for wound healing assessment. Photonics Research 9, 272. https://doi.org/10.1364/prj.410168.

Leal-Junior, A., Theodosiou, A., Frizera-Neto, A., Pontes, M.J., Shafir, E., Palchik, O., Tal, N., Zilberman, S., Berkovic, G., Antunes, P., André, P., Kalli, K., Marques, C., 2018. Characterization of a new polymer optical fiber with enhanced sensing capabilities using a Bragg grating. Optics Letters 43, 4799. https://doi.org/10.1364/OL.43.004799.

Liu, W., Guo, T., Wong, A.C.l., Tam, H.Y., He, S., 2010. Highly sensitive bending sensor based on Er^{3+}-doped dbr fiber laser. Optics Express 18, 17834. https://doi.org/10.1364/OE.18.017834.

Lynch, K.M., Marchuk, N., Elwin, M.L., 2016. Sensors. In: Embedded Computing and Mechatronics with the PIC32. Elsevier, pp. 317–340.

Ma, H., Jen, A.K., Dalton, L.R., 2002. Polymer-based optical waveguides: materials, processing, and devices. Advanced Materials 14, 1339–1365. https://doi.org/10.1002/1521-4095(20021002)14:19<1339::AID-ADMA1339>3.0.CO;2-O.

Makino, K., Kado, T., Inoue, A., Koike, Y., 2012. Low loss graded index polymer optical fiber with high stability under damp heat conditions. Optics Express 20, 12893–12898. https://doi.org/10.1364/Oe.20.012893.

Markos, C., Stefani, A., Nielsen, K., Rasmussen, H.K., Yuan, W., Bang, O., 2013. High-tg topas microstructured polymer optical fiber for fiber Bragg grating strain sensing at 110 degrees. Optics Express 21, 4758–4765. https://doi.org/10.1364/OE.21.004758.

Markos, C., Travers, J.C., Abdolvand, A., Eggleton, B.J., Bang, O., 2017. Hybrid photonic-crystal fiber. Reviews of Modern Physics 89, 1–55. https://doi.org/10.1103/RevModPhys.89.045003.

Mohammed, H.A., Rashid, S.A., Abu Bakar, M.H., Ahmad Anas, S.B., Mahdi, M.A., Yaacob, M.H., 2019. Fabrication and characterizations of a novel etched-tapered single mode optical fiber ammonia sensors integrating pani/gnf nanocomposite. Sensors and Actuators. B, Chemical 287, 71–77. https://doi.org/10.1016/j.snb.2019.01.115.

Oliveira, R., Bilro, L., Nogueira, R., 2015. Bragg gratings in a few mode microstructured polymer optical fiber in less than 30 seconds. Optics Express 23, 10181. https://doi.org/10.1364/OE.23.010181.

Prado, A., Leal-Junior, A., Marques, C., Leite, S., De Sena, G., Machado, L., Frizera, A., Ribeiro, M., Pontes, M., 2017. Polymethyl methacrylate (pmma) recycling for the production of optical fiber sensor systems. Optics Express 25. https://doi.org/10.1364/OE.25.030051.

Shan, D., Zhang, C., Kalaba, S., Mehta, N., Kim, G.B., Liu, Z., Yang, J., 2017. Flexible biodegradable citrate-based polymeric step-index optical fiber. Biomaterials 143, 142–148. https://doi.org/10.1016/j.biomaterials.2017.08.003.

Smith, F.G., King, T.A., Dawes, D.L., 2001. Optics and photonics: an introduction. American Journal of Physics 69, 236–237. https://doi.org/10.1119/1.1336840.

Stefani, A., Nielsen, K., Rasmussen, H.K., Bang, O., 2012. Cleaving of TOPAS and PMMA microstructured polymer optical fibers: core-shift and statistical quality optimization. Optics Communications 285, 1825–1833. https://doi.org/10.1016/j.optcom.2011.12.033.

Talataisong, W., Ismaeel, R., Marques, T.H., Abokhamis Mousavi, S., Beresna, M., Gouveia, M.A., Sandoghchi, S.R., Lee, T., Cordeiro, C.M., Brambilla, G., 2018. Mid-IR hollow-core microstructured fiber drawn from a 3D printed PETG preform. Scientific Reports 8, 1–8. https://doi.org/10.1038/s41598-018-26561-8.

Tricker, R., 2002. The history of fiber optics. Optoelectronics and Fiber Optic Technology, 1–35. https://doi.org/10.1016/b978-075065370-1/50003-9.

Veber, A., Lu, Z., Vermillac, M., Pigeonneau, F., Blanc, W., Petit, L., 2019. Nano-structured optical fibers made of glass-ceramics, and phase separated and metallic particle-containing glasses. Fibers 7, 105. https://doi.org/10.3390/fib7120105.

Webb, D.J., 2015. Fibre Bragg grating sensors in polymer optical fibres. Measurement Science and Technology 26, 092004. https://doi.org/10.1088/0957-0233/26/9/092004.

Woyessa, G., Fasano, A., Markos, C., Rasmussen, H., Bang, O., 2017. Low loss polycarbonate polymer optical fiber for high temperature fbg humidity sensing. IEEE Photonics Technology Letters 29, 575–578. https://doi.org/10.1109/LPT.2017.2668524.

Woyessa, G., Fasano, A., Stefani, A., Markos, C., Nielsen, K., Rasmussen, H.K., Bang, O., 2016. Single mode step-index polymer optical fiber for humidity insensitive high temperature fiber Bragg grating sensors. Optics Express 24, 1253–1260. https://doi.org/10.1364/OE.24.001253.

Yuan, W., Khan, L., Webb, D.J., Kalli, K., Rasmussen, H.K., Stefani, A., Bang, O., 2011. Humidity insensitive topas polymer fiber Bragg grating sensor. Optics Express 19, 19731–19739. https://doi.org/10.1364/OE.19.019731.

Zhao, Q., Tian, F., Yang, X., Li, S., Zhang, J., Zhu, X., Yang, J., Liu, Z., Zhang, Y., Yuan, T., Yuan, L., 2017. Optical fibers with special shaped cores drawn from 3D printed preforms. Optik 133, 60–65. https://doi.org/10.1016/j.ijleo.2017.01.002.

Ziemann, O., Krauser, J., Zamzow, P.E., Daum, W., 2008. POF Handbook. Springer, Berlin, Heidelberg.

Zubia, J., Arrue, J., 2001. Plastic optical fibers: an introduction to their technological processes and applications. Optical Fiber Technology 7, 101–140. https://doi.org/10.1006/ofte.2000.0355.

Chapter 5

Optical fiber materials

5.1 Optically transparent materials

The widespread of applications in optical fiber technology, especially for sensors systems, resulted in the development of novel optical fibers with customized or optimized parameters with a myriad of transparent materials. Actually, optical fibers are fabricated with different materials and approaches as summarized in many literature reviews (Luo et al., 2017; Broadway et al., 2019; Liu et al., 2019). In fact, solid core and microstructured optical fiber fabrication methods cover from the conventional draw towers (Nguyen et al., 2011) to novel 3D printing technologies (Luo et al., 2020), including simple single step approaches (Cordeiro et al., 2020) and materials recycling (Prado et al., 2017), as also indicated in Section 4.5. In addition, there are also advances on optical fiber doping doped with different materials (optically active or not) for applications in photonics devices, especially lasers and amplifiers as well as sensing approaches (Liu et al., 2019), including optical fibers doped with nanoparticles for enhanced Rayleigh scattering, which find important applications in distributed optical fiber sensors systems (Veber et al., 2019; Beisenova et al., 2019; Silveira et al., 2020). Generally, fiber materials are divided into glass (or silica) optical fibers and POFs, where each class of material has its subdivisions and optical fiber types (Liu et al., 2019). Silica still is the preferred material when telecommunications applications are concerned due to its lower optical losses, which guarantee an optical signal transmission in long distances (kilometer range), making them preferable for remote sensing applications as well (Tejedor et al., 2017). The silica optical fibers are also widely employed in medical applications, especially in invasive sensors systems due to their biocompatibility and small sizes (Roriz et al., 2013).

Considering the sensors applications, the material's physical properties play an important role in the material selection, since most of the sensors are based on mechanical and thermal variations in the optical fibers. Conventional silica optical fibers have high glass transition temperature (Tg), which leads to high temperatures on the extrusion in the fiber fabrication, circa 1200 °C. In soft glass materials, such as chalcogenide glass, it needs lower temperatures on its drawing, which can be as low as 200 °C (Shiryaev, 2015), similar to the ones of POFs (Luo et al., 2017). Regarding the mechanical properties, a small strain limit of below 5% makes silica fibers fragile and applicable only to small strains on the fiber. Moreover, its elastic modulus is close to 70 GPa, whereas the soft glasses

Optical Fiber Sensors for the Next Generation of Rehabilitation Robotics
https://doi.org/10.1016/B978-0-32-385952-3.00014-7

have elastic modulus in the range of 10 GPa to 20 GPa, thereby being known as "soft glasses." These elastic moduli values of silica fibers are at least ten times higher than the ones commonly reported in POF (generally around a few GPa). Moreover, commercial POFs present strain limits as high as 10%, indicating a higher strain range when compared with silica fibers. It is important to mention that the additional fabrication methods for POFs (discussed in Chapter 4) enable the fabrication of POFs with much higher mechanical flexibility can be achieved, where there are reports of POFs with elastic modulus in the range MPa (Leal-Junior et al., 2018b) and even kPa (Guo et al., 2017). In this case, the POF is created from a mixture of monomers in the light polymerization spinning as well as UV-curing or thermosetting resins fabrication methods discussed in Chapter 4. These methods can result in a POF not only with elastic modulus in MPa or kPa range, but also with strain limits higher than 200% (Leal-Junior et al., 2018b). As the majority of sensors used on wearable robots instrumentation are applied on the measurement of physical, strain/stress-related parameters, the mechanical properties are key parameters in the sensor sensitivity, resolution, and even linearity, since the elastic modulus, for example, is important for the stress sensitivity of the sensors, as discussed in Chapter 6. Similarly, the material elastic limit influences on the linear dynamic range of the sensor for the measurement of stress/strain parameters. A tradeoff between optical attenuation (important for optical signal transmission in longer lengths) and elastic modulus (related to the sensor's stress sensitivity) can be inferred in the Ashby chart shown in Fig. 5.1. The range with lowest optical losses for each material was chosen, i.e., visible wavelength range for POFs and near infrared for silica fibers, since the optical losses vary on the wavelength range. In the material selection, it is desired low optical losses and low elastic modulus, where the importance of each parameter depends on the sensor application field as shown in the optical attenuation curve in Fig. 5.1 inset.

POFs present higher variation of the materials used, among these many materials options, PMMA is yet the most employed material in POF production, despite its low glass transition temperature (Tg) compared to some of the other polymers used on POF development, such as polycarbonate (PC). Such feature can limit its application at higher temperatures (Fasano et al., 2016), since the operation temperature of POFs is generally below its Tg. Furthermore, the higher moisture absorption capability of PMMA can harm its application in temperature or strain sensing where a humidity cross-sensitivity is undesirable (Yuan et al., 2011). These limitations motivated the development of novel POFs to address the aforementioned issues and cross-sensitivities to environmental variations, e.g., to mitigate the humidity cross-sensitivity. POFs made of cyclic olefin copolymers (COC) were proposed throughout the years. The COCs used on POF fabrication include TOPAS grade 8007 (Johnson et al., 2011) and 5013 (Markos et al., 2013). In addition, POF fabrication was also proposed with cyclic olefin homopolymer such as Zeonex 480R (Woyessa et al., 2017). The optical fiber fabricated with these materials presented a humidity

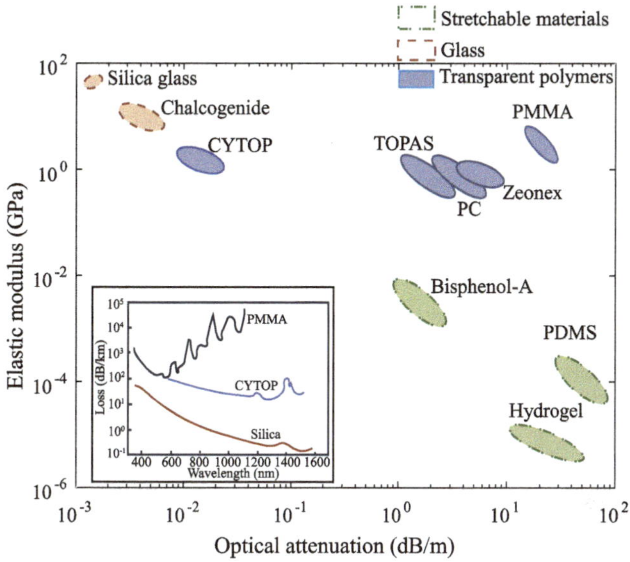

FIGURE 5.1 Ashby chart of optical attenuation and elastic modulus of optical fiber commonly used materials. Figure inset shows the optical attenuation curve.

sensitivity at least 30 times lower than the one of PMMA. However, the glass transition temperature shows higher variation among the different COC grades, e.g., TOPAS 8007 presents Tg of only 78 °C, whereas TOPAS 5013 has a Tg of 134 °C (Markos et al., 2013). Another polymer material employed for high temperature and strain sensing is Zeonex 480R, which has a Tg of 138 °C with the additional advantage of superior drawability (Woyessa et al., 2017). An even higher Tg is obtained in PC optical fibers. In this case, PC has a Tg of 145 °C, which is higher than all aforementioned POFs (Fasano et al., 2016) with higher strain limits as well. The aforementioned POFs, i.e., TOPAS, Zeonex, and PC, have attenuation curves similar to the one of the PMMA shown in Fig. 5.1 inset. For this reason, these POFs have high optical attenuation in the 1550 nm region, which limits the optical signal transmission to a few centimeters. Thus optical fiber sensors in POFs, especially fiber Bragg gratings, are used in the 850 nm wavelength region with some progress on the 600 nm wavelength region as well (Min et al., 2018), which is the preferred wavelength region for intensity variation-based sensors. In order to tackle the limitations imposed by the higher attenuation of POFs, graded-index CYTOP fibers have been developed (Koike and Asai, 2009), where carbon-fluoride bonds replace carbon-hydrogen in the polymer structure, which lead to lower losses in near infrared (especially in 1550 nm) when compared with acrylate-based POFs (Koike and Asai, 2009), as also shown in the Fig. 5.1 inset. In addition, this material also has low dispersion (Koike and Asai, 2009) and these advantages have led to the rapid widespread of CYTOPs as commercial solutions for POFs.

Due to the variety of materials used on POF fabrication, Fig. 5.2 presents a collection of spider diagrams for the comparison of different parameters in the aforementioned POFs, where the POFs are categorized with their temperature, strain, and humidity sensitivities as well as their attenuation and Young's modulus the functionalities, which provide guidelines for the materials choice, depending on the desired application. It is possible to observe in Fig. 5.2 that POFs made of highly stretchable materials have higher sensitivity than the other materials for temperature and humidity. In addition, the low Young's modulus of such fibers also enables the development of force sensors with higher sensitivity than the other fibers. However, these materials provide higher optical attenuation than the other POFs. It is also important to observe that their low glass transition temperature limits the temperature operation range.

5.2 Viscoelasticity overview

The mechanical, thermal, and attenuation properties of an optical fiber under flexion, torsion, and elongation have been studied throughout the years. Since the POFs are made of plastic materials, it is expected that the Young modulus of the POFs be lower than silica fibers. However, there is not an exact value for the Young modulus on polymers; these values lie in the range between 1.6 and 5 GPa for PMMA POF (Peters, 2011). The differences in the results are caused by the polymer anisotropy, internal stresses generated in the manufacture process and differences between the experimental approaches, which include different strain rates for the Young modulus estimation, and testing conditions (Peters, 2011). Yang et al. (2004) employed a commercial elongation testing machine to enhance the fidelity of the tests on a polymer optical fiber manufactured on their own laboratory. The stain-stress curve was obtained with extension speed of 1 mm/min. The sampling rate was 20 points before and 4 points after the elongation of 2.5 mm, which results on a Young modulus of 2.747 GPa on 6% of the tensile strain. In order to evaluate the Young modulus deviation with the strain rate and strain range, Kiesel et al. (2007) measure a static Young modulus for three different strain rates (0.01, 0.3, 0.6 mm–1) with the strain range between 1% and 5.5%. The static Young modulus lies between 3.83 GPa and 5.04 GPa. Furthermore, the static Young modulus is applied for small strains (lower than 1%) in static tests. In strains larger than 1%, the nonlinear effects became relevant and must be considered in Young modulus estimations. Therefore the nonlinear effects of the polymeric material, which was not considered on most estimations, could cause the deviation between the Young modulus values in strain ranges above 1%.

Stefani et al. (2012) considered the polymers viscoelasticity, and for this reason, a dynamic mechanical analysis (DMA) is performed. A current driven shaker is employed to perform the DMA, the shaker can provide a peak-to-peak displacement of 6 mm in a frequency range of 0 to 18 kHz and it is comprised of a shaker within a climatic chamber. The fiber is fixed at one end

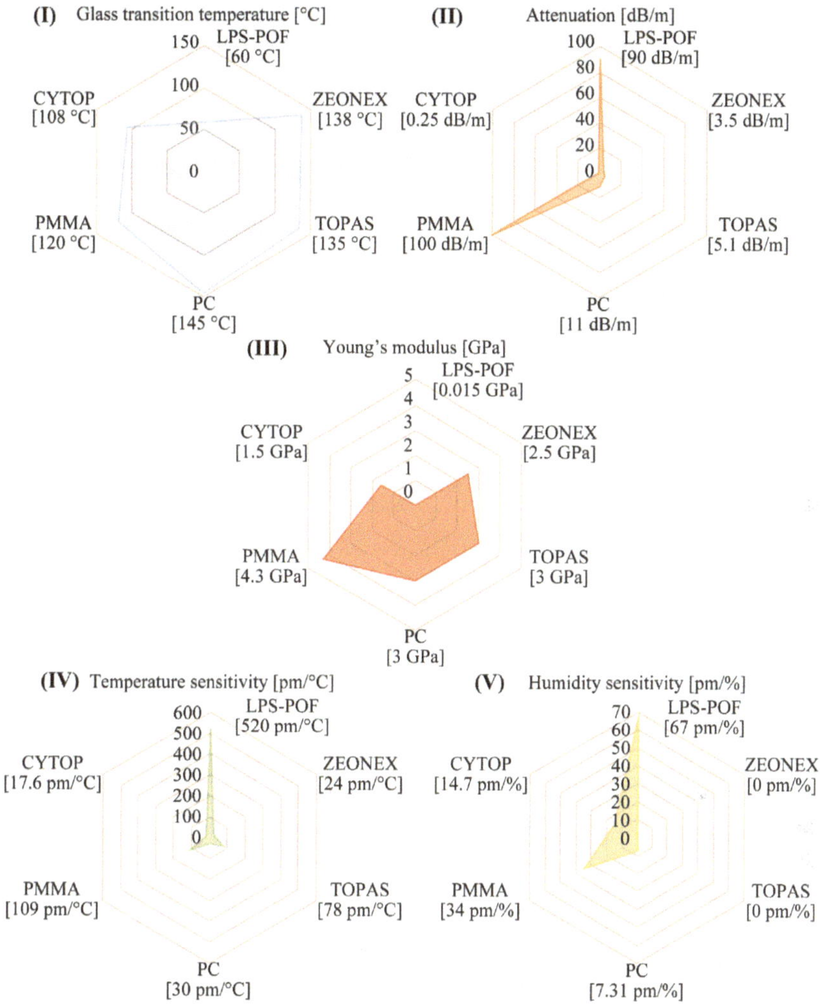

FIGURE 5.2 Properties of different POF materials. (I) Glass transition temperature, (II) optical attenuation, (III) Young's modulus, (IV) temperature sensitivity (considering FBGs), and (V) humidity sensitivity (considering FBGs).

of the shaker, which have an accelerometer beneath it. There is a force gauge on the other side of the testing machine. The shaker is driven by a waveform generator, which generates a sine wave on a fixed frequency. The data acquisition card monitors the accelerometer, force gauge, and wave input. Fig. 5.3 shows the comparison between a force for a generic elongation input with constant frequency and strain. Results show the typical phase lag of the viscoelastic materials. Moreover, the phase lag increases when the frequency is increased. Almost all parameters, which defines the dynamic Young modulus, depend on the phase lag, as it is shown in Eq. (5.1):

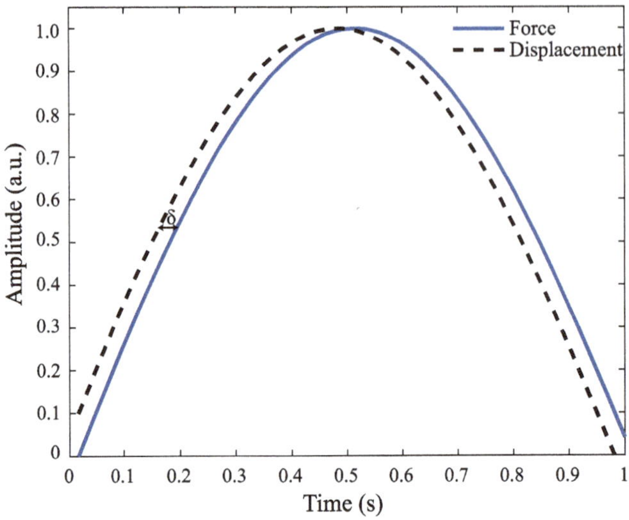

FIGURE 5.3 Force and periodic displacement for a viscoelastic material.

$$E^* = E_0 \cos(\delta) + i E_0 \sin(\delta), \tag{5.1}$$

where E^* is the dynamic Young modulus, E_0 is the static Young modulus, and δ is the phase lag. Since there is no phase lag in elastic materials, the dynamic Young modulus is the static Young modulus, which can be obtained from the stress-strain curve inclination calculation in the linear region.

There are several models to describe the viscoelastic behavior of a material, which may be divided in three major types: integral models, differential models, and molecular models (Riande et al., 2000). Among the differential models, some different approaches exist as well. However, all the approaches consider the stress or strain response of the material as a combination of springs and dashpots. The spring behavior is the behavior of an elastic material and the dashpot represents the behavior of a viscous material. Since the response of a viscoelastic material is a combination of the elastic and viscous response, the combination between springs and dashpots result on the response of the material (Lakes, 2009). The differential models are Maxwell's model, Kelvin–Voigt model, three-element standard solid, Burgers model among others. The differences between these models are the configuration between the spring and dashpot and the number of such elements (Riande et al., 2000). For example, the Maxwell model has one spring and one dashpot connected in series, whereas the three-element standard solid model has a spring in series with a spring parallel with a dashpot (Fig. 5.4). The parameters of the model can be obtained experimentally. Although models with higher number of components result in response closer to the real one, the number of parameters to determine also increase, which needs a more complex experimental setup and additional viscoelastic characterization. Nevertheless, in some cases, the Maxwell model can predict a POF response with good approximation degree as shown by the DMA made in Stefani et al. (2012).

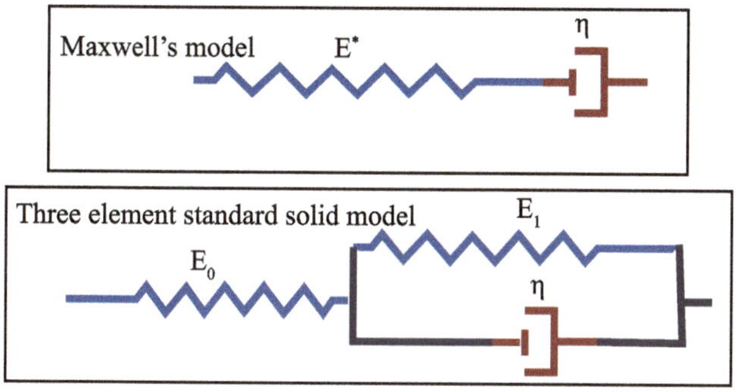

FIGURE 5.4 Maxwell model and three-element standard linear solid model.

Besides the viscoelastic response and characteristics, optical fiber mechanical tests and models can also provide information about stress-optical effects, namely the change of the refractive index with the stress variation. Zubia et al. (1997) presented theoretical and experimental analyses of a torsion-induced optical effect in a POF. There were also presented the effects of a fiber torsion, uniaxial stress, fiber bending, and the stress transmitted to the cladding. A refractive index variation equation was obtained and exhibited a torsion-induced anisotropy slightly smaller than the same effect provoked by tensile and bending stresses.

As presented in Zubia et al. (1997), bending effects can be significant to refractive index variation. Moreover, flexion is the most common condition in optical fibers both in telecommunication application and in sensing purposes. Therefore research effort is toward the experimental and analytical models of the losses due to optical fiber bending, which is divided into two types: macrobending and microbending. Although the both types of bending create losses in an optical fiber, the losses are lower in a microbend. Furthermore, the microbending is more difficult to be measured and controlled. Nevertheless, macrobending can be measured, controlled, and several POF based sensors apply bending losses to measure the desired parameter. Arrue and Zubia (1996) made an extensive investigation over the curvature losses of a polymer optical fiber. It was investigated the reflection coefficient deviation with incident angles close to the critical incident angles for PMMA and PS POFs. These results are the references for the measurement of the power loss under different conic sections. The conic sections investigated were hyperbola, circumference, ellipse with lower eccentricity (closer to zero), and ellipse with higher eccentricity (closer to 1). Results show that bigger critical angles result in lower total attenuation for sharp sections and the circumference was the curve that yields higher attenuation. Therefore for a curvature based POF sensor, a circumferential curve for bending is more suitable than the hyperbola and ellipse. Moreover, the circumference is the easiest curve to achieve. Zhang et al. (2009) presented a theoretical analysis of the surface scattering losses in a curvature optical fiber sensor. However, experimental results showed that the analytical model is not accurate enough. In order to find the signal attenuation and bandwidth on bending POFs in strained and unstrained mode, Losada et al. (2004) proposed an experimental setup, which comprises of a path with several posts, the fiber makes a 90° turn at each post. This configuration permits the evaluation of a strained and unstrained fiber with different lengths (0 to 100 m). It was proven that a higher attenuation occurs on a strained fiber (0.26 dB/m for strained fiber and 0.16 dB/m for the unstrained one). The bandwidth for unstrained fiber is wider than for the strained POF for short lengths. However, if the fiber length is higher than 40 m, the bandwidth remains the same for the two conditions.

Although the findings reported earlier for the POF curvature are important for the development of a new experimental setup for a curvature sensor, all experiments were made on static or quasistatic conditions. Therefore the viscoelasticity effects are not considered. An experimental apparatus to evaluate successive bending on a POF is presented in Ziemann et al. (2008). However, the experimental purpose is to evaluate the transmission ratio with the number of bending cycles as a measure of the functionality and serviceable life expectancy of the optical fiber. Hence, the transmission is measured at intervals after a repeated bending cycle. Moreover, the transmission is measured only after a relaxation period of about 60 s, which eliminates almost all viscoelastic effects. Fig. 5.5 shows the results for different bending radii for a controlled temperature of 23 °C.

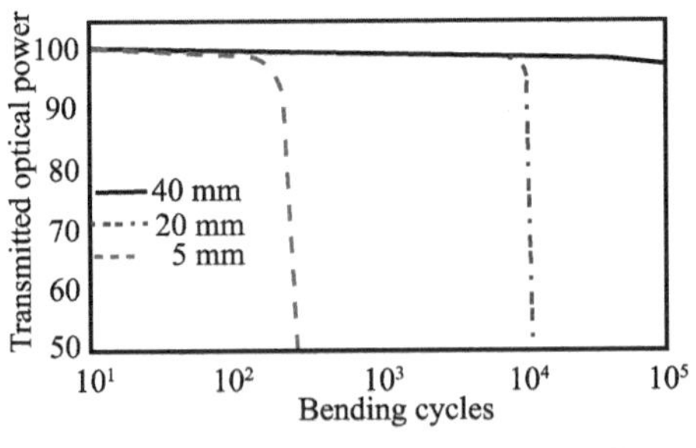

FIGURE 5.5 Repeated bending analysis. Transmission and number of repeated bending cycles for different curvature radii and fixed temperature of 23 °C.

As shown in Fig. 5.5, the temperature has to be controlled because it may cause deviations in transmission results. However, the temperature change in polymers are limited due to glass transition. The glass transition temperature for PMMA is about 100 °C (Peters, 2011). Aware of the temperature limitation, Ziemann et al. (2008) presented an experimental analysis to evaluate the attenuation with a cyclic temperature variation over the time. The temperature varies from −40 °C to 90 °C over 6 hours, the temperature remains at the 90 °C during 1h and the same happens at −40 °C, this cyclic variation occurs during 1000 h. The humidity is fixed at 90% after 1 h of test. Results show a 10% decrease of transmission after 700 h, which represents a very low decrease with time. Kovacevic and Djordjevich (2009) analyzed some optical parameters variations with temperature and POF length. Although the temperature range was quite large (−60 °C to 100 °C), the parameters show little variation. The refractive index maximum variation was 0.03. Bandwidth and mode dispersion have appreciable variation with temperature only after 10 m length. The mode dispersion decreased with the temperature and bandwidth slightly increased when the temperature rises. Silva-López et al. (2005) investigated the strain and temperature sensitivity for POFs, the results show a strain and temperature sensitivity greater in magnitude than that of the silica fibers. Moreover, the thermooptic is negative, which presents the possibility for temperature compensation in strain sensors (Peters, 2011). Bilro et al. (2011) evaluate the power loss due to temperature variation from 25 °C to 70 °C in a POF; however, there was not an appreciable power loss in this temperature range since it was used a compensation method in the characterization. Additionally, the power loss with the relative humidity variation was evaluated in a range of 50% to 90% in a fixed temperature of 30 °C. There was also not a considerable variation with relative humidity. However, all experiments presented were made on static or quasistatic conditions, thus the viscoelastic effects were suppressed. Since the viscoelastic material is affected by temperature, it is possible that a dynamic mechanical analysis with temperature variation achieves a variation in the response curve.

PMMA is a thermorheologically simple material (Lakes, 2009). In this type of polymer the effect of lowering or raising the temperature simply shifts the viscoelastic response horizontally without change the curve shape. This horizontal shift is the time-temperature shift factor; this factor depends on the polymer relaxation time, reference temperature, and actual temperature (Roylance, 2001). Fig. 5.6 shows the logarithmic scale plot of the time-temperature shift factor in a typical response curve, where a_T is the time-temperature shift factor, τ is the polymer relaxation time, C_{crp} is the strain response to a unit stress input.

Eq. (5.2) shows the time-temperature shift factor calculation, where C_1 and C_2 are material constants, which depends on the reference temperature chosen. Universal values for C_1 and C_2 are 17.4 and 51.6, respectively. The reference temperature for these values is the glass transition temperature of the polymer (Roylance, 2001):

$$\log(a_T) = \frac{-C_1(T - T_{ref})}{C_2 + (T - T_{ref})}. \tag{5.2}$$

5.3 Dynamic mechanical analysis in polymer optical fibers

In order to characterize the mechanical properties and polymers viscoelastic response, dynamic characterization tests are performed. The dynamic character-

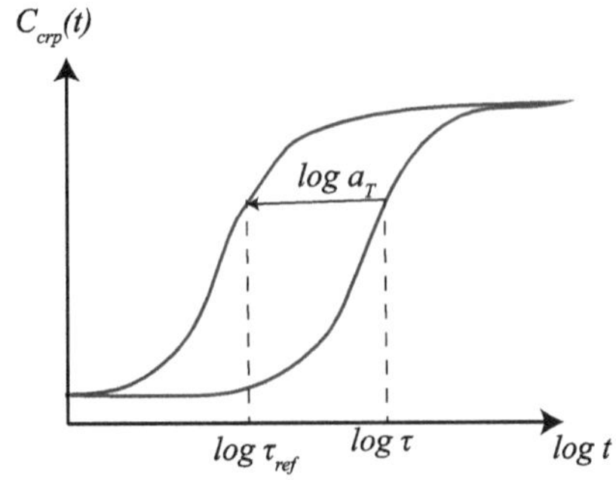

FIGURE 5.6 Time-temperature shift factor.

ization tests are widely employed for polymers in biological applications (Lin et al., 2009), automotive (Ropers et al., 2016), aircraft (Henry et al., 2015), industrial (Celentano et al., 2015) among others. Dynamic tests on polymers can be made by means of nanoindentation techniques (Huang et al., 2004) and dynamic mechanical analysis (DMA) (Badawi, 2015). Nanoindentation techniques are more suitable for the evaluation of thin films and microstructures. Moreover, it is also applicable to the assessment of local properties of the polymer (Huang et al., 2004). Therefore, the DMA technique is a suitable technique due to the fiber dimensions on the order of millimeters. DMA applies an oscillatory load on the polymer with specific temperature and frequency ranges. It is a well-established method to evaluate the polymer glass transition, storage modulus, loss modulus, and stress relaxation (Badawi, 2015), which are sufficient parameters to characterize the POF viscoelastic response.

These parameters include the storage modulus (E'), loss modulus (E''), and relaxation time (τ), which are sufficient to determine the viscoelastic behavior of the material (Lakes, 2009). The combination of the storage and loss modulus is the dynamic Young's modulus (E^*) of the polymer (see Eq. (5.3)).

$$E^* = E' + i E''. \tag{5.3}$$

The ratio between the storage and loss modulus is the loss factor ($\tan(\delta)$) defined in Eq. (5.4). This is a ratio between the dissipated energy and the storage energy per cycle of applied load:

$$\tan(\delta) = \frac{E''}{E'}. \tag{5.4}$$

The analysis of the loss factor presents some advantages over the analysis of the loss modulus. The Young's modulus is divided into the loss and storage modulus due to the duality of a viscoelastic response, which is the combination of the elastic and viscous responses of the polymer. The loss modulus refers to the energy loss due to the viscous effect, whereas the storage modulus refers to the energy storage due to the elastic effect. Therefore, the loss modulus may be understood as the relationship between the viscous and elastic components of the polymer response. For this reason, the loss factor of an elastic material is zero ($\tan(\delta) = 0$) and Young's modulus is referred as a static Young's modulus (E_0).

The dynamic mechanical analyzer employed on the tests of this chapter is the DMA 8000 (Perkin Elmer, USA), where the tests for Young's modulus estimation were made following ISO 527-1:2012, which is the standard recommended for Young's modulus estimation in polymers (ISO, 2012). In addition, the dynamic characterizations were made following ASTM D4065 standard for DMA testing in polymers (International, 2012). The length and width of the DMA's clamps (see Fig. 5.7) are about 22 mm and 5 mm, whereas mass of the clamps is about 6 g. The sample length on the tests is about 20 mm. The analyzer also has a climatic chamber with a heater to isothermal tests and tests with temperature increase. The fiber fixation is shown in Fig. 5.7.

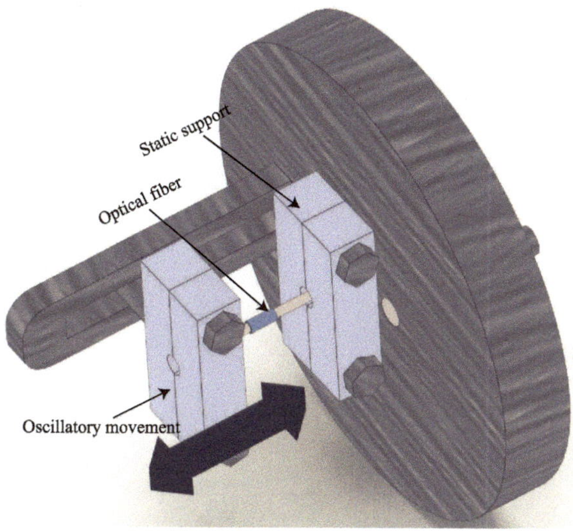

FIGURE 5.7 Schematic view of the optical fiber sample positioning in the DMA.

5.3.1 DMA on PMMA solid core POF

The dynamic analysis of the PMMA POF includes the temperature variation from 0 to 130 °C to evaluate the storage modulus variation as a function of this parameter and the glass transition temperature of the optical fiber. The lower bound of the temperature variation is below the material's Tg, whereas the upper bound is higher than the Tg to demonstrate the material's behavior within the temperature operation range for POF sensors, i.e., below the materials' Tg. Furthermore a constant frequency of 1 Hz is employed and the maximum displacement is 0.05 mm to guarantee the stress-strain curve in the linear region, where the displacement is dynamically applied on the sample following a periodic function with 1 Hz and 0.05 mm amplitude, as shown in Fig. 5.8, which also shows the storage modulus and loss factor with respect to the temperature for the PMMA sample. In this case, it is possible to observe Tg in temperatures higher than 100 °C, characterized by the peak on the loss factor curve, i.e., the storage modulus is low (below the analyzer resolution) and the loss factor is high (above 1.9). The results presented in Fig. 5.8 indicate a temperature range of 24 °C to 60 °C as the suitable temperature operation limit for dynamic sensors applications as it coincides with two transitions in the PMMA, i.e., α-transition (about 20 °C) and β-transition at 65 °C. Since $\tan(\delta)$ is the ratio between the storage and loss modules, as presented in Eq. (5.4), the storage modulus is almost ten times higher than the loss modulus on temperatures between 5 °C and 25 °C (see Fig. 5.8). Such difference is caused by the lower movement of the polymer molecules, leading to the decrease of friction between them that causes the increase of the storage energy (Lakes, 2009), which in this case is almost ten times higher than the dissipated energy.

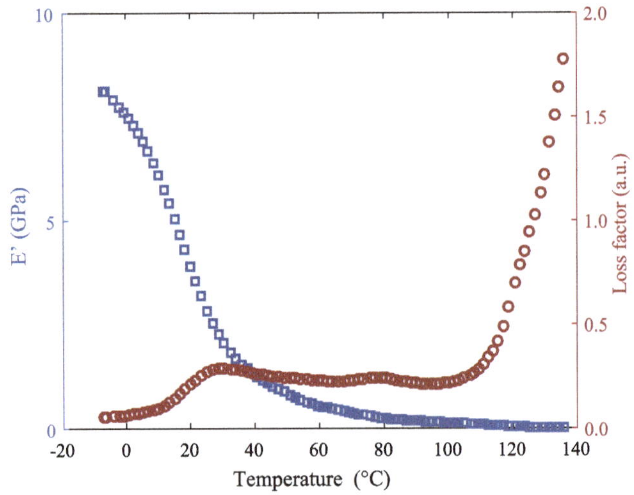

FIGURE 5.8 Storage modulus and loss factor as a function of the temperature for PMMA POF.

In addition, the frequency is also varied on different temperature conditions scan (25 °C to 45 °C with steps of 5 °C) to evaluate the variation of polymer viscoelastic properties as a function of both parameters, i.e., frequency and temperature. This analysis can indicate the possibility of applying the TTS principle, depicted in Eq. (5.2). The frequency range is between 0.01 Hz and 21 Hz, where each frequency sweep is performed at a constant temperature conditions. In this case, the temperatures used on frequency variation tests are 25 °C to 45 °C in 5 °C steps, which are within the temperature operation limits of the PMMA POF. These temperatures are also within the range of thermal comfort of human skin, as depicted in Moreno et al. (2008). The lower frequency bound (0.01 Hz) is limited by the analyzer resolution, whereas the upper bound of 21 Hz is sufficiently large to cover the frequency of movement for applications of wearable sensors in human joints (Kirtley, 2006). The combined analysis of the temperature and frequency effects enable the application of the TTS principle. According to this principle, there is an interchangeability between the variation of the temperature and frequency over the viscoelastic parameters (Menard, 1999). Fig. 5.9 shows the storage modulus variation as a function of the frequency employed on the oscillatory loads in different temperature conditions, leading to five curves, one for each temperature condition. Since there is only an offset as the temperature increases, the PMMA can be regarded as a thermorheological simple material and it is possible to apply the TTS principle (Lakes, 2009).

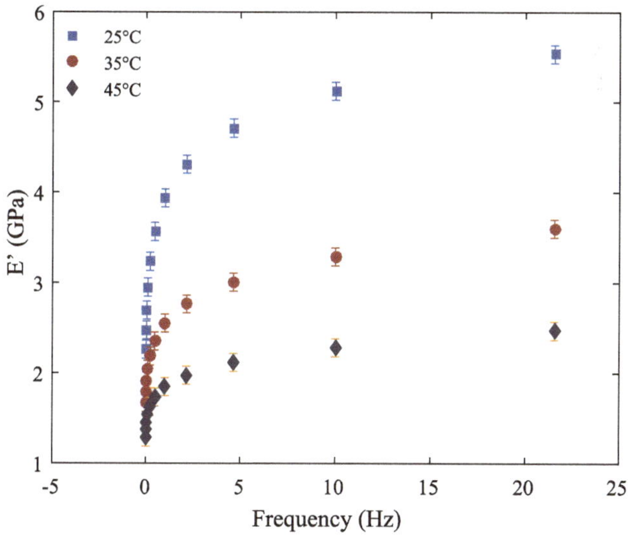

FIGURE 5.9 Storage modulus for frequencies between 0.01 Hz and 21 Hz with different temperatures.

Finally, creep recovery experiments are performed in order to obtain the POF relaxation time and the stress rate dependency of the PMMA optical fiber. The test comprises of applying a constant stress (with constant frequency and temperature) on the sample, which enable the assessment of the polymer relaxation time. In order to evaluate the relaxation time dependency with the stress or strain rate, the creep recovery test is made with different load conditions, where four loads were applied, between 0.5 N and 2 N with 0.5 N steps. The frequency and temperature employed in the creep tests are 1 Hz and 25 °C, respectively. In order to enable the analysis of the relaxation, the input and output are normalized with respect to its mean value, which enable the possibility of obtaining the relaxation time for each test. The difference between the maximum and minimum of the normalized strain measured on the test with 0.5 N is 0.058. This difference increases as the force applied on the test increases until it reaches 0.17 on the test with 2 N load. Fig. 5.10 shows the results for the test with different loads, where the error bars are related to the precision of the DMA sensors.

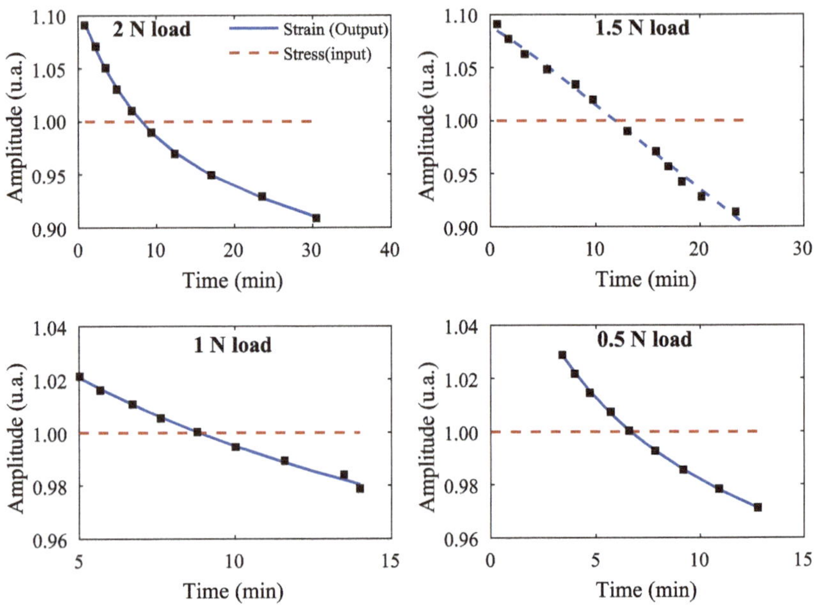

FIGURE 5.10 Creep recovery tests in PMMA solid-core POF using loads of 0.5 N to 2 N in 0.5 N steps.

AAs presented in Fig. 5.10, an exponential fit is made on each curve of the creep recovery tests and the POF relaxation time is the time constant obtained on these exponential approximations. Since there are two different materials, polyethylene (in the POF coating) and PMMA (in the optical fiber core/cladding), the performed exponential regression is a sum on two exponentials, one may be related to the PMMA relaxation and the other to the polyethylene, as presented in Eq. (5.5):

$$\epsilon = y_1 e^{\tau_1 t} + y_2 e^{\tau_2 t}, \tag{5.5}$$

where ϵ is the strain, t is the time, τ_1 and τ_2 are time constants, and y_1 and y_2 are the weights of each exponential. This approach is similar to the Prony series, which is employed for the estimation of the polymer relaxation (Lakes, 2009).

The first relaxation time (τ_1) obtained is 4.89±1.61 seconds, whereas the second relaxation time (τ_2) is 273.46±202.54 seconds. Since it is closer to the one obtained in the literature, the first relaxation time may be related to the PMMA (Stefani et al., 2012) and the second to the polyethylene (jacket of the PMMA POF). However, the relaxation time is not constant either. However, the relaxation time is not constant in many viscoelastic materials, as they follow a range of relaxation times instead of a unique value of this parameter. This behavior of the polymer also can be a source of hysteresis in viscoelastic materials. The polyethylene presents higher relaxation time than the PMMA and will also present a range of relaxation times proportionally higher, which can explain the higher variation of the polyethylene relaxation time. In addition, the temperature also influences the relaxation time of some polymers (Lakes, 2009), which can be the case of the polyethylene, since there is a variation of about 0.5 °C on the tests with 0.5 N and 1.5 N.

As the PMMA has moisture absorption properties, the last set of tests is the humidity sensitivity assessment. These tests present some operational limitations, since the humidity needs to be kept constant for some minutes in order to enable the polymer moisture absorption. The relative humidity (RH) range of the test is from 65% to 95%, where each humidity step is maintained constant for about 30 minutes to enable the water absorption by the polymer. Fig. 5.11 presents the storage modulus variation of the PMMA POF with respect to the RH, where it is possible to observe that the PMMA mPOF presented variations in the storage modulus as the humidity increases. Thus PMMA POFs are suitable for humidity sensing, and on the other hand, compensation for humidity effects are needed in temperature, strain, angle, and force sensors.

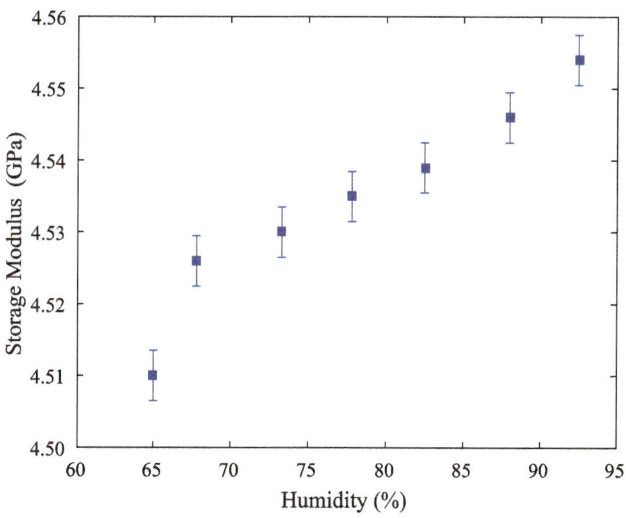

FIGURE 5.11 PMMA POF storage modulus variation as a function of the relative humidity.

5.3.2 Dynamic characterization of CYTOP fibers

In the CYTOP characterization, the DMA was performed in separate regions of the same CYTOP fiber: one in the region of the bare fiber and the other in the region where a femtosecond (fs) laser was focused. Each sample has 7 mm length and the tests are made with a maximum displacement of 0.5 mm. For the dynamic characterization of the polymer viscoelasticity, the displacement was about 0.05 mm, and for the temperature characterization of the storage modulus, the temperature varies from 27 °C to 125 °C at a constant strain cycle frequency was 1 Hz. In contrast, the storage modulus characterizations as a function of the strain cycle frequency are performed at constant temperatures, where the cycle frequency varies from 0.01 Hz to 100 Hz. The frequency characterization was performed at four different temperatures, namely 30 °C, 40 °C, 50 °C, and 60 °C.

In order to verify the Young's modulus of each sample, stress-strain cycles were applied on the samples with and without laser incidence. First, a stress-strain cycles with strain up to 6% is applied on the CYTOP without fs-laser incidence in order to evaluate its response with larger strains. Then, stress-strain cycles are applied on both samples (with and without fs-laser incidence) with strains up to 0.25%, since this region (0–0.25%) is the one recommended by ISO 527-1:2012 standard for Young's modulus estimation in polymers (ISO, 2012), where the Young's modulus is estimated through the slope of the stress-strain curve.

The stress-strain test with higher range for the CYTOP sample is presented in Fig. 5.12(a), whereas the tests with smaller range for Young's modulus estimation of both regions of the CYTOP are shown in Fig. 5.12(b). The stress-strain curve of Fig. 5.12(a) shows a linear response until about 2.5% strain with a yield stress of 70 MPa for the CYTOP fiber (without fs-laser incidence) and the strain region for the Young's modulus evaluation are also presented. In the evaluation of the Young's modulus, depicted in Fig. 5.12(b), it is possible to see a linear stress-strain relation in which there is an increase of the stress-strain curve slope for the sample with fs-laser incidence. Thus the sample with laser incidence has a higher Young's modulus than the bare fiber. However, the difference between both samples is about 0.22 GPa, which represents a relative Young's modulus variation higher than 10%. It is noteworthy that this relative variation is higher than the standard deviation between the three performed tests.

The CYTOP material properties are also evaluated under different temperature and frequency of oscillatory movements. In order to estimate the CYTOP moduli at higher frequency range, the TTS principle is applied, which enables the creation of master curves for estimation of the material viscoelastic properties within ranges higher than the ones of the DMA equipment (Menard, 1999). As the principle requires a linear viscoelastic regime, Fig. 5.13(a) shows the storage modulus of the sample as a function of the temperature, whereas Fig. 5.13(b) depicts the frequency response of the material within the temperature range 30 °C to 60 °C.

In the analyzed range, the temperature leads to a linear shift (a_T) on the storage modulus—frequency curves, which is in agreement with the first requirement for the TTS principle and define the material as thermorheological simple in this temperature range (Leal-Junior et al., 2018a). The second require-

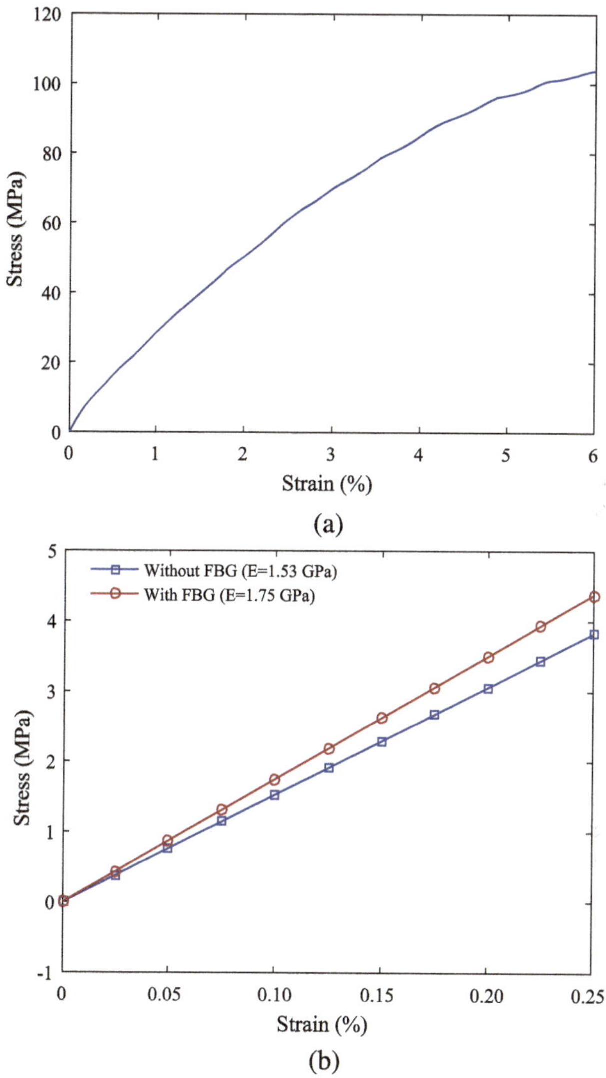

FIGURE 5.12 (a) Stress-strain curve for the CYTOP in a strain range of 0%–6%. (b) Stress-strain curves for Young's modulus calculation for CYTOPs with and without fs-laser incidence.

ment is that there is no physical changes or phase transition on the material on the considered temperature range (Menard, 1999). In order to verify this assumption, the loss factor, tan(δ), of the material is analyzed, as Fig. 5.14 depicts. In this case, the tan(δ) peak (at 122 °C) is associated with the material glass transition temperature (Tg), which limits the TTS principle application in temperature higher than (or close to) the Tg. Nevertheless, the results of Fig. 5.14

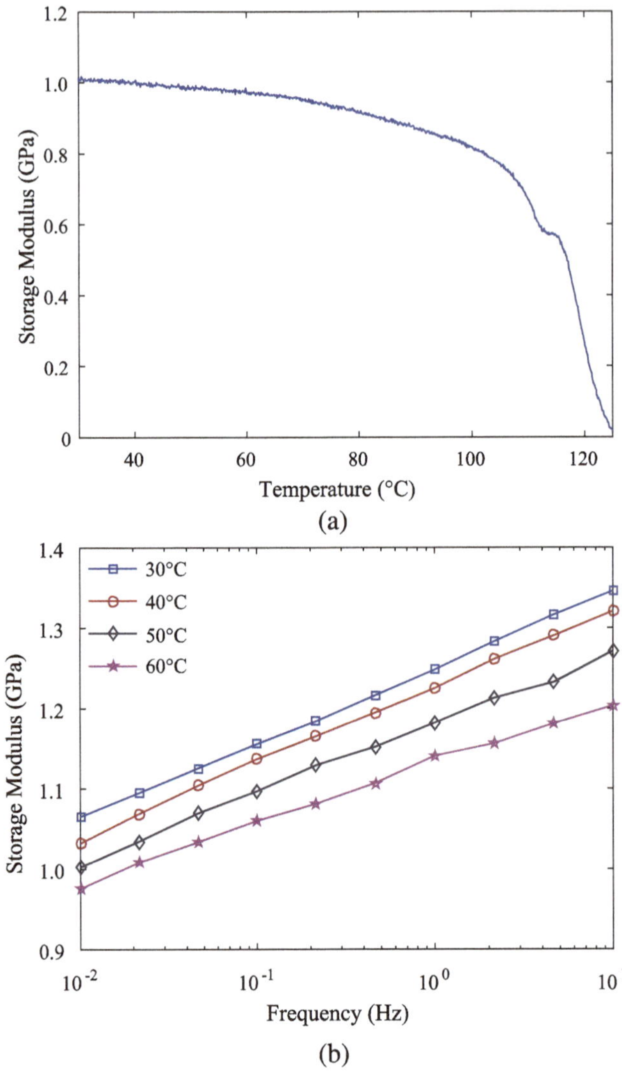

FIGURE 5.13 (a) CYTOP's storage modulus as a function of the temperature. (b) Frequency response in 4 different temperatures (30 °C, 40 °C, 50 °C, and 60 °C).

also show that the material does not undergo any phase transition in the analyzed temperature range (30 °C to 60 °C).

Hence, Figs. 5.13 and 5.14 provide sufficient background to the application of the TTS principle on the frequency and temperature results. In this case, it is possible to extent the estimation of the material viscoelastic properties to a frequency range higher than the DMA operation range.

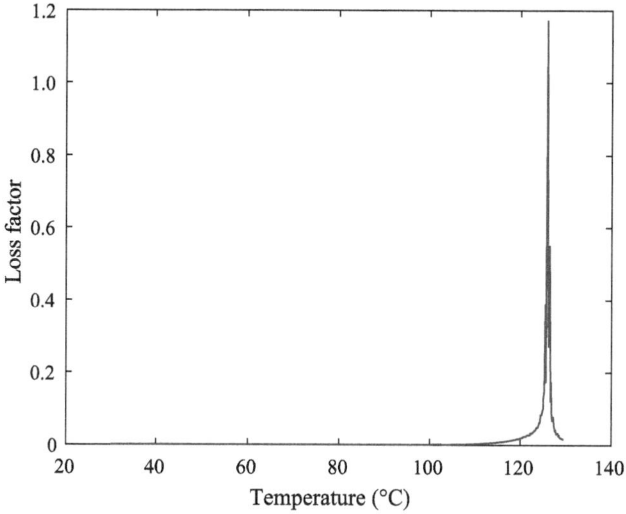

FIGURE 5.14 Loss factor on the range of 27 °C to 130 °C.

5.4 Influence of optical fiber treatments on polymer properties

As discussed in Sections 5.2 and 5.3 the polymer viscoelasticity can influence the material responses, and consequently, the sensors' behavior, especially due to the possibility of significant hysteresis on the sensor's responses to dynamic loading conditions, which can be reduced through by compensation techniques on the signal processing (Leal-Junior et al., 2018c), will be discussed throughout this book or with the proper thermal treatments on the optical fiber (Yuan et al., 2011). A commonly applied thermal treatments in optical fibers, especially in POFs, is the annealing, which comprises of keeping the POF on a temperature higher than the polymer β-transition for some time (generally some hours), which results in a relaxation of the internal stress created on its manufacturing process of the optical fiber (Pospori et al., 2017), especially on the extrusion processes (discussed in Chapter 4). Furthermore, the fiber annealing can be mediated by predefined solutions, such as a methanol solution that allows the annealing at room temperature (Fasano et al., 2017). Another variant for the annealing treatment is to perform it under water (or high humidity conditions) that reduces the annealing time to some minutes, instead of hours (Pospori et al., 2017). The fiber annealing can also influence the sensor sensitivity to strain and stress as reported in Pospori et al. (2017). In addition, such thermal treatment can influence the behavior of fiber Bragg gratings (presented in Chapter 6), as it can reduce the gratings inscription time when it is made before the grating inscription (Marques et al., 2017) or it can lead to blue shift of the Bragg wavelength, especially when it is made at high humidity conditions for the case of the fiber annealing after the grating inscription (Woyessa et al., 2016). Another commonly employed post-fabrication treatment in optical fiber is the etching, a chemical treatment that generally comprises of positioning the optical fiber in a container filled with a solution that reacts with the optical fiber material, e.g., acetone for PMMA POFs. Such chemical treatment can lead to variation of the polymer properties due to the molecular chain relaxation, related to the solvent absorption (Bhowmik et al., 2016). The chemical reaction between the PMMA POF and the acetone leads to a diameter reduction of the fiber, which provide lower response times for POF-based humidity sensors (Rajan et al., 2013). In addition, the fiber etching can provide higher reflectivity in fiber Bragg gratings and lower exposure times on the gratings inscription (Bhowmik et al., 2016).

In order to evaluate the influences of each treatment on a PMMA POF, different samples were fabricated and tested. There are five PMMA POF samples, one without any post-fabrication treatment, two annealed samples, and two etched samples. The annealing at low humidity is performed with a constant temperature of 70 °C and relative humidity of 10%, the annealing time is 24 hours. The annealing temperature is higher than the PMMA β-transition temperature (around 50 °C), which makes it suitable for molecular rearrangement. For comparison purposes, the annealing under water is made with the same annealing temperature and time (i.e., 70 °C and 24 hours). In order to eliminate the influence of any water absorbed by the fiber, the POFs annealed under water are positioned outside the water recipient for about 24 hours after the annealing. For the etching process, the samples are placed at the vertical position in a container filled with pure acetone, where one set of five samples is etched for 4 minutes and the other, for 7 minutes. The etching is made in about 10 mm of the fiber length, which resulted in a diameter reduction to 0.92 mm and 0.88 mm for the etching times of 4 and 7 minutes, respectively. Such reduction represents a rate of about 0.012 mm/min for the etching made with pure acetone. In addition, a deviation of about 0.01 mm was obtained in all etched samples.

The stress-strain cycles are performed on each sample to indicate the effect of each thermal and chemical treatment on the PMMA POF mechanical and thermal properties, namely the Young's modulus and thermal expansion coefficient depicted in Table 5.1. As the annealing leads to a relaxation of the fiber stress generated at its fabrication process, it is expected a reduction of the PMMA samples Young's modulus. In addition, the annealing at high humidity conditions leads to a higher relaxation of the polymer, which result in an even lower Young's modulus. Comparing with the fiber without thermal and chemical treatments, the annealing at low humidity resulted in a Young's modulus reduction of about 32.5%, which is further reduced if the annealing is made at high humidity condition (under water). In this case, a further Young's modulus reduction of 12.8% was obtained when compared with the Young's modulus of the POF annealed at low humidity (2.96 GPa).

Regarding the chemical etching, the Young's modulus presented a reduction proportional to the etching time. However, it is possible that the Young's modu-

TABLE 5.1 Measured thermal expansion coefficient and Young's modulus for each PMMA POF sample.

Thermal/Chemical Treatment	Thermal Expansion Coefficient (K^{-1})	Young's modulus (GPa)
No treatment	$(7.14 \pm 0.07) \times 10^{-5}$	4.38 ± 0.50
Annealing at low humidity	$(7.28 \pm 0.07) \times 10^{-5}$	2.96 ± 0.32
Annealing under water	$(7.46 \pm 0.09) \times 10^{-5}$	2.58 ± 0.23
Etching (4 minutes)	$(7.40 \pm 0.12) \times 10^{-5}$	2.64 ± 0.28
Etching (7 minutes)	$(7.38 \pm 0.13) \times 10^{-5}$	2.16 ± 0.20

lus only reduces until a time period due to a saturation on the solvent absorption by the polymer chains, which can be related to the polymer absorption time. Nevertheless, this saturation time is higher than 7 minutes for PMMA POFs, since the Young's modulus still presents reduction at this etching time.

The effect of thermal and chemical treatments is also verified with respect to the POF thermal expansion coefficient, where the obtained results are presented in Table 5.1, where the results show a small variation of the thermal expansion coefficient with the thermal and chemical treatments. The fiber annealing leads to an increase of the thermal expansion coefficient, especially when it is made under water, which also happens with the etching treatment. However, results show the thermal expansion coefficient starts to decrease when the etching is made after the annealing for etching times higher than 4 minutes, where the thermal expansion coefficient is approximately the same as the one of the fiber without treatment (considering the experimental uncertainty). However, it is important to mention that only minor variations of the thermal expansion coefficient were found in all analyzed cases. These relative differences are smaller than the ones of the Young's modulus, which indicate that the thermal and chemical treatments in the optical fiber present higher influence in the mechanical properties than in the thermal parameters.

In the following tests, the PMMA samples are positioned on the DMA for the temperature and frequency variation tests. In the temperature characterization, the storage modulus variation of the samples are analyzed as a function of the temperature, where the temperature increases in a range 2 °C/min until the temperature of 50 °C is reached. Fig. 5.15(a) shows storage modulus variation with respect to the temperature for each sample, i.e., without treatments, annealed at low humidity, annealed at high humidity, etched for 4 minutes, and etched for 7 minutes. The PMMA POF without heat and chemical treatments presented the highest variation of the storage modulus, whereas the annealed POFs show a storage modulus decrease with almost constant slope, where the POF annealed under water has the lowest variation among the ones submitted to heat treatment. Regarding the etched fibers, the etching time of 7 minutes provided the lowest storage modulus variation among all the tested samples.

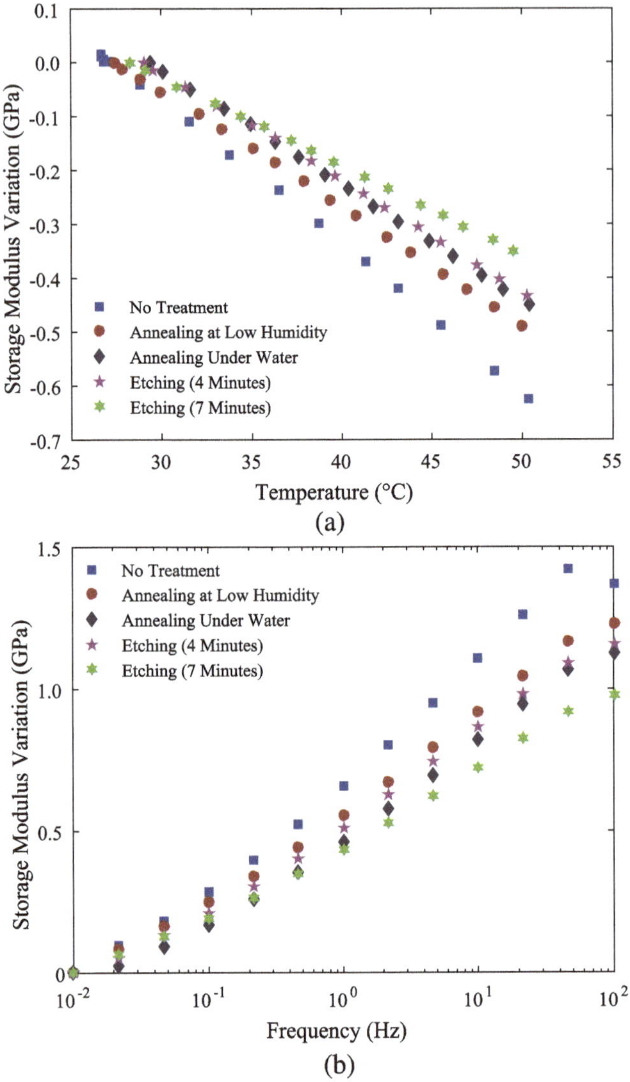

FIGURE 5.15 Storage modulus variation as a function of (a) temperature and (b) frequency for the annealed and etched POFs.

In Fig. 5.15(b), the storage modulus variation is evaluated with respect to the strain cycle frequency to explore the thermal and chemical treatments. The results for the nontreated, etched, and annealed fibers are presented in Fig. 5.15(b), where the higher variation of the PMMA POF sample without thermal treatment indicates that this sample presents high cross-sensitivity with the movement frequency on strain applications that requires oscillatory movement such as angle measurements on flexion and extension cycles, as depicted throughout this book, especially on Chapters 7 and 8. In such applications, the POF with etching time of 7 minutes may be preferred, since it presents lower storage modulus variation. However, the etching treatment increases the optical fiber brittleness, which makes them unsuitable for applications that involve high deformation, such as the ones in wearable sensors and robots. In addition, the samples without thermal and chemical treatments present a decrease of the modulus variation in frequencies close to 100 Hz as presented in Stefani et al. (2012), which was not observed in the annealed and etched samples. Thus the results indicate that the annealing treatment can provide significant advantages for sensors applications, especially for wearable applications due to the Young's modulus reduction, which leads to higher stress sensitivity as well as smaller Young's modulus variation with frequency and temperature when compared with the sample without thermal and chemical treatments, which that indicate smaller cross-sensitivity with the movement dynamics and environmental conditions for a sensor for mechanical parameters (e.g., pressure, angle, or force sensor).

References

Arrue, J., Zubia, J., 1996. Analysis of the decrease in attenuation achieved by properly bending plastic optical fibres. IEE Proceedings. Optoelectronics 143, 135. https://doi.org/10.1049/ip-opt:19960181.

Badawi, A., 2015. Characterization of the optical and mechanical properties of cdse qds/pmma nanocomposite films. Journal of Materials Science: Materials in Electronics 26, 3450–3457. https://doi.org/10.1007/s10854-015-2854-1.

Beisenova, A., Issatayeva, A., Iordachita, I., Blanc, W., Molardi, C., Tosi, D., 2019. Distributed fiber optics 3d shape sensing by means of high scattering np-doped fibers simultaneous spatial multiplexing. Optics Express 27, 22074. https://doi.org/10.1364/oe.27.022074.

Bhowmik, K., Peng, G.D., Luo, Y., Ambikairajah, E., Lovric, V., Walsh, W.R., Rajan, G., 2016. Etching process related changes and effects on solid-core single-mode polymer optical fiber grating. IEEE Photonics Journal 8, 1–9. https://doi.org/10.1109/JPHOT.2016.2524210. http://ieeexplore.ieee.org/lpdocs/epic03/wrapper.htm?arnumber=7397839.

Bilro, L., Oliveira, J.G., Pinto, J.L., Nogueira, R.N., 2011. A reliable low-cost wireless and wearable gait monitoring system based on a plastic optical fibre sensor. Measurement Science and Technology 22, 045801. https://doi.org/10.1088/0957-0233/22/4/045801.

Broadway, C., Min, R., Leal-Junior, A.G., Marques, C., Caucheteur, C., 2019. Toward commercial polymer fiber Bragg grating sensors: review and applications. Journal of Lightwave Technology 37, 2605–2615. https://doi.org/10.1109/JLT.2018.2885957.

Celentano, D., Wimmer, D., Colabella, L., Cisilino, A.P., 2015. Viscoelastic mechanical characterization of a short-fiber reinforced polyethylene tube: experiments and modelling. International Journal of Pressure Vessels and Piping 134, 82–91. https://doi.org/10.1016/j.ijpvp.2015.08.007.

Cordeiro, C.M., Ng, A.K., Ebendorff-Heidepriem, H., 2020. Ultra-simplified single-step fabrication of microstructured optical fiber. Scientific Reports 10, 9678. https://doi.org/10.1038/s41598-020-66632-3.

Fasano, A., Woyessa, G., Janting, J., Rasmussen, H.K., Bang, O., 2017. Solution-mediated annealing of polymer optical fiber Bragg gratings at room temperature. IEEE Photonics Technology Letters 29, 687–690. https://doi.org/10.1109/LPT.2017.2678481.

Fasano, A., Woyessa, G., Stajanca, P., Markos, C., Stefani, A., Nielsen, K., Rasmussen, H.K., Krebber, K., Bang, O., 2016. Fabrication and characterization of polycarbonate microstructured polymer optical fibers for high-temperature-resistant fiber Bragg grating strain sensors. Optical Materials Express 6, 649. https://doi.org/10.1364/OME.6.000649. http://www.osapublishing.org/viewmedia.cfm?uri=ome-6-2-649&seq=0&html=true.

Guo, J., Niu, M., Yang, C., 2017. Highly flexible and stretchable optical strain sensing for human motion detection. Optica 4, 1285. https://doi.org/10.1364/optica.4.001285.

Henry, T.C., Bakis, C.E., Smith, E.C., 2015. Viscoelastic characterization and self-heating behavior of laminated fiber composite driveshafts. Materials and Design 66, 346–355. https://doi.org/10.1016/j.matdes.2014.10.083.

Huang, G., Wang, B., Lu, H., 2004. Measurements of viscoelastic functions of polymers in the frequency-domain using nanoindentation. Mechanics of Time-Dependant Materials 8, 345–364. https://doi.org/10.1007/s11043-004-0440-7.

International, A., 2012. Astm d4065 – standard practice for plastics: dynamic mechanical properties: determination and report of procedures. Annual Book of ASTM Standards i, 1–7. https://doi.org/10.1520/D4065-12.1.8.

ISO, 2012. https://www.iso.org/standard/56045.html.

Johnson, I., Yuan, W., Stefani, A., Nielsen, K., Rasmussen, H., Khan, L., Webb, D., Kalli, K., Bang, O., 2011. Optical fibre Bragg grating recorded in TOPAS cyclic olefin copolymer. Electronics Letters 47, 271. https://doi.org/10.1049/el.2010.7347.

Kiesel, S., Peters, K., Hassan, T., Kowalsky, M., 2007. Behaviour of intrinsic polymer optical fibre sensor for large-strain applications. Measurement Science & Technology 18, 3144–3154. https://doi.org/10.1088/0957-0233/18/10/s16.

Kirtley, C., 2006. Clinical Gait Analysis: Theory and Practice. Elsevier, Philadelphia.

Koike, Y., Asai, M., 2009. The future of plastic optical fiber. NPG Asia Materials 1, 22–28. https://doi.org/10.1038/asiamat.2009.2.

Kovacevic, M., Djordjevich, A., 2009. Temperature dependence analysis of mode dispersion in step-index polymer optical fibers 116, 649–651. https://doi.org/10.12693/APhysPolA.116.649.

Lakes, R., 2009. Viscoelastic Materials. Cambridge University Press, Cambridge.

Leal-Junior, A., Marques, C., Frizera, A., Pontes, M., 2018a. Dynamic mechanical analysis on a polymethyl methacrylate (pmma) polymer optical fiber. IEEE Sensors Journal 18. https://doi.org/10.1109/JSEN.2018.2797086.

Leal-Junior, A., Theodosiou, A., Frizera-Neto, A., Pontes, M.J., Shafir, E., Palchik, O., Tal, N., Zilberman, S., Berkovic, G., Antunes, P., André, P., Kalli, K., Marques, C., 2018b. Characterization of a new polymer optical fiber with enhanced sensing capabilities using a Bragg grating. Optics Letters 43, 4799. https://doi.org/10.1364/OL.43.004799.

Leal-Junior, A.G.A., Frizera, A., Pontes, M.M.J., 2018c. Dynamic compensation technique for POF curvature sensors. Journal of Lightwave Technology 36, 1112–1117. https://doi.org/10.1109/JLT.2017.2752361. http://ieeexplore.ieee.org/document/8037980/.

Lin, I.K., Ou, K.S., Liao, Y.M., Liu, Y., Chen, K.S., Zhang, X., 2009. Viscoelastic characterization and modeling of polymer transducers for biological applications. Journal of Microelectromechanical Systems 18, 1087–1099. https://doi.org/10.1109/JMEMS.2009.2029166.

Liu, Z., Zhang, Z.F., Tam, H.Y., Tao, X., 2019. Multifunctional smart optical fibers: materials, fabrication, and sensing applications. Photonics 6, 1–24. https://doi.org/10.3390/photonics6020048.

Losada, M.A., Mateo, J., Garcés, I., Zubía, J., Casao, J.A., Pérez-Vela, P., 2004. Analysis of strained plastic optical fibers. IEEE Photonics Technology Letters 16, 1513–1515. https://doi.org/10.1109/LPT.2004.826780.

Luo, Y., Canning, J., Zhang, J., Peng, G.D., 2020. Toward optical fibre fabrication using 3d printing technology. Optical Fiber Technology 58, 102299. https://doi.org/10.1016/j.yofte.2020.102299.

Luo, Y., Yan, B., Zhang, Q., Peng, G.D., Wen, J., Zhang, J., 2017. Fabrication of polymer optical fibre (pof) gratings. Sensors 17, 511. https://doi.org/10.3390/s17030511.

Markos, C., Stefani, A., Nielsen, K., Rasmussen, H.K., Yuan, W., Bang, O., 2013. High-tg topas microstructured polymer optical fiber for fiber Bragg grating strain sensing at 110 degrees. Optics Express 21, 4758–4765. https://doi.org/10.1364/OE.21.004758.

Marques, C.A., Pospori, A., Demirci, G., Çetinkaya, O., Gawdzik, B., Antunes, P., Bang, O., Mergo, P., André, P., Webb, D.J., 2017. Fast Bragg grating inscription in PMMA polymer optical fibres: impact of thermal pre-treatment of preforms. Sensors (Switzerland) 17, 1–8. https://doi.org/10.3390/s17040891.

Menard, K., 1999. Dynamic Mechanical Analysis: A Practical Introduction. CRC Press.

Min, R., Ortega, B., Leal-Junior, A., Marques, C., 2018. Fabrication and characterization of Bragg grating in CYTOP POF at 600-nm wavelength. IEEE Sensors Letters 2, 1–4. https://doi.org/10.1109/LSENS.2018.2848542. https://ieeexplore.ieee.org/document/8387446/.

Moreno, J.C., Bueno, L., Pons, J.L., Baydal-Bertomeu, J.M., Belda-Lois, J.M., Prat, J.M., Barber, R., 2008. Wearable robot technologies. In: Wearable Robots. John Wiley & Sons, Ltd, Chichester, UK, pp. 165–200. From Duplicate 2 (Wearable robot technologies – Moreno J.C., Bueno L., Pons J.L., Baydal-Bertomeu J.M., Belda-Lois J.M., Prat J.M., Barber R.).

Nguyen, L., Bhaktha, B.N.S., Sebbah, P., Blanc, W., Pal, B.P., Dussardier, B., Upr, C., Gregory, B., Lucioles, L., 2011. Fabrication of rare Earth-doped transparent glass ceramic optical fibers by modified chemical vapor deposition. Journal of the American Ceramic Society 2318, 2315–2318. https://doi.org/10.1111/j.1551-2916.2011.04672.x.

Peters, K., 2011. Polymer optical fiber sensors—a review. Smart Materials and Structures 20, 013002. https://doi.org/10.1088/0964-1726/20/1/013002. From Duplicate 2 (Polymer optical fiber sensors—a review – Peters Kara).

Pospori, A., Marques, C.A.F., Sáez-Rodríguez, D., Nielsen, K., Bang, O., Webb, D.J., 2017. Thermal and chemical treatment of polymer optical fiber Bragg grating sensors for enhanced mechanical sensitivity. Optical Fiber Technology 36, 68–74. https://doi.org/10.1016/j.yofte.2017.02.006. http://www.sciencedirect.com/science/article/pii/S106852001730007X.

Prado, A., Leal-Junior, A., Marques, C., Leite, S., De Sena, G., Machado, L., Frizera, A., Ribeiro, M., Pontes, M., 2017. Polymethyl methacrylate (pmma) recycling for the production of optical fiber sensor systems. Optics Express 25. https://doi.org/10.1364/OE.25.030051.

Rajan, G., Noor, Y.M., Liu, B., Ambikairaja, E., Webb, D.J., Peng, G.D., 2013. A fast response intrinsic humidity sensor based on an etched singlemode polymer fiber Bragg grating. Sensors and Actuators. A, Physical 203, 107–111. https://doi.org/10.1016/j.sna.2013.08.036. https://linkinghub.elsevier.com/retrieve/pii/S0924424713004202.

Riande, E., Diaz-Calleja, R., Prolongo, M., Masegosa, R., Salom, C., 2000. Polymer Viscoelasticity: Stress and Strain in Practice. Marcel Dekker, New York.

Ropers, S., Kardos, M., Osswald, T.A., 2016. A thermo-viscoelastic approach for the characterization and modeling of the bending behavior of thermoplastic composites. Composites. Part A, Applied Science and Manufacturing 90, 22–32. https://doi.org/10.1016/j.compositesa.2016.06.016.

Roriz, P., Frazão, O., Lobo-Ribeiro, A.B., Santos, J.L., Simões, J.A., 2013. Review of fiber-optic pressure sensors for biomedical and biomechanical applications. Journal of Biomedical Optics 18, 050903. https://doi.org/10.1117/1.JBO.18.5.050903.

Roylance, D., 2001. Engineering viscoelasticity, pp. 1–37.

Shiryaev, V.S., 2015. Chalcogenide glass hollow-core microstructured optical fibers. Frontiers in Materials 2, 1–10. https://doi.org/10.3389/fmats.2015.00024.

Silva-López, M., Fender, A., Macpherson, W.N., Barton, J.S., Jones, J.D.C., Zhao, D., Dobb, H., Webb, D.J., Zhang, L., Bennion, I., 2005. Strain and temperature sensitivity of a single-mode polymer optical fiber. Optic Letters 30, 3129–3131. https://doi.org/10.1364/OL.30.003129.

Silveira, M., Frizera, A., Leal-Junior, A., Ribeiro, D., Marques, C., Blanc, W., Camilo, C.A., 2020. Transmission–reflection analysis in high scattering optical fibers: a comparison with single-mode optical fiber. Optical Fiber Technology 58, 102303. https://doi.org/10.1016/j.yofte.2020.102303.

Stefani, A., Andresen, S., Yuan, W., Bang, O., 2012. Dynamic characterization of polymer optical fibers. IEEE Sensors Journal 12, 3047–3053. https://doi.org/10.1109/JSEN.2012.2208951.

Tejedor, J., Macias-Guarasa, J., Martins, H.F., Pastor-Graells, J., Corredera, P., Martin-Lopez, S., 2017. Machine learning methods for pipeline surveillance systems based on distributed acoustic sensing: a review. Applied Sciences (Switzerland) 7, 1–26. https://doi.org/10.3390/app7080841.

Veber, A., Lu, Z., Vermillac, M., Pigeonneau, F., Blanc, W., Petit, L., 2019. Nano-structured optical fibers made of glass-ceramics, and phase separated and metallic particle-containing glasses. Fibers 7, 105. https://doi.org/10.3390/fib7120105.

Woyessa, G., Fasano, A., Markos, C., Stefani, A., Rasmussen, H.K., Bang, O., 2017. Zeonex microstructured polymer optical fiber: fabrication friendly fibers for high temperature and humidity insensitive Bragg grating sensing. Optical Materials Express 7, 286. https://doi.org/10.1364/OME.7.000286.

Woyessa, G., Nielsen, K., Stefani, A., Markos, C., Bang, O., 2016. Temperature insensitive hysteresis free highly sensitive polymer optical fiber Bragg grating humidity sensor. Optics Express 24, 1206. https://doi.org/10.1364/OE.24.001206.

Yang, D.X., Yu, J., Tao, X., Tam, H., 2004. Structural and mechanical properties of polymeric optical fiber. Materials Science and Engineering A 364, 256–259. https://doi.org/10.1016/j.msea.2003.08.025.

Yuan, W., Khan, L., Webb, D.J., Kalli, K., Rasmussen, H.K., Stefani, A., Bang, O., 2011. Humidity insensitive TOPAS polymer fiber Bragg grating sensor. Optics Express 19, 19731–19739. https://doi.org/10.1364/OE.19.019731. http://www.ncbi.nlm.nih.gov/pubmed/21996915.

Zhang, J., Liu, H., Wu, X., 2009. Curvature optical fiber sensor by using bend enhanced method. Frontiers of Optoelectronics in China 2, 204–209. https://doi.org/10.1007/s12200-009-0032-x.

Ziemann, O., Krauser, J., Zamzow, P.E., Daum, W., 2008. POF Handbook. Springer, Berlin, Heidelberg.

Zubia, J., Arrue, J., Mendioroz, a., 1997. Theoretical analysis of the torsion-induced optical effect in a plastic optical fiber. Optical Fiber Technology 3, 162–167. https://doi.org/10.1006/ofte.1997.0212.

Chapter 6

Optical fiber sensing technologies☆

Optical fiber sensors (OFS) can be applied in several fields of remote sensing, since they require low electrical power and have small size, i.e., they can be easily fit in small areas (Othonos and Kalli, 1999). They present immunity to electromagnetic interference, and for this reason can be used in places with high electromagnetic field interference, such as inside motors (Alwis et al., 2016) or magnetic resonance imaging equipment (Dziuda et al., 2012). Since there are only optical signals in the sensor head, it is suitable for classified areas. Furthermore, the advantages of OFS over conventional sensors are the fact of being ideal for microwave and harsh environments since OFS can monitor wide range of physical and chemical parameters. It consists of thin, low-loss waveguide made of glass or polymer with higher refractive index than its surrounding region, which allows light to propagate through single or multimode fiber (Othonos and Kalli, 1999). The worldwide consumption value of OFS systems is estimated to increase from \$3.56 billion to \$7.95 billion from 2017 to 2027 (Photonic sensors market, 2020). The flexibility and small dimensions of optical fiber sensors in their many variants enable the embedment in different structures with a variety of geometries and configurations (Liu ct al., 2019). In order to study and monitor the structural health, optical fibers have been embedded in concrete and composite structures in the last few years (Kinet et al., 2014). Similar to monitor leakages, such sensors are integrated in dams or diaphragm construction walls (Ukil et al., 2012). Moreover, the optical fibers are embedded in pipelines for their structural monitoring as well as the assessment of leakages (Tejedor et al., 2017).

6.1 Intensity variation sensors

Intensity variation-based sensors employ the optical power variation (transmitted or reflected) along the optical fiber as the sensing principle. Generally, these sensors use the optical fiber attenuation mechanisms and their relation with environmental variations and mechanical loadings to perform the estimation of a predefined parameter. Among their many variants, intensity variation sensors are commonly based on the macrobending losses in an optical fiber and the

☆ This chapter is carried out with the participation of Leticia Munhoz de Avellar.

https://doi.org/10.1016/B978-0-32-385952-3.00015-9

light coupling principle between a light source and the waveguide. In addition, quasidistributed sensing systems using intensity variation sensors were recently proposed and the operation principle of this approach is also discussed in this section.

6.1.1 Macrobending sensors

Intensity variation-based sensors have been used in different sensing fields, including healthcare applications (Leal-Junior et al., 2019). In general, these sensors consist of multimode polymer optical fibers (POFs) and take advantage of the optical fiber radiation loss due the macrobend. Considering that the fiber is fixed on a rotating joint, as the angle of the joint change the output power also changes, i.e., if a positive bend occurs the output power decreases. On the other hand, if the bend occurs on the opposite direction, the output power increases (Leal Junior et al., 2017). Therefore this type of sensor can measure strain or stress on both directions (flexion and extension).

A typical POF macrobending-based sensor generally consists of a POF with a lateral section. The lateral section creates a sensitive zone, which increases the sensor sensitivity and linearity of the signal attenuation when the fiber is under a bending. A conventional POF has a circular cross-section with three layers: core, cladding, and coating. The core section is where the most part of the optical signal is transmitted due to its higher refractive index than the cladding and the jacket provides only mechanical and chemical protection. The light losses are due to absorption and frustrated total internal reflection when in contact with the surface (Bilro et al., 2012). When the bending occurs, the incident angle increases and creates a variation on the transmitted signal. When the sensitive zone of the fiber is bending, there are higher losses due to this POF region that has no cladding, increasing the radiation losses. Another source of loss is the surface scattering caused by the coupling between higher and lower guided modes (Liu et al., 2006). Fig. 6.1 presents the description of a fiber geometry considered in the POF sensor modeling. The sensitive zone is represented by the section length given by c and the section depth of removed material on the fiber core denoted by p. The optical fiber length in Fig. 6.1 is given by L, meanwhile the optical fiber diameter is d, and the curvature radius is R.

The optical power of the POF sensor with lateral section is modeled using geometrical optics concepts, as considered by (Kovacević et al., 2006), and presented in Eq. (6.1). In the following equation, (dP) is an element power radiated into a solid angle subtended by a section (dS) at an angle $(d\theta)$. In addition, L_0 represents the light source radiance. The integration of Eq. (6.1) over the cross-sectional area gives the input power of the model, where a is the optical fiber core radius, and θ_c is the critical angle, as shown in Eq. (6.2):

$$dP = 2\pi L_0 \cos(\theta)\sin(\theta)d\theta dS, \tag{6.1}$$

$$P_{in} = \pi^2 L_0 a \sin^2(\theta_c). \tag{6.2}$$

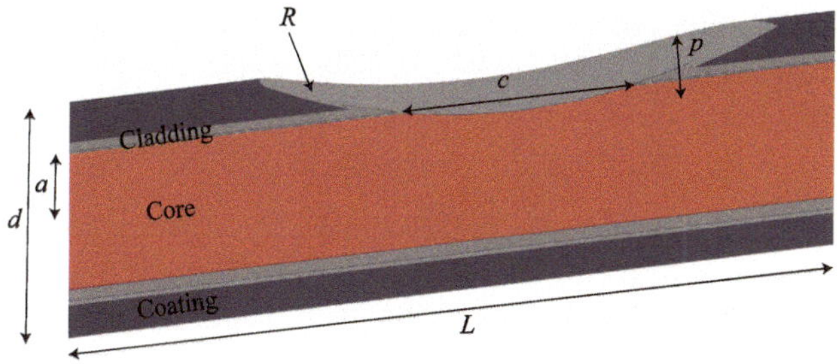

FIGURE 6.1 Schematic representation of the sensitive zone in a macrobending sensor.

Assuming a constant radiance of the light source, i.e., an uniform mode distribution, the difference of the fiber core cross-sectional area (S_c) and the maximum area of removed material (S_o) can be defined (Kovacević et al., 2006) from the parameters of Fig. 6.1, as presented in Eq. (6.3):

$$S_c - S_o = \frac{\pi a^2}{2} + a^2 \arcsin\left(\frac{a-p}{a}\right) + (a-p)\sqrt{a^2 - (a-p)^2}. \quad (6.3)$$

The evaluation of the optical fiber sensitive zone power (P_s) depends of the cross-section of the sensitive zone, which is defined in Eq. (6.3). P_s can also be defined as the remaining power when it enters the sensitive zone, and can be calculated by Eq. (6.4):

$$P_s = \pi L_0 (S_c - S_o) \sin^2(\theta_c). \quad (6.4)$$

Furthermore, a loss of the transmitted power can occur when the light meets the side-polished interface (Kovacević et al., 2006), and due to it is a dielectric media, the losses can be calculated through the Fresnel's equations and the reflection coefficient (r_T) for an unpolarized light is calculated by Eq. (6.5). In Eq. (6.5), (r_{PP}) represents the reflection of the perpendicular rays and (r_{PL}) represent the reflection of parallel rays, and both are defined in Eq. (6.6) and Eq. (6.7), respectively. In addition, θ_p is the angle of propagation and n_c is the POF core refractive index,

$$r_T = \frac{r_{PP}^2 + r_{PL}^2}{2}, \quad (6.5)$$

$$r_{PP} = \frac{\sin(\theta_p) - \sqrt{\frac{1}{n_c^2} - \cos^2(\theta_p)}}{\sin(\theta_p) + \sqrt{\frac{1}{n_c^2} - \cos^2(\theta_p)}}, \quad (6.6)$$

$$r_{PL} = \frac{\frac{1}{n_c^2}\sin(\theta_p) - \sqrt{\frac{1}{n_c^2} - \cos^2(\theta_p)}}{\frac{1}{n_c^2}\sin(\theta_p) + \sqrt{\frac{1}{n_c^2} - \cos^2(\theta_p)}}. \quad (6.7)$$

Assuming multiple reflections inside the sensitive zone of the fiber, the number of internal reflections (N) can be calculated as the closest integer number obtained by

$$N = \frac{c\tan(\theta_p)}{2(2a - p)}. \tag{6.8}$$

The combination of the power losses caused by the lateral section and the losses caused by the reflected rays results in an output power (P_o) defined in Eq. (6.9). It accounts the output power when the fiber is not under a bending. When a bending occurs, the integral upper limit θ_c has to be corrected to evaluate the effect of the fiber bending by a certain angle. In addition, the fiber has a decrease on its numerical aperture and the output signal decreases. The angle corrected by the fiber bending (θ_b) for a concave sensitive zone can be written as in Eq. (6.10):

$$P_o = \pi L_0(S_c - S_o)\int_0^{\theta_c} r_T^N \cos(\theta)\sin(\theta)d\theta. \tag{6.9}$$

$$\theta_b = \theta_c\sqrt{1 - \frac{2a}{R\theta_c^2}}. \tag{6.10}$$

Eq. (6.11) presents the ratio between the input and output power before and after bending the fiber with respective parameters and characteristics. Analyzing Eq. (6.11) in addition to Eqs. ((6.3) and (6.10)), it is possible to observe that only the curvature radius (R) will increase or decrease depending on a positive or negative bending and the bending angle. Therefore such expression relies only on the curvature radius, and may not be an adequate approach for dynamic scenario in which the sensor is under stress or strain condition:

$$\frac{P_o}{P_i} = \frac{(S_c - S_o)\sin^2(\theta_b)}{S_c\sin^2(\theta_c)}. \tag{6.11}$$

In the aforementioned model, no effect of the load on the fiber is included. When the fiber is under a bending, there is a stress tensor of symmetric second-rank with six independent variables acting on it (Zubia et al., 1997). Moreover, due to the viscoelasticity of the POF material, the elements of the tensor are not constant. A viscoelastic material has a time-dependent relationship between stress and strain due to its molecular rearrangement, in which part of the accumulated energy is dissipated leading to a time varying tensor (Lakes, 2009). Thus the viscoelastic effect can not be neglected on the stress-optical analysis in POFs.

In the stress-optical analysis, some issues must be discussed. First, the effects due to temperature changes are neglected. Second, the model is valid only for step-index POF, which is considered transparent, homogeneous, and isotropic (Zubia et al., 1997). Moreover, there are minor differences between the optical behavior along the fiber and perpendicular to it due to the anisotropy induced on the fabrication processes. However, these differences for a Poly (Methyl Methacrylate) (PMMA) fiber are small and difficult to measure (Dugas et al., 1994). Since most of the POFs have its core made of PMMA, the assumption of isotropy is acceptable.

Eq. (6.12) defines the stress-optical relation described by a second-rank tensor (ΔB_i) to represent the changes of coefficients in optical indicatrix under the action of the bending stress (Zubia et al., 1997). In Eq. (6.12), $q_{i,j}$ represents the stress-optical tensor, which is a fourth-rank tensor with 36 components. However, as the assumption of isotropy is acceptable the independent variables are reduced to only two (q_{11} and q_{12}). The values of the elements of the stress-optical tensor are related to the material of the fiber core and it can be measured through the experimental setup presented in Szczurowski et al. (2010).

$$\Delta B_i = q_{i,j}\sigma_i. \tag{6.12}$$

Eq. (6.13) shows the stress-optical tensor under symmetry conditions assuming the material isotropy:

$$\begin{bmatrix} q_{11} & q_{12} & q_{12} & 0 & 0 & 0 \\ q_{12} & q_{11} & q_{12} & 0 & 0 & 0 \\ q_{12} & q_{12} & q_{11} & 0 & 0 & 0 \\ 0 & 0 & 0 & q_{11}-q_{12} & 0 & 0 \\ 0 & 0 & 0 & 0 & q_{11}-q_{12} & 0 \\ 0 & 0 & 0 & 0 & 0 & q_{11}-q_{12} \end{bmatrix} \tag{6.13}$$

Since the most of the stress has an axial component, the stress tensor can be simplified by applying a correction on the Young's modulus of the material to cover the effects of the shear stress (Riande et al., 2000). In Eq. (6.14) the expression for the corrected Young modulus for shear stress is presented:

$$E^* = E_a^*\left[1 + \frac{6}{5}(2 + \nu^*)\left(\frac{d}{L}\right)^2\right]. \tag{6.14}$$

In Eq. (6.14), E^* is the corrected dynamic Young modulus, E_a^* is the uncorrected Young modulus, ν^* is the material Poisson ratio, d and L are defined in Fig. 6.1 and represent the fiber diameter and length, respectively. Since the fiber length generally is some orders of magnitude higher than the fiber diameter, the quadratic term in Eq. (6.14) can be neglected. The Poisson ratio of the materials typically lie on the range 0.3 to 0.5 and for this reason the second term of the summation continues to be close to zero. Thus the corrected Young modulus is close to the uncorrected one. Moreover, the stress tensor depends on the load condition on the fiber, such as the fiber under a bending stress with the configuration presented in Fig. 6.2.

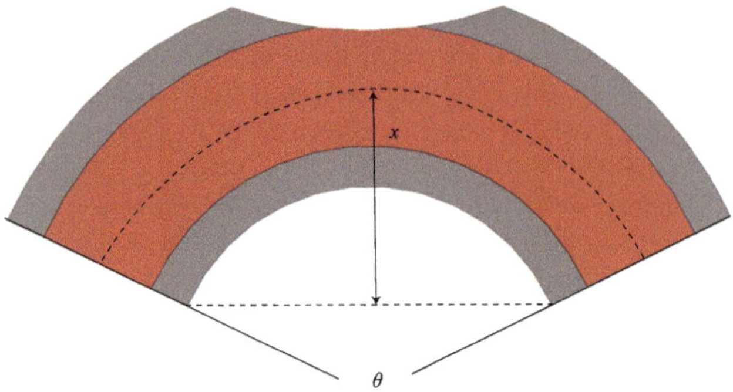

FIGURE 6.2 Polymer optical fiber under the bending stress condition.

In this configuration, x represents the distance between the bending part to the neutral line, and θ represents the bending angle. For this stress condition, the only component of the tensor different from zero is the axial stress on the z direction. The multiplication of the stress tensor and the stress-optical tensor gives the following results for ΔB_i presented in Eqs. (6.15)–(6.18):

$$\Delta B_1 = q_{12}\sigma(t), \tag{6.15}$$

$$\Delta B_2 = q_{12}\sigma(t), \tag{6.16}$$

$$\Delta B_3 = q_{11}\sigma(t), \tag{6.17}$$

$$\Delta B_4 = \Delta B_5 = \Delta B_6 = 0. \tag{6.18}$$

In Eq. (6.19), B is defined as presented in Zubia et al. (1997).

$$B = \frac{1}{n_c^2}. \tag{6.19}$$

By combining Eqs. ((6.15)–(6.19)), it is possible to obtain the variation of the refractive indexes Δn_x, Δn_y, and Δn_z for the bending condition, presented in Eqs. (6.20)–(6.22):

$$\Delta n_x = \frac{n_c^3 q_{12}\sigma(t)}{2}, \tag{6.20}$$

$$\Delta n_y = \frac{n_c^3 q_{12}\sigma(t)}{2}, \tag{6.21}$$

$$\Delta n_z = \frac{n_c^3 q_{11}\sigma(t)}{2}. \tag{6.22}$$

The stress component is time-dependent due to the viscoelasticity of the material. The viscoelastic behavior of a material can be described by several models, which may be divided in three major types: integral models, differential models, and molecular models (Riande et al., 2000). Among the differential models, there are different approaches as well. However, all the differential models consider the stress or strain response of the material as a combination of springs and dashpots, as thoroughly discussed in Chapter 5.

6.1.2 Light coupling-based sensors

The light coupling principle is based on the alignment difference between two optical fibers. In general, POFs with 1 mm core diameter are used due to their larger acceptance cone, related to the numerical aperture, as discussed in Chapter 4, leading to a larger measurement range. The setup consists of two fibers, where one POF is connected on the light source, called "illuminated" fiber. The another POF is connected to the photodetector, called "non-illuminated" fiber, as shown in Fig. 6.3. The output of the illuminated fiber has a cone of light, known as the acceptance cone. If the nonilluminated fiber input is within the acceptance cone boundaries of the illuminated fiber output, the light will be transmitted to the nonilluminated fiber and the output power is measured by the photodetector. In addition, variations of the alignment (angular, axial or lateral) between the fibers lead to variation on the light coupling efficiency, which can be measured with the photodetector.

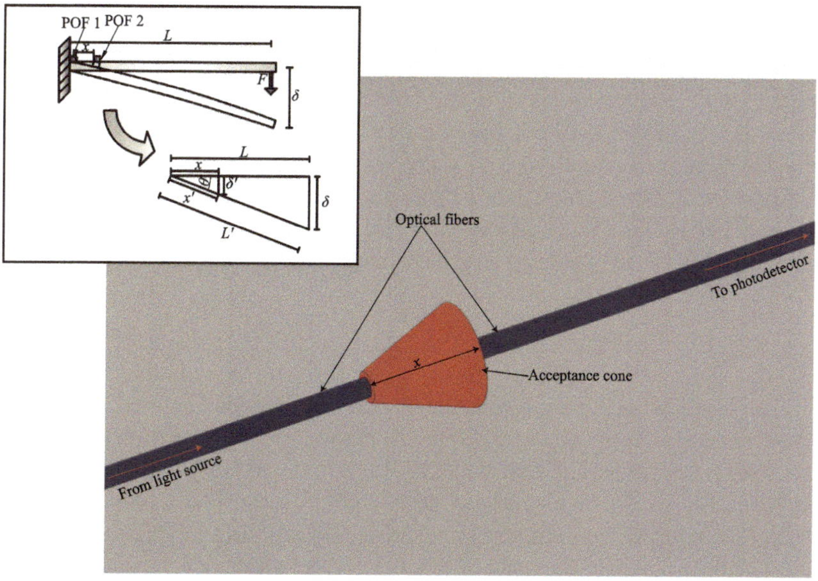

FIGURE 6.3 Operation principle of light coupling based sensors with polymer optical fibers. Figure inset shows application of the light coupling based optical fiber sensor on a cantilever beam.

If the sensor presented in Fig. 6.3 is positioned on a cantilever beam under flexural stress, it is possible to measure strain of the structure, as shown in Fig. 6.3 inset, in which the structure is represented by a metallic beam. The stress causes a deflection on the beam, creating an alignment difference between the two POFs, which leads to a power attenuation when the beam is under deflection. Thus it is possible apply the light coupling principle between two POFs to measure deflection or strain of a structure. In Fig. 6.3 inset the solid lines represent the initial condition of the system, where there is no deflection on the beam. In this case, L represents the distance between the beam support and the region that a force F is applied. In addition, x is the distance between the illuminated and nonilluminated POF (POF 1 and POF 2, respectively), and the dashed lines represent the beam deflection when a force F is applied, which is deflected by δ. The inset of Fig. 6.3 also shows the geometrical changes of the alignment between the fibers when the beam is under deflection.

By considering that the only variation on the POF response is due to the light coupling between two fibers the POF power variation described by $(\frac{P}{P_i})$, where P is coupled power into the nonilluminated fiber and P_i is the incident light power, will be equal to the coupling efficiency between the illuminated and nonilluminated POFs. Eq. (6.23) describes the power variation detected by the photodetector, for a multimode step index POF considering an uniform power distribution (Antunes et al., 2013):

$$\frac{P}{P_i} = \left[1 - \frac{\theta n_0}{\pi NA}\right]\left[1 - \frac{xNA}{4an_0}\right]\left[\frac{\pi}{2}\left(\cos^{-1}\frac{l}{2a}\right) - \frac{l}{2a}\sqrt{1 - \frac{l}{2a}^2}\right]. \tag{6.23}$$

In Eq. (6.23), n_0 represents the medium refractive index, whereas NA and a are the POF numerical aperture and core radius, respectively. In addition, the lateral displacement between the two POFs is represented by l, whereas the angular displacement and axial gap between the fibers are represented by θ and x, respectively. The deflection of a beam under a force presented in Fig. 6.3 can be estimated by Eq. (6.24), as presented in Ashby (2005), where A is the cross-sectional area of the beam and E is the Young modulus of the beam material:

$$\delta = \frac{FL}{AE}. \tag{6.24}$$

The beam deflection represents the lateral misalignment between the fibers; however, in the region where the force is applied, the deflection increases. For this reason, the deflection that the POF 2 is submitted is δ', which can be observed in Eq. (6.25). The axial gap between POF 1 and POF 2 after the beam deflection, represented by x, can be calculated by applying the Pythagorean theorem on the triangle of the inset of Fig. 6.3 and the difference between x and x' gives the variation of the axial gap between the illuminated and non-illuminated fiber. Moreover, the angular misalignment (θ) caused by the beam deflection can also be calculated from geometry, as presented in Eq. (6.26),

$$\delta' = \frac{x}{L}\delta, \tag{6.25}$$

$$\theta = \tan^{-1}\frac{\delta}{L}. \tag{6.26}$$

By substituting the results obtained by the longitudinal, lateral, and angular misalignments on Eq. (6.23), the variation of the nonilluminated POF power response with respect to the applied momentum on the beam can be estimated.

Different applications have already used the principle of light coupling between two fibers to measure different parameters, among them acceleration (Antunes et al., 2013), spine bending angle (Zawawi et al., 2013) and displacement (Vallan et al., 2012). This sensor presented some advantages, such as the low cost, ease of implementation, and simplicity on signal processing, since it is an intensity variation based sensor (Bilro et al., 2012). In addition, the proposed strain gauge does not have temperature sensitivity. Since the employed POF has 1 mm of diameter, different supports to position the fiber on the structure can be designed without the need of glues or resins. Moreover, there is no stress directly applied on the fiber, and for this reason, the viscoelastic behavior of the POF will not lead to errors on the sensor strain measurement.

6.1.3 Multiplexed intensity variation sensors

A technique of multiplexing intensity variation sensors is based on the schematic representation presented in Fig. 6.4. This technique consists of side-coupling of the light source to lateral sections made on the fiber. Two photodetectors are positioned on each end facet of the fiber. In addition, each lateral section on the fiber presented in Fig. 6.4 represents one sensor, where each sensor has a side-coupled LED. A microcontroller controls the activation of the LED array, which are activated with a predefined sequence and frequency one activated at a time, i.e., there is not the simultaneous activation of two or more light sources. Moreover, a microcontroller is responsible for the signal acquisition of both photodetectors at the moment in which each LED is activated. Thus a matrix is obtained with each column representing the POF power acquired by one photodetector when a predefined LED is activated and the rows representing the acquired samples. This matrix is presented in Fig. 6.4, where the number of columns is 2n (two photodetectors with n LEDs). By analyzing and comparing each vector of the matrix, it is possible to obtain the response of n sensors without the influence of light source deviations.

The presented technique is limited by the maximum number of sensors to be multiplexed in one fiber. Since there is a power attenuation at each sensor's lateral section, a high number of sensors in the fiber may decrease the output power to be detected. In addition, the sensors sensitivities also lead to a decrease of the power, which need to be considered in the design of the sensor arrays. The initial

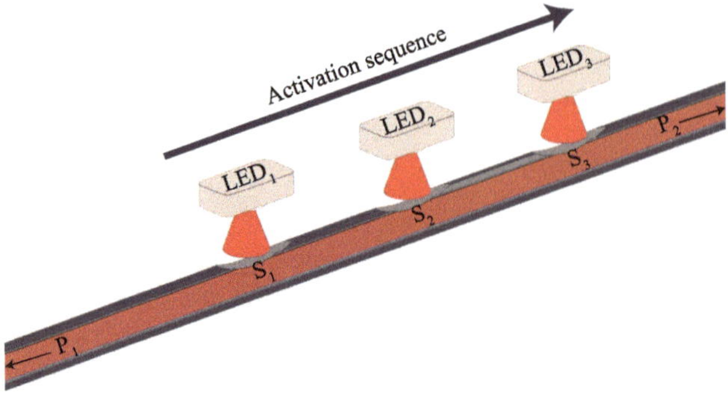

FIGURE 6.4 Schematic representation of the proposed multiplexing technique for intensity variation-based sensors.

power attenuation caused by the lateral section and sensor sensitivity depends on two different parameters related to the lateral section: the length and depth of each optical fiber sensor (Fu et al., 2010; Leal-Junior et al., 2018d). In addition, the optical power estimation is based on the power attenuation principle, and for this reason, the sensors sensitivities and dynamic range are considered in the calculation. Taking these parameters into account, Eq. (6.27) presents the estimation of the aforementioned power attenuation, where N is the number of sensors, α is the fiber attenuation coefficient, $sens_i$ and r_i are the sensitivity and dynamic range of each sensor, respectively,

$$\frac{P_0}{P_{in}} = \frac{(S_c - N S_0)}{S_c} - \alpha L - \sum sens_i r_i. \tag{6.27}$$

Since the fiber attenuation coefficient and the loss due to the lateral section only lead to an offset in the sensor response, and only the last term of Eq. (6.27) is considered. Eq. (6.28) presents the simulation of the sensors responses, where i is the sensor ($i = 1, 2, 3$ in this case) and j is the photodetector ($j = 1, 2$),

$$\left(\frac{P_0}{P_{in}}\right)_{i,j} = \sum_{i=1}^{n} sens_i r_i, \tag{6.28}$$

where i is the sensor ($i = 1, 2, 3$ in this case) and j is the photodetector ($j = 1, 2$)

Eq. (6.28) shows the response calculation of each sensor based on the sum of the previous sensors, i.e., the power of Sensor 3 consists of the sum of the responses of Sensors 1, 2, and 3, whereas Sensor 2 is the sum of 1 and 2. Therefore in order to obtain the response of a single sensor without the influence of other sensors, it is necessary to compensate the response of the other sensors. In this case, Eq. (6.28) is rewritten as Eq. (6.29). Thus it is noticeable the need of characterization of each sensor individually, prior to their applications for simultaneous measurements,

$$r_i = \frac{P_i - P_{i-1}}{sens_i}. \tag{6.29}$$

6.2 Interferometers

There are different types of interferometric sensors, such as Mach–Zehnder, Fabry–Perot, Sagnac and Michelson interferometers, as summarized in Lee et al. (2012). The great advantage of this sensor approach is related to the simplicity on the fabrication process and the low cost of the setup when compared to fiber Bragg gratings (FBGs) (Rodríguez et al., 2016). Moreover, optical interferometers have high sensitivity, high accuracy, and capability of multiparameters sensing (Castellani et al., 2017).

Fabry–Perot interferometers (FPIs) are regarded the ones with the higher fabrication flexibility or simplicity. The FPI principle is based on two reflecting surfaces separated by a known distance in another medium also known as cavity (Lee et al., 2012). In general, the FPI response is acquired in the reflection (instead of transmission), which facilitate the application of these sensors in harsh environments and areas of difficult access, in addition to enable a compact sensing element. Thus FPIs have been used on the measurement of parameters such as strain (Frazão et al., 2015), temperature (Tian et al., 2018), pressure (Zhu et al., 2017), liquid level (Lü and Yang, 2007), relative humidity (Oliveira et al., 2018), and even as edge-filter for low-cost FBG interrogators (Díaz et al., 2017).

The possibility of fabricating structures with many different approaches as summarized in Rajibul Islam et al. (2014) is one of the attractive features of FPIs. There are different methods in the FPIs fabrication, such as the use of photonic crystal fibers (Tian et al., 2018), silica tubes (Frazão et al., 2015), reflecting surfaces (Cui et al., 2019), air bubbles in fusion splicing (Rajibul Islam et al., 2014), femtosecond laser micromachining (Theodosiou et al., 2018a), and even using the periodic voids created on the optical fiber after the catastrophic fuse effect (Antunes et al., 2014). A novel FPI fabrication method was proposed recently in (Oliveira et al., 2018) in which ultraviolet (UV) curable resins were used to create the FPI cavity. The advantages of this method are the low relative cost, the no need of specialized equipment, and the control of the cavity dimensional/geometrical parameters as further explored in Oliveira et al. (2019).

A possible sensor configuration consists of two single mode fibers (SMFs) separated by a distance L connected by UV-curable resin. Thus the end facet of each fiber acts as a low reflectivity mirror, as a Fizeau interferometer, which is a FPI with a low reflectivity cavity (Lee et al., 2016). Fig. 6.5 presents the sensor configuration.

In this configuration, the sensor presents some key aspects such as the optical path difference (OPD). Eq. (6.30) shows the OPD estimation, where n is the refractive index of the cavity's medium and L is the cavity length.

The sensor configuration is comprised of two single mode fibers (SMFs) separated by a distance L with the UV-curable resin in between. In this configuration, the end facet of each fiber acts as a low reflectivity mirror. This feature makes the sensor as a Fizeau interferometer, which is a FPI with a low reflectivity cavity (Lee et al., 2016). The sensor basic configuration is presented in Fig. 6.5.

In this configuration, the sensor presents some key aspects which can be understood such as the optical path difference (OPD). Eq. (6.30) shows the OPD estimation, where n is the refractive index of the cavity's medium and L is the cavity length,

$$OPD = 2nL. \tag{6.30}$$

Then the optical phase difference can be estimated from the free space wavelength (λ) as

$$\phi = \frac{2\pi}{\lambda} OPD, \tag{6.31}$$

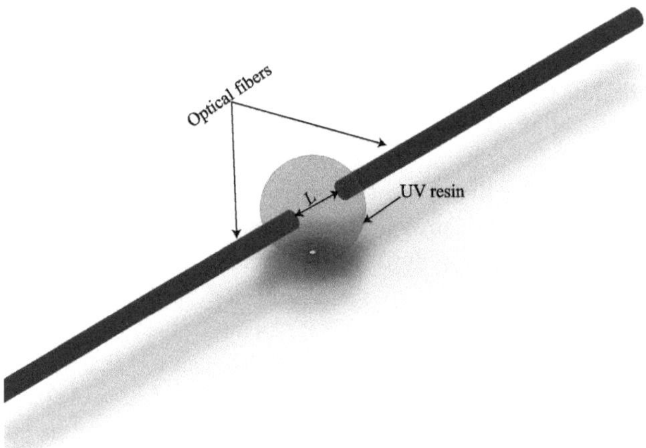

FIGURE 6.5 FPI configuration using UV curable resins.

In this configuration, it is possible to infer the intensity of the interference signal by considering the reflectivity of each surface. In order to simplify, both surfaces are considered with the same reflectivity, resulting in the same reflected intensity (R). Eq. (6.32) presents the interference intensity calculation:

$$I(\lambda) = 2R\big(1 + \cos(\phi)\big). \tag{6.32}$$

The phase of a peak in the spectrum depends on an integer m and the wavelength at a peak (or dip) can be expressed as (Oliveira et al., 2018):

$$\lambda_m = \frac{2nL}{m}. \tag{6.33}$$

In addition, the free spectral range (FSR), i.e., distance between two adjacent peaks is defined in Eq. (6.34):

$$\Delta\lambda_{FSR} = \frac{\lambda^2}{2nL}. \tag{6.34}$$

Eq. (6.34) was presented in Oliveira et al. (2019), in which the feasibility of this equation was confirmed by comparing the FSR measured to the one estimated by Eq. (6.34). Hence, the equation was able to use to estimate the cavity length of the FPI.

An FPI sensor is based on the wavelength shift of the spectrum related to variations in cavity length and/or in the refractive index of the cavity medium, as presented in Eq. (6.35),

$$\Delta\lambda = \lambda_m\left(\frac{\Delta n}{n} + \frac{\Delta L}{L}\right). \tag{6.35}$$

The first analysis is to consider the influence of temperature variation on the sensor response. The temperature change leads to a difference on two aspects of the sensor; first, it may lead to a variation of the refractive index due to the thermooptic effect, and second, it may lead to changes on the cavity length due to thermal expansion of the UV-curable resin. Thus Eq. (6.35) presents the calculation of Eq. (6.34) when the temperature variation is considered. In Eq. (6.35), ΔT is the temperature variation, α is the linear thermal expansion coefficient of the resin and ξ is the thermooptic coefficient of the medium,

$$\Delta\lambda(T) = \lambda_m\left(\frac{\xi\Delta T}{n} + \frac{\alpha\Delta T}{L}\right). \tag{6.36}$$

In an analysis of an FPI under uniaxial strain, there is a cavity length variation due to the strain on the UV-curable resin and there is also the refractive index variation due to the stress-optic effect. In addition, since the resin is a polymer it has a viscoelastic behavior and it needs to be addressed in the analysis. A viscoelastic behavior consists of a viscous and an elastic components. Thus a way to model this behavior is by combining springs (for the elastic response) and dashpots (for the viscous response) (Leal et al., 2018). The Maxwell's model is one of the simplest models used for viscoelastic materials, in which the material response is estimated as the spring connected in series with a dashpot, resulting in the differential equation (Lakes, 2009) presented in Eq. (6.37), where E is the Young's modulus, σ is the stress, ϵ is the strain, and η is the viscosity,

$$\frac{d\epsilon}{dt} = \frac{1}{E}\frac{d\sigma}{dt} + \frac{\sigma}{\eta}. \tag{6.37}$$

Considering a constant strain rate applied on the UV-curable resin from Eq. (6.37), and solving the differential equation, the result is Eq. (6.38). Moreover, in Eq. (6.38), the stress depends on the time (t) as well as on the material elasticity and viscosity,

$$\sigma(t) = \sigma_0\left[1 + exp\left(\frac{-E}{\eta}t\right)\right]. \tag{6.38}$$

The σ_0 is a time-independent component of the stress, which in the case of uniaxial strain, is defined as presented in Eq. (6.39), where ν is the material Poisson's ratio,

$$\sigma_0 = \epsilon\frac{E(1-\nu)}{(1+\nu)(1-2\nu)}. \tag{6.39}$$

The variation on the refractive index is result of the stress on the UV-curable resin caused by the uniaxial strain. Eq. (6.40) presents the variation on the refractive index following the stress-optic effect, where q is the stress-optic coefficient,

$$\Delta n = \frac{-n^3 q\sigma(t)}{2}, \tag{6.40}$$

where q is the stress-optic coefficient.

Fig. 6.6 presents the response of an FPI simulated spectra under different uniaxial strain conditions, where the simulation was performed considering the aforementioned equations for the FPI intensity and phase as well as the strain effects on the cavity length and refractive index. Table 6.1 presents the parameters used in the simulation.

TABLE 6.1 Parameters employed on the FPI simulation.

Symbol	Parameter	Value
n	Refractive index	1.56
L	Cavity length	50 μm
q	Stress-optic coefficient	$10^{-11}\,\mathrm{Pa}^{-1}$
ν	Poisson's ratio	0.40
E	Young's modulus	170 MPa
η	Viscosity	12000 m^2/s

As can be observed in Eq. (6.40) the FPI response with respect to mechanical loadings will also depend on the material properties. Thus it is expected not only different response time, but also different sensor sensitivities depending on the material in the FPI cavity. In addition, as a viscoelastic material well-known property, the moduli of these resins will also depend on the temperature, where such modulus variation was demonstrated in Leal et al. (2018) for PMMA fibers. For this reason, the temperature variation can lead to a change of the sensor sensitivity due to the thermal effect on the material properties (Leal-Junior et al., 2018c).

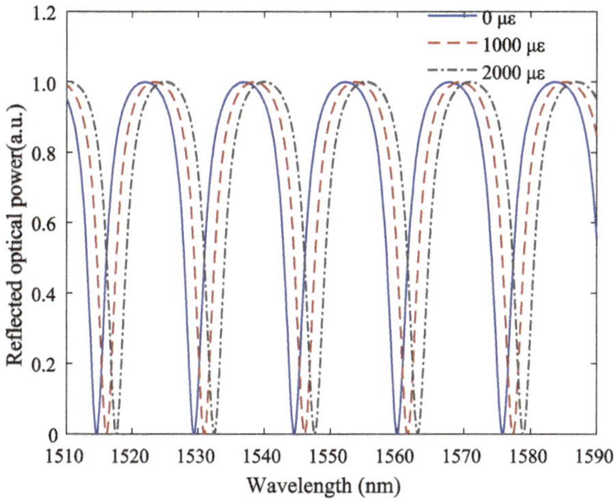

FIGURE 6.6 Simulation of FPI spectra under different strain conditions.

6.3 Gratings-based sensors

Fiber Bragg gratings (FBGs) and long period gratings (LPGs) are created by periodic modulations of the fiber core's refractive index, in which the period of LPGs is much higher than the one of FBGs. LPGs present periods with factors of hundred microns, whereas FBGs present period less than one micron. However, the modulation can be achieved with the interference of two laser beams (Hill and Meltz, 1997), where there is a photo-degradation of the polymer when the grating is inscribed (Sáez-Rodríguez et al., 2014), and this photo-degradation is related to the wavelength intensity of the light source employed (Pospori et al., 2017). Therefore the rate of photo-degradation is different depending on which UV source is employed. Due to the ablation issues reported in Peng et al. (1999), the 248 nm laser was not applied on FBG inscription for years. However, the reason for the ablation reported with the 248 nm laser was the high energy density employed on the grating inscription. Nevertheless, if low energy density and repetition rate is employed it is possible to obtain a refractive index modification without ablation, as presented in Wochnowski et al. (2005), which was used energy density of 40 mJ/cm^2 and repetition rate of 5 Hz. It is also possible to observe in the demonstration of a crosslinking between ester side chains of two polymer molecules with a 248 nm irradiation that can result in the increase of the refractive index. It can be related to the lower inscription time of the 248 nm laser when compared to the 325 nm laser. In addition, the polymer starts to photo-degradate as the 248 UV irradiation continues making the material less dense, leading to a reduction of the refractive index (Pospori et al., 2017). As the photo-degradation continues, there is the polymer ablation, which can be faster if higher energy densities are employed. In Oliveira et al. (2015) a low energy density with a repetition rate of 1 Hz was employed to obtain a

FBG in POFs (POFBGs) without material ablation and with lower inscription time. Such pulsed laser parameters enable the inscription of POFBGs at undoped polymethyl methacrylate (PMMA) with only 30 seconds, which is a lot faster than the lower inscription time for undoped PMMA fiber obtained with the 325 nm continuous laser (7 minutes) (Bundalo et al., 2014). For this reason, such repetition rate was employed in the grating inscription presented on other sensors. Since the optimal energy density depends on the polymer properties (Pospori et al., 2017), this parameter was experimentally optimized for the different POF materials.

In FBG-based sensors, the shift of the reflected wavelength as a function of the monitored parameter is evaluated. The FBG is intrinsically sensitive to temperature and strain, following Eq. (6.41) (Cusano et al., 2009):

$$\frac{\Delta\lambda_B}{\lambda_B} = (1 - P_e)\epsilon + (\alpha_T + \xi)\Delta T, \tag{6.41}$$

where $\Delta\lambda_B$ is the Bragg wavelength shift, λ_B is the Bragg wavelength, P_e is the photoelastic constant, α is the thermal expansion coefficient of the fiber, ξ is the thermooptic coefficient, ΔT is the temperature variation, and ϵ is the strain. Thus historically the typical applications of POFBGs were generic and based on the fundamental parameters of strain and temperature. However, FBGs can be embedded in different structures to sense other parameters besides strain and temperature as discussed below.

In terms of FBG applications, Stefani et al. (2012) proposed an accelerometer in a PMMA mPOFBG for SHM, displaying higher sensitivity than a silica counterpart. Theodosiou et al. (2016) demonstrated mode shape capturing using an array of CYTOP FBGs embedded in a metal beam. When compared to a silica counterpart, the authors demonstrated that the POFBG array had a sensitivity up to 6 times greater. A similar application was presented by the same group for a rear helicopter rotor blade, measuring imbalance on the cantilever (Theodosiou et al., 2018b). The sensors were used to measure the vibration due to imbalances created by loadings positioned at different regions of the cantilever.

POFBGs have also been embedded into patches produced using 3D printing, which is highly relevant for plug-and-play devices (Zubel et al., 2016). POFBGs embedded within a silicone diaphragm were also used for liquid level measurement, showing a system delivering multi-level measurement (Marques et al., 2015). A comparable silica sensor had a sensitivity 5 times lower than the POF sensor. Then the sensor was used to compensate some issues associated with conventionally employed sensors, including fuel density sensitivity, temperature, and g-force variations (Marques et al., 2016). This work could lead toward similar applications in the oil and gas industry where liquid level, temperature, pressure, and flow are important monitoring parameters (Qiao et al., 2017). The pressure response of POFBGs inscribed in CYTOP fibers was demonstrated in (Ishikawa et al., 2017), where again POF outperformed silica fibers in terms of sensitivity

Curvature sensors have been proposed by Chen et al. (2010), who demonstrated angle measurements using an FBG in eccentric core PMMA fiber. This type of sensor can be suitable for gait measurements, where both the angular range and frequencies are within the range for the lower limbs. In addition, gait and movement analysis has been extensively considered in recent years, with another viable example in Vilarinho et al. (2017). In this instance, an array of 5 FBGs were embedded in an insole for plantar pressure monitoring and subjected to static and dynamic testing. The response was in good agreement with the calibration system and the expected behavior during use.

Humidity sensing has been much more explored, as an inherent property of some polymers as characterized in Leal-Junior et al. (2018b). For POFBGs, the water affinity of the fiber introduces a significant Bragg wavelength shift, and as a result, the wavelength-encoded response is immune to fluctuations in light source and transmission loss. POFBGs do not need any coating or corrugation for humidity sensing and are robust, compact, and cost-effective. However, water absorption is very slow and response time remains a key issue for practical applications. It has been shown that etching the fiber (i.e., reducing its diameter) with acetone lowers the response time below the 7-minute threshold (Zhang and Webb, 2014), where the lowest response time reported is below 10 seconds (Rajan et al., 2013).

Fig. 6.7 presents an experimental setup that can be employed on the FBG inscription, where an excimer laser operating at UV wavelength with a pulse duration in the order of nanoseconds can be employed (with the possibility of also employ a continuous wave laser), generally, the beam profile can present a rectangular Tophat function with dimensions predefined dimensions and divergence. In addition, the setup shows mirrors, focal lens, and a slit with 4.5 mm width to positioning the UV beam on the phase mask employed, which is focused onto the fiber core through a plano-convex cylindrical lens. Moreover, 3D translational stages are applied to obtain the correct positioning of the fiber, in addition to avoid the fiber bending during the inscription.

This setup can be employed to the inscription of different types of FBGs, such as uniform FBGs, chirped FBGS and PS-FBGs. The difference of these FBGs are related to refractive index variation along the fiber length (z-axis), in which in the uniform FBG the period (Λ) is constant, whereas for the chirped FBG there are a linear aperiodic grating with the period variation along the z-axis, and the PS-FBG there is a phase shift that modifies the period in a certain length of the fiber along the z-axis. Fig. 6.8 presents the period of uniform FBGs with respect to the fiber x-axis.

Oliveira et al. (2015) employed a similar setup with the 248 nm KrF pulsed laser to inscribe uniform POFBGs at 1550 nm region for a PMMA POF. Moreover, the setup presented in Fig. 6.7 was employed to inscribe POFBGs in different POF materials at the wavelength region of 850 nm and the obtained inscription time was lower than the one presented in previous reports for the PMMA POF and the inscription times were lower than the ones obtained with

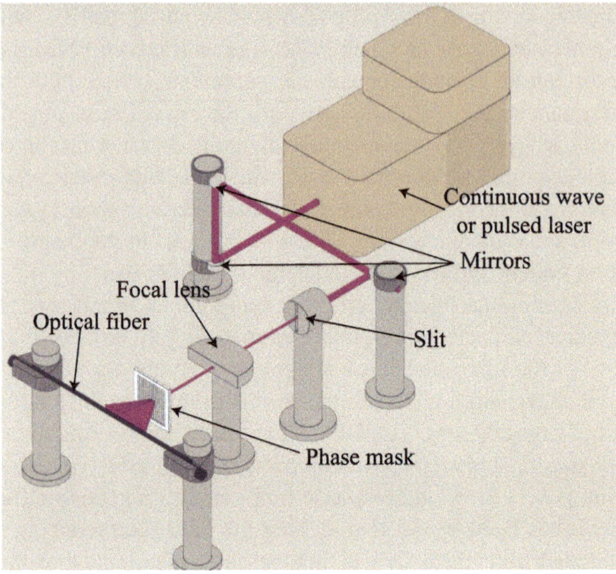

FIGURE 6.7 Inscription setup with the continuous wave or pulsed laser for FBGs inscription.

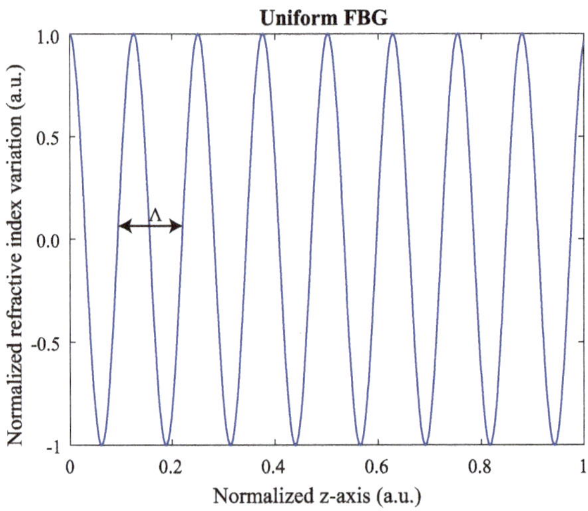

FIGURE 6.8 Periodic variation of the refractive index with respect to the fiber z-axis (fiber length) for uniform FBGs.

the 325 nm continuous laser for TOPAS high and low-Tg, PC, and Zeonex fibers as further discussed in Marques et al. (2018).

Different types of POFBGs have been inscribed using the setup presented in Fig. 6.7. The first chirped POFBG on undoped fiber was inscribed with a chirped phase mask, in which its responses of the temperature, strain, and pressure were characterized a chirped phase mask; the chirp parameters of the grating are limited to the ones of the phase mask and it necessary different chirped phased masks to obtain different chirp parameters, which can make the technology very expensive. One technique using chirped FBGs consists of tapering the POF to obtain the chirp, which is obtained by the fiber etching on a container filled with acetone. This process provides an improvement of the flexibility on chirped POFBGs manufacturing. In addition, a BDK-doped POF was employed and the chirped POFBG was obtained with a single laser pulse using a uniform phase mask. Fig. 6.9 shows the reflection spectrum of the chirped POFBG inscribed using the method described. Fig. 6.9 also presents the same chirped POFBG when a strain of 0.3% is applied in order to show the chirped POFBG tuning with strain.

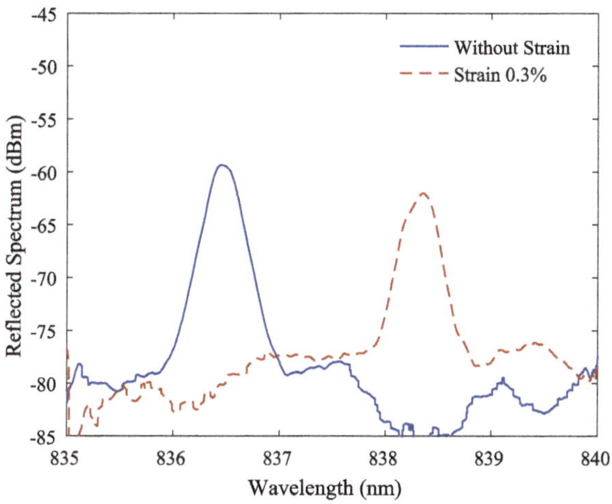

FIGURE 6.9 Chirped POFBG inscribed in POFs under different strain conditions.

The KrF 248 nm pulsed laser was also employed in the inscription of PS-FBGs, in which a metal wire divides the laser beam, leading to the creation of a phase shift on the Bragg grating (Pereira et al., 2017). It also demonstrated the possibility of creating multiple phase shifts on the grating with additional metal wires in the laser beam. However, due to the low dimensions of the laser beam and phase mask, it can be difficult to position the metal wires correctly.

Another important issue FBG applications is the cost of the interrogation equipment, which generally present high cost. This issue is especially undesired for health monitoring applications, in which the user's health condition or activities can be monitored from home as a smart healthcare solution for patients monitoring (Majumder et al., 2017). In addition to the low cost, the compactness and flexibility of optical fiber sensors are desirable features for this application, due to the possibility of embedment of the sensors in different materials, or the sensor positioning on the human skin for joint angle (Leal-Junior et al., 2018a) or physiological parameters (Leal-Junior et al., 2019) monitoring. The biocompatibility and electromagnetic field immunity of optical fibers are additional advantages of these sensors when applications involving invasive monitoring or wearable robotics are concerned. In this context, wavelength-based sensors (such as FBGs, optical interferometers, and LPGs) have been proposed (mostly using FBGs) on the assessment of joint angles (Pant et al., 2018), plantar pressure (Tavares et al., 2018), arterial pulse (Díaz et al., 2017), breathing and heart rates (Bonefacino et al., 2018). However, in these applications, commercial FBG interrogators were used, leading to a high size device which reduces the sensor system's portability (Pant et al., 2018). Although the arterial pulse application uses the edge filter interrogation technique (Díaz et al., 2017), only one FBG can be assessed with each filter and the FBG spectrum cannot be reconstructed, since only the correlation between the wavelength and optical power is analyzed. Therefore the edge-filter technique can result in information loss, especially for displacement (angle, strain, or torsion) sensors, since that the analysis of the whole FBG spectrum can provide additional information about the mechanical loadings on the grating, where it can be possible to obtain a 3D displacement sensor by this analysis, as presented in Leal Junior et al. (2018e).

Considering this issue, there is development and characterization of portable interrogators for dynamic measurements using tunable filters. In general, these interrogators have two stages: (i) optical stage that includes tunable filter, light source, photodetector, and optical circulator, and (ii) electronic stage with the microcontrollers and single-board computers, which can also have an embedded Wi-Fi module, enabling remote health monitoring. In order to obtain a fast and high-resolution sweep on the filter optical spectrum the microcontrollers tune the filter. In addition, the convolution between the filter and the wavelength-based sensor is acquired by the photodetectors for the spectral reconstruction, which results in a system with lower cost when compared with commercially available interrogators.

6.4 Compensation techniques and cross-sensitivity mitigation in optical fiber sensors

In general, optical fiber sensors measure strain (or stress) transmitted from a structure to the optical fiber. This transmitted strain can be converted to power variations (in case of intensity variation-based sensors) or wavelength shifts (in case of FBGs or FPIs). Thus the variation on the OFS spectral features and the stress applied on the structure are correlated in which the optical fiber is embedded and/or positioned (Leal Junior et al., 2018e). Fig. 6.10(a) shows an optical fiber embedded in a diaphragm, where there is also the influence of the optical fiber material in the sensor response. When silica and PMMA POF responses are compared, as shown in Fig. 6.10(a), it is possible to observe a five-fold increase on the sensitivity when POFs are used, due to its smaller elastic modulus when compared to silica fibers. Considering the example of the optical fiber embedded in a polymer diaphragm (or other polymer structure), it is important to note that the pressure response of diaphragm-embedded optical fiber sensors is correlated to the strain transmission between diaphragm and the optical fiber, which can be related to the material properties and dimensions of each component (Leal-Junior et al., 2020). Since the optical fiber elastic modulus is some orders of magnitude higher than the ones of diaphragms, the diaphragm displacement under pressure is limited and it can lead to variations on the strain transmission between materials (Diaz et al., 2018). Fig. 6.10(a) presents the strain variation on the diaphragm for a pressure of 10 kPa.

Since most optical fiber sensors are temperature sensitive, it is important to consider such parameter when developing the sensor system. The sensors embedded in the polymer structures (especially FBGs), in general, present higher temperature sensitivity when compared to the similar sensors without embedment (Marques et al., 2015). It is due to the diaphragm material, which has its own thermal expansion coefficient. Thus as the temperature increases the strain is thermally induced on the polymer, which is transmitted to the optical fiber, and hence, provoking additional variation on the spectral features (Leal-Junior et al., 2018a). This thermally induced strain is simulated in a polyurethane diaphragm via FEM, and presented in inset of Fig. 6.10(b) for 80 °C. Also, Fig. 6.10(b) shows the relation between the wavelength shift and the temperature of a FBG before and after the embedment in polyurethane diaphragm; it is noticeable there is a two-fold increase of the temperature sensitivity, related to the diaphragm thermal expansion. In this case, a TPU diaphragm was analyzed, however, if a material with higher thermal expansion coefficient (such as nitrile rubber) is analyzed, an even higher temperature sensitivity increase is expected (Leal-Junior et al., 2020).

The fiber embedment with polymer materials improves the temperature responses in optical fiber sensors, as previously noted. However, it can also harm the detection of strain-related parameters such pressure, liquid level, acceleration among others due to the temperature cross-sensitivity. In order to overcome this issue, different compensation techniques have been proposed for years in order to achieve a temperature-insensitive sensor system. A simple and most used technique is relating the same sensor under two different scenarios: one subjected to variations of both parameters (temperature and the desired parameter) and the other subjected only to temperature variations. Thus it is possible to obtain the direct difference between the responses. This technique has several proposed configurations, such as dual FBGs, FBG, and FPI (or other interferometers), dual interferometers. In this context, there are different fabrication and assembly methods proposed to achieve the strain isolation in the sensor for the temperature assessment. A simple approach is to separate a temperature sensor out of the diaphragm and the other sensor embedded in a structure (for stress-related parameters assessment), since the other is not embedded. Since the FBGs do not present high stress sensitivity without the embedment in a transducer mechanism, such as a diaphragm, they are commonly used in this

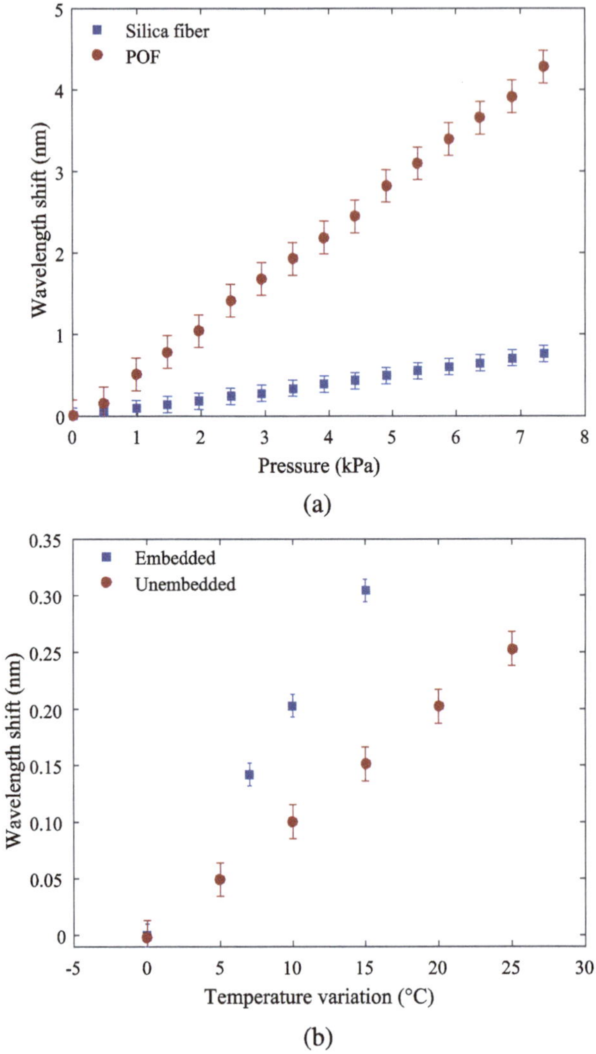

FIGURE 6.10 (a) Pressure response of diaphragm-embedded FBGs with silica fiber and POFs. (b) Temperature response of unembedded FBG in silica fiber and the same FBG embedded in a polyurethane diaphragm.

approach. In addition, metallic housing can also be employed on the FBG for temperature assessment to guarantee the stress insensitivity. There are other approaches, which consists of adding a metallic sheet in a region of the diaphragm at which the FBG for temperature compensation is positioned. Moreover, by positioning of two FBGs in the same diaphragm has also been explored for the development of temperature-compensated systems. An aforementioned approach for temperature compensation consisted of characterizing the sensor's sensitivities of pressure (or other strain-related parameter) and temperature, and hence, to subtract the sensor responses considering the sensitivities for each parameter. Eq. (6.42) shows this principle, based on a linear system with the same number of equations (responses of sensors 1 and 2) and unknown parameters (temperature and pressure, for example). In Eq. (6.42), ΔP and ΔT represent pressure and temperature variations, respectively, whereas K_{1-2}, P and K_{1-2}, T represent the sensitivities of sensors 1 and 2 as a function of the pressure and temperature, respectively. In addition, parameters $\Delta\lambda_{B,1-2}$ are the wavelength (or optical power in case of intensity variation sensors) variations for sensors 1 and 2,

$$\begin{bmatrix} \Delta P \\ \Delta T \end{bmatrix} = \begin{bmatrix} K_{1,P} & K_{1,T} \\ K_{2,P} & K_{2,T} \end{bmatrix} \cdot \begin{bmatrix} \Delta\lambda_{B,1} \\ \Delta\lambda_{B,2} \end{bmatrix}. \tag{6.42}$$

Another approach to compensate the temperature influence is using chirped FBGs, which can present large variations on the bandwidth according to the strain (or temperature) profile applied on the grating region due to its cascaded configuration. In this approach, one part of the grating region is embedded on the diaphragm. The other one can be embedded in a metallic surface (to avoid strain) or in a retainer ring. In this configuration, there are variations in the wavelength shift and in the bandwidth when the grating is under pressure and subjected to temperature variations. Thus to decouple the pressure and temperature it is necessary a method similar to the one shown in Eq. (6.42), where the wavelength shift and bandwidth are analyzed as a function of temperature and pressure (as a linear system with two equations and two unknowns).

It is possible to observe in Fig. 6.10(b) that the temperature sensitivity of the embedded optical fiber sensor is higher than the not embedded one. Hence, the embedment of two FBGs in the same structure leads to a similar temperature sensitivity, which in theory, can result in a temperature insensitivity for the sensor system. Moreover, the optical fiber can be embedded in different positions of the diaphragm. Fig. 6.11 shows the FBGs embedded in different positions in which the pressure on the diaphragm leads to compression in one optical fiber sensor and extension in the other. Therefore, the analysis of the responses difference of each sensor leads to an increase in sensitivity and resolution.

The materials of diaphragm and optical fiber are important parts in any temperature compensation approach. As discussed in Chapter 5, polymer materials have a viscoelastic behaviors, which present a time and temperature dependence on their mechanical properties. Thus when using polymer structures or optical fibers, they may present variations in their mechanical properties (including the Young's modulus) depending on the system's dynamics or the temperature variation. For this reason, a sensor embedded in polymer structure can present different sensitivities in pressure tests depending on the temperature (as shown in Fig. 6.11) due to variations in the material properties. Since the strain on the diaphragm depends on its material, the modulus reduction provoked by the

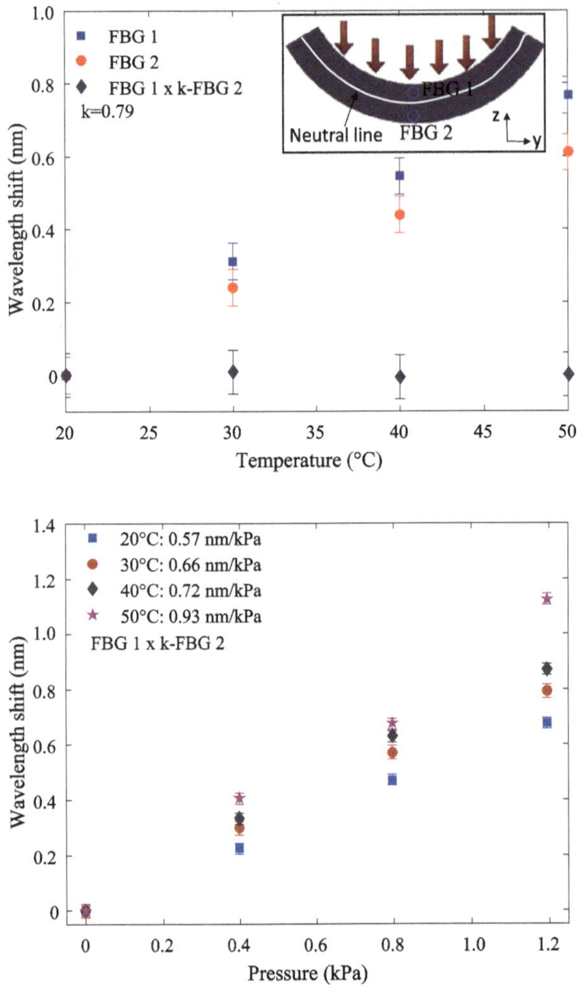

FIGURE 6.11 Compression and extension regions of a diaphragm under transverse stress. Figure also shows the differences in the sensors responses based on their positioning and temperature cross-sensitivities.

temperature increase leads to a higher transmitted strain to the optical fiber. It results in higher variations on the spectral features under a constant pressure, but in different temperature conditions, as presented in Fig. 6.11. It presented a comparison of the temperature cross-sensitivity in different diaphragm materials, in which the dependence between the temperature, the material composition, and the material type is demonstrated, and it can also occur with polymer fibers. In these cases, it is necessary to consider the material properties variation in the temperature compensation, since they are correlated and for higher temperature variations there will be changes in the sensitivity of coefficients in Eq. (6.42).

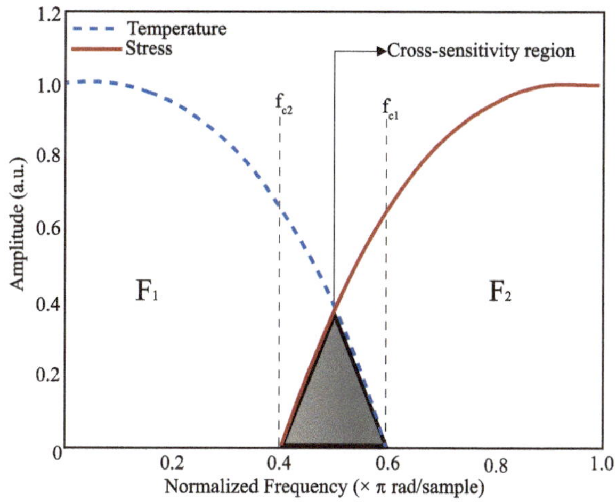

FIGURE 6.12 Schematic representation of the normalized frequency regions for temperature (F1) and pressure (F2) responses.

It may result in errors on the pressure estimation under different temperatures. Therefore the stress (or force and pressure) sensitivity variation as a function of the temperature is considered in the temperature compensations, in which this sensitivity variation can be estimated from pressure characterization experiments at different temperatures.

In several dynamic applications, the temperature changes slower than the strain-related parameters. In addition, depending on the sensors' dimensions and material, the dynamics of heat convection between the environment and the diaphragm are faster than the one of heat conduction inside the sensor structure. Both in the two conditions, i.e., strain-related parameters (such as pressure, force, displacement) with faster and slower temperature thermal dynamics in the embedded structure, lead to the possibility of decoupling the temperature and strain responses by analyzing the frequency components of these responses. Thus the temperature presents a low frequency components and the strain (or pressure) presents high frequency components. They can be separated using filtering technique and schematically represented in Fig. 6.12 for the case of dynamic pressure and temperature variations. By using a high-pass filter, the resultant signal has only the high frequency components, related to the strain-related parameter response. Such technique is especially used for vibration and acoustic sensing in which there are high frequencies signals due to dynamics much faster than the ones of temperature variations.

References

Alwis, L., Sun, T., Grattan, V.K.T., Grattan, V.K.T., 2016. Developments in optical fibre sensors for industrial applications. Optics and Laser Technology 78, 62–66. https://doi.org/10.1016/j.

optlastec.2015.09.004. From Duplicate 2 (Developments in optical fibre sensors for industrial applications – Alwis, L., Sun, T., Grattan, V.K.T.).

Antunes, P.F.C., Domingues, M.F.F., Alberto, N.J., André, P.S., 2014. Optical fiber microcavity strain sensors produced by the catastrophic fuse effect. IEEE Photonics Technology Letters 26, 78–81. https://doi.org/10.1109/LPT.2013.2288930.

Antunes, P.F.C., Varum, H., Andre, P.S., 2013. Intensity-encoded polymer optical fiber accelerometer. IEEE Sensors Journal 13, 1716–1720. https://doi.org/10.1109/JSEN.2013.2242463.

Ashby, M.F., 2005. Materials Selection in Mechanical Design. Elsevier, Cambridge.

Bilro, L., Alberto, N., Pinto, J.L., Nogueira, R., 2012. Optical sensors based on plastic fibers. Sensors (Switzerland) 12, 12184–12207. https://doi.org/10.3390/s120912184.

Bonefacino, J., Tam, H.Y., Glen, T.S., Cheng, X., Pun, C.F.J., Wang, J., Lee, P.H., Tse, M.L.V., Boles, S.T., 2018. Ultra-fast polymer optical fibre Bragg grating inscription for medical devices. Light: Science & Applications 7, 17161. https://doi.org/10.1038/lsa.2017.161. http://www.nature.com/doifinder/10.1038/lsa.2017.161.

Bundalo, I.L., Nielsen, K., Markos, C., Bang, O., 2014. Bragg grating writing in pmma microstructured polymer optical fibers in less than 7 minutes. Optics Express 22, 5270. https://doi.org/10.1364/OE.22.005270.

Castellani, C.E.S., Ximenes, H.C.B., Silva, R.L., Frizera-neto, A., Ribeiro, M.R.N., Pontes, M.J., 2017. Multi-parameter interferometric sensor based on a reduced diameter core axial offseted fiber. IEEE Photonics Technology Letters 29, 239–242.

Chen, X., Zhang, C., Webb, D.J., Peng, G.D., Kalli, K., 2010. Bragg grating in a polymer optical fibre for strain, bend and temperature sensing. Measurement Science and Technology 21. https://doi.org/10.1088/0957-0233/21/9/094005.

Cui, Q., Thakur, P., Rablau, C., Avrutsky, I., Cheng, M.M.C., 2019. Miniature optical fiber pressure sensor with exfoliated graphene diaphragm. IEEE Sensors Journal PP, 5621–5631. https://doi.org/10.1109/JSEN.2019.2904020.

Cusano, A., Cutolo, A., Albert, J., 2009. Fiber Bragg Grating Sensors: Market Overview and New Perspectives. Bentham Science Publishers, Potomac.

Diaz, C.A.R., Leal-Junior, A.G., Andre, P.S.B., Antunes, P.F.d.C., Pontes, M.J., Frizera-Neto, A., Ribeiro, M.R.N., 2018. Liquid level measurement based on FBG-embedded diaphragms with temperature compensation. IEEE Sensors Journal 18, 193–200. https://doi.org/10.1109/JSEN.2017.2768510. http://ieeexplore.ieee.org/document/8093646/.

Díaz, C., Leitão, C., Marques, C., Domingues, M., Alberto, N., Pontes, M., Frizera, A., Ribeiro, M., André, P., Antunes, P., 2017. Low-cost interrogation technique for dynamic measurements with fbg-based devices. Sensors 17, 2414. https://doi.org/10.3390/s17102414.

Dugas, J., Pierrejean, I., Farenc, J., Peichot, J.P., 1994. Birefringence and internal stress in polystyrene optical fibers. Applied Optics 33, 3545–3548. https://doi.org/10.1364/AO.33.003545.

Dziuda, L., Lewandowski, J., Skibniewski, F., Nowicki, G., 2012. Fibre-optic sensor for respiration and heart rate monitoring in the mri environment. Procedia Engineering 47, 1291–1294. https://doi.org/10.1016/j.proeng.2012.09.391.

Frazão, O., Kobelke, J., Schuster, K., Aichele, C., Bierlich, J., Santos, J.L., Roriz, P., Wondraczek, K., Ferreira, M.S., 2015. Fabry–Perot cavity based on silica tube for strain sensing at high temperatures. Optics Express 23, 16063. https://doi.org/10.1364/oe.23.016063.

Fu, Y., Di, H., Liu, R., 2010. Light intensity modulation fiber-optic sensor for curvature measurement. Optics and Laser Technology 42, 594–599. https://doi.org/10.1016/j.optlastec.2009.10.009.

Hill, K.O., Meltz, G., 1997. Fiber Bragg grating technology fundamentals and overview. IEEE Journal of Lightwave Technology 15, 1263–1276. https://doi.org/10.1109/50.618320.

Ishikawa, R., Lee, H., Lacraz, A., Theodosiou, A., Kalli, K., Mizuno, Y., Nakamura, K., 2017. Pressure dependence of fiber Bragg grating inscribed in perfluorinated polymer fiber. IEEE Photonics Technology Letters 29, 2167–2170. https://doi.org/10.1109/LPT.2017.2767082.

Kinet, D., Mégret, P., Goossen, K., Qiu, L., Heider, D., Caucheteur, C., 2014. Fiber Bragg grating sensors toward structural health monitoring in composite materials: challenges and solutions. Sensors 14, 7394–7419. https://doi.org/10.3390/s140407394.

Kovacević, M., Djordjevich, A., Nikezić, D., 2006. An analytical approach and optimization of curvature gauge. Journal of Physics. Conference Series 48, 850–858. https://doi.org/10.1088/1742-6596/48/1/161.

Lakes, R., 2009. Viscoelastic Materials. Cambridge University Press, Cambridge.

Leal, A.G., Marques, C., Frizera, A., Pontes, M.J., 2018. Dynamic mechanical analysis on a polymethyl methacrylate (pmma) polymer optical fiber. IEEE Sensors Journal 18, 2353–2361. https://doi.org/10.1109/JSEN.2018.2797086.

Leal Junior, A., Frizera, A., Pontes, M., 2017. Analytical model for a polymer optical fiber under dynamic bending. Optics and Laser Technology 93. https://doi.org/10.1016/j.optlastec.2017.02.009.

Leal-Junior, A., Casas, J., Marques, C., Pontes, M., Frizera, A., 2018a. Application of additive layer manufacturing technique on the development of high sensitive fiber Bragg grating temperature sensors. Sensors 18, 4120. https://doi.org/10.3390/s18124120.

Leal-Junior, A., Frizera, A., Pontes, M.J., Fasano, A., Woyessa, G., Bang, O., Marques, C.A.F., 2018b. Dynamic mechanical characterization with respect to temperature, humidity, frequency and strain in mpofs made of different materials. Optical Materials Express 8, 804. https://doi.org/10.1364/OME.8.000804.

Leal-Junior, A., Theodosiou, A., Marques, C., Pontes, M.M.J., Kalli, K., Frizera, A., 2018c. Compensation method for temperature cross-sensitivity in transverse force applications with fbg sensors in pofs. Journal of Lightwave Technology 36, 3660–3665. https://doi.org/10.1109/JLT.2018.2848704.

Leal-Junior, A.G., Frizera, A., José Pontes, M., 2018d. Sensitive zone parameters and curvature radius evaluation for polymer optical fiber curvature sensors. Optics & Laser Technology 100, 272–281. https://doi.org/10.1016/j.optlastec.2017.10.006.

Leal Junior, A.G., Theodosiou, A., Diaz, C., Marques, C., Pontes, M.J., Kalli, K., Frizera, A., 2018e. Simultaneous measurement of axial strain, bending and torsion with a single fiber Bragg grating in CYTOP fiber. Journal of Lightwave Technology 8724, 971–980. https://doi.org/10.1109/JLT.2018.2884538. https://ieeexplore.ieee.org/document/8558545/.

Leal-Junior, A.G., Díaz, C.R., Leitão, C., Pontes, M.J., Marques, C., Frizera, A., 2019. Polymer optical fiber-based sensor for simultaneous measurement of breath and heart rate under dynamic movements. Optics and Laser Technology 109, 429–436. https://doi.org/10.1016/j.optlastec.2018.08.036.

Leal-Junior, A.G., Frizera, A., Marques, C., 2020. Thermal and mechanical analyses of fiber Bragg gratings-embedded polymer diaphragms. IEEE Photonics Technology Letters 32, 623–626. https://doi.org/10.1109/LPT.2020.2988554. https://ieeexplore.ieee.org/document/9069892/.

Lee, B.H., Kim, Y.H., Park, K.S., Eom, J.B., Kim, M.J., Rho, B.S., Choi, H.Y., 2012. Interferometric fiber optic sensors. Sensors 12, 2467–2486. https://doi.org/10.3390/s120302467.

Lee, C.L., You, Y.W., Dai, J.H., Hsu, J.M., Horng, J.S., 2016. Hygroscopic polymer microcavity fiber fizeau interferometer incorporating a fiber Bragg grating for simultaneously sensing humidity and temperature. Sensors and Actuators. B, Chemical 222, 339–346. https://doi.org/10.1016/j.snb.2015.08.086.

Liu, R., Fu, Z., Zhao, Y., Cao, Q., Wang, S., 2006. Operation principle of a bend enhanced curvature optical fiber sensor. In: IEEE International Conference on Intelligent Robots and Systems, pp. 1966–1971.

Liu, Z., Zhang, Z.F., Tam, H.Y., Tao, X., 2019. Multifunctional smart optical fibers: materials, fabrication, and sensing applications. Photonics 6, 1–24. https://doi.org/10.3390/photonics6020048.

Lü, T., Yang, S., 2007. Extrinsic Fabry–Perot cavity optical fiber liquid-level sensor. Applied Optics 46, 3682. https://doi.org/10.1364/ao.46.003682.

Majumder, S., Mondal, T., Deen, M., 2017. Wearable sensors for remote health monitoring. Sensors 17, 130. https://doi.org/10.3390/s17010130.

Marques, C.A.F., Min, R., Leal Junior, A., Antunes, P., Fasano, A., Woyessa, G., Nielsen, K., Rasmussen, H.K., Ortega, B., Bang, O., 2018. Fast and stable gratings inscription in POFs made of different materials with pulsed 248 nm KrF laser. Optics Express 26. https://doi.org/10.1364/OE.26.002013, 2013. https://www.osapublishing.org/abstract.cfm?URI=oe-26-2-2013.

Marques, C.A.F., Peng, G.D., Webb, D.J., 2015. Highly sensitive liquid level monitoring system utilizing polymer fiber Bragg gratings. Optics Express 23, 6058–6072. https://doi.org/10.1364/OE.23.006058.

Marques, C.A.F., Pospori, A., Saez-Rodriguez, D., Nielsen, K., Bang, O., Webb, D.J., 2016. Aviation fuel gauging sensor utilizing multiple diaphragm sensors incorporating polymer optical fiber Bragg gratings. IEEE Sensors Journal 16, 6122–6129. https://doi.org/10.1109/JSEN.2016.2577782.

Oliveira, R., Bilro, L., Nogueira, R., 2015. Bragg gratings in a few mode microstructured polymer optical fiber in less than 30 seconds. Optics Express 23, 10181. https://doi.org/10.1364/OE.23.010181.

Oliveira, R., Bilro, L., Nogueira, R., 2018. Fabry–Pérot cavities based on photopolymerizable resins for sensing applications. Optical Materials Express 8, 2208. https://doi.org/10.1364/ome.8.002208.

Oliveira, R.F., Bilro, L., Nogueira, R., Rocha, A.M.M., 2019. Adhesive based Fabry–Pérot hydrostatic pressure sensor with improved and controlled sensitivity. Journal of Lightwave Technology 8724, 1909–1915. https://doi.org/10.1109/jlt.2019.2894949.

Othonos, A., Kalli, K., 1999. Fiber Bragg gratings: fundamentals and applications in telecommunications and sensing. In: Fiber Bragg Gratings: Fundamentals and Applications in Telecommunications and Sensing. Artech House Inc., Norwood.

Pant, S., Umesh, S., Asokan, S., 2018. Knee angle measurement device using fiber Bragg grating sensor. IEEE Sensors Journal 18, 10034–10040. https://doi.org/10.1109/JSEN.2018.2875564.

Peng, G., Xiong, Z., Chu, P., 1999. Photosensitivity and gratings in dye-doped polymer optical fibers. Optical Fiber Technology 5, 242–251. https://doi.org/10.1006/ofte.1998.0298.

Pereira, L.M., Pospori, A., Antunes, P., Domingues, M.F., Marques, S., Bang, O., Webb, D.J., Marques, C.A.F., 2017. Phase-shifted Bragg grating inscription in pmma microstructured pof using 248-nm uv radiation. Journal of Lightwave Technology 35, 5176–5184. https://doi.org/10.1109/JLT.2017.2771436.

Photonic sensors market – growth, trends, and forecasts, 2020. https://www.marketresearchfuture.com/reports/photonic-sensors-market-2644.

Pospori, A., Marques, C.A.F., Bang, O., Webb, D.J., André, P., 2017. Polymer optical fiber Bragg grating inscription with a single uv laser pulse. Optics Express 25, 9028–9038. https://doi.org/10.1364/OE.25.009028.

Qiao, X., Shao, Z., Bao, W., Rong, Q., 2017. Fiber Bragg grating sensors for the oil industry. Sensors 17, 429. https://doi.org/10.3390/s17030429.

Rajan, G., Noor, Y.M., Liu, B., Ambikairaja, E., Webb, D.J., Peng, G.D., 2013. A fast response intrinsic humidity sensor based on an etched singlemode polymer fiber Bragg grating. Sensors and Actuators. A, Physical 203, 107–111. https://doi.org/10.1016/j.sna.2013.08.036. From Duplicate 1 (A fast response intrinsic humidity sensor based on an etched singlemode polymer fiber Bragg grating – Rajan Ginu, Noor Yusof Mohd, Liu Bing, Ambikairaja Eliathamby, Webb David J., Peng, Gang-Ding).

Rajibul Islam, M., Mahmood Ali, M., Lai, M.H., Lim, K.S., Ahmad, H., 2014. Chronology of Fabry-Perot interferometer fiber-optic sensors and their applications: a review. Sensors (Switzerland) 14, 7451–7488. https://doi.org/10.3390/s140407451.

Riande, E., Diaz-Calleja, R., Prolongo, M., Masegosa, R., Salom, C., 2000. Polymer Viscoelasticity: Stress and Strain in Practice. Marcel Dekker, New York.

Rodríguez, C.A., Castellani, C.E.S., Frizera-Neto, A., Pontes, M.J., Ribeiro, M.R.N., 2016. Envelope-based technique for liquid level sensors using an in-line fiber Mach–Zehnder interferometer. Applied Optics 55, 9803. https://doi.org/10.1364/ao.55.009803.

Stefani, A., Andresen, S., Yuan, W., Herholdt-Rasmussen, N., Bang, O., 2012. High sensitivity polymer optical fiber-Bragg-grating-based accelerometer. IEEE Photonics Technology Letters 24, 763–765. https://doi.org/10.1109/LPT.2012.2188024.

Szczurowski, M.K., Martynkien, T., Gabriela, S.B., Urbanczyk, W., Khan, L., Webb, D.J., Statkiewicz-Barabach, G., Urbanczyk, W., Khan, L., Webb, D.J., 2010. Measurements of stress-optic coefficient in polymer optical fibers. Optics Letters 35, 2013. https://doi.org/10.1364/OL.35.002013.

Sáez-Rodríguez, D., Nielsen, K., Bang, O., Webb, D.J., 2014. Photosensitivity mechanism of undoped poly(methyl methacrylate) under uv radiation at 325 nm and its spatial resolution limit. Optics Letters 39, 3421–3424. https://doi.org/10.1364/OL.39.003421/r10.1117/122026796.

Tavares, C., Domingues, M., Frizera-Neto, A., Leite, T., Leitão, C., Alberto, N., Marques, C., Radwan, A., Rocon, E., André, P., Antunes, P., 2018. Gait shear and plantar pressure monitoring: a non-invasive ofs based solution for e-health architectures. Sensors 18, 1334. https://doi.org/10.3390/s18051334.

Tejedor, J., Macias-Guarasa, J., Martins, H.F., Pastor-Graells, J., Corredera, P., Martin-Lopez, S., 2017. Machine learning methods for pipeline surveillance systems based on distributed acoustic sensing: a review. Applied Sciences (Switzerland) 7, 1–26. https://doi.org/10.3390/app7080841.

Theodosiou, A., Polis, M., Lacraz, A., Kalli, K., Komodromos, M., Stassis, A., 2016. Comparative study of multimode cytop graded index and single-mode silica fibre Bragg grating array for the mode shape capturing of a free-free metal beam. p. 988600. https://doi.org/10.1117/12.2230792.

Theodosiou, A., Hu, X., Caucheteur, C., Kalli, K., 2018a. Bragg gratings and Fabry–Perot cavities in low-loss multimode cytop polymer fibre. IEEE Photonics Technology Letters 30, 857–860. https://doi.org/10.1109/LPT.2018.2820381.

Theodosiou, A., Komodromos, M., Kalli, K., 2018b. Carbon cantilever beam health inspection using a polymer fiber Bragg grating array. Journal of Lightwave Technology 36, 986–992. https://doi.org/10.1109/JLT.2017.2768414. From Duplicate 2 (Carbon antilever beam health inspection using a polymer fiber Bragg grating array – Theodosiou Antreas, Komodromos Michael, Kalli Kyriacos).

Tian, J., Jiao, Y., Fu, Q., Ji, S., Li, Z., Quan, M., Yao, Y., 2018. A Fabry–Perot interferometer strain sensor based on concave-core photonic crystal fiber. Journal of Lightwave Technology 36, 1952–1958. https://doi.org/10.1109/JLT.2018.2797104.

Ukil, A., Braendle, H., Krippner, P., 2012. Distributed temperature sensing: review of technology and applications. IEEE Sensors Journal 12, 885–892. https://doi.org/10.1109/JSEN.2011.2162060.

Vallan, A., Casalicchio, M.L., Olivero, M., Perrone, G., 2012. Assessment of a dual-wavelength compensation technique for displacement sensors using plastic optical fibers. IEEE Transactions on Instrumentation and Measurement 61, 1377–1383. https://doi.org/10.1109/TIM.2011.2180975.

Vilarinho, D., Theodosiou, A., Leitão, C., Leal-Junior, A.A., Domingues, M., Kalli, K., André, P., Antunes, P., Marques, C., de Fátima Domingues, M., Kalli, K., André, P., Antunes, P., Marques, C., 2017. Pofbg-embedded cork insole for plantar pressure monitoring. Sensors 17, 2924. https://doi.org/10.3390/s17122924.

Wochnowski, C., Eldin, M.A., Metev, S., 2005. Uv-laser-assisted degradation of poly(methyl methacrylate). Polymer Degradation and Stability 89, 252–264. https://doi.org/10.1016/j.polymdegradstab.2004.11.024.

Zawawi, M.A., O'Keeffe, S., Lewis, E., 2013. Plastic optical fibre sensor for spine bending monitoring with power fluctuation compensation. Sensors (Basel, Switzerland) 13, 14466–14483. https://doi.org/10.3390/s131114466.

Zhang, W., Webb, D.J., 2014. Humidity responsivity of poly(methyl methacrylate)-based optical fiber Bragg grating sensors. Optics Letters 39, 3026. https://doi.org/10.1364/OL.39.003026.

Zhu, J., Wang, M., Chen, L., Ni, X., Ni, H., 2017. An optical fiber Fabry–Perot pressure sensor using corrugated diaphragm and angle polished fiber. Optical Fiber Technology 34, 42–46. https://doi.org/10.1016/j.yofte.2016.12.004.

Zubel, M.G., Sugden, K., Saez-Rodriguez, D., Nielsen, K., Bang, O., 2016. 3d printed sensing patches with embedded polymer optical fibre Bragg gratings. p. 99162E. https://doi.org/10.1117/12.2237495.

Zubia, J., Arrue, J., Mendioroz, a., 1997. Theoretical analysis of the torsion-induced optical effect in a plastic optical fiber. Optical Fiber Technology 3, 162–167. https://doi.org/10.1006/ofte.1997.0212.

Part III

Optical fiber sensors in rehabilitation systems

Chapter 7

Wearable robots instrumentation

7.1 Optical fiber sensors on exoskeleton's instrumentation

The robotic therapy for rehabilitation and assistance experienced a large widespread in the last few years, as thoroughly discussed in Chapter 1 and in the first part of this book. The wearable and/or assistive robots are employed on functional rehabilitation of lower and upper limbs as well as in the assistance for the activities of daily living, especially on the gait assistance. In this case, it is possible to use novel control paradigms and strategies for human-robot physical and cognitive interaction aiming at a higher transparency between the user and the robot, where the robot does not inhibit the user's natural movement and does not apply excessive torques on the user's joints. In order to obtain such natural interface between the human and wearable robot, the sensor system of the robotic device needs to be able of accurately detect many parameters, which include the joint angles in the robotic device and the interaction forces between user and robot. Furthermore, as the wearable robots have supports or pads in direct contact with the user, it is necessary to estimate the microclimate conditions, i.e., temperature and humidity, in the interface region between the exoskeleton and its user.

Conventionally, electronic or electromechanic sensors are employed on the assessment of the aforementioned parameters. However, their main drawback in this case is the electromagnetic field sensitivity, which is especially undesirable for wearable robots since their actuation is generally performed with electric motors (Huo et al., 2016). Thus the electromagnetic interference caused by the constant activation of the electric motors can harm the performance of the electronic sensors, especially if they are not properly shielded. Moreover, the potentiometers and encoders for joint angle assessment have sensitivity to misalignments, which leads to the necessity of developing mechanical supports accurately attached to the robot, reducing the system compactness (Leal-Junior et al., 2018g). As another commonly used technology for joint angles assessment, inertial measurement units (IMUs) have the advantage of providing the three dimensional joint kinematics with a compact system. However, they need frequent calibration (El-Gohary and McNames, 2012) and the majority of wearable robots are limited to the sagittal plane (Huo et al., 2016). Regarding to the most used technology for interaction force assessment, the electronic strain

https://doi.org/10.1016/B978-0-32-385952-3.00017-2

gauges used to evaluate human-robot force/torque interaction also have sensitivity to electromagnetic fields and share some drawbacks of the electronic technologies previously mentioned for angle assessment, i.e., necessity of frequent calibration and careful attachment on the exoskeleton structure. Similar drawbacks occur with electronic sensor for microclimate conditions assessment, which are not widely used in commercially available exoskeletons.

A feasible approach for the mitigation (and even the elimination, in some cases) of the limitations presented by the electronic sensors in exoskeleton's sensor systems, the optical fiber sensors present some advantages in such application, which include compactness, lightweight, chemical stability, multiplexing capabilities, and electromagnetic fields immunity (Peters, 2011). As discussed in Chapter 6, there are two major types of optical fibers: silica optical fibers (SOFs) and polymer optical fibers (POFs). The former is widely used in telecommunication purposes and present low optical losses suitable for long distance communications and remote sensing, whereas the latter present higher flexibility and fracture toughness, which is suitable for sensing large forces and deflections. In addition, the smaller Young's modulus of POFs make them more sensitive to stress-related parameters than SOFs. Thus the measurement of joint angles, human-robot interaction forces and microclimate conditions are discussed in this chapter mainly using POFs due to their favorable mechanical features for these applications. In addition, two sensors approaches are presented in this chapter: (i) intensity variation-based sensors due to their low cost and portability (Leal-Junior et al., 2019b); (ii) Fiber Bragg Grating (FBG) sensors due to their wavelength-encoded data, which generally leads to higher accuracy on parameters estimation, and their multiplexing capabilities, where numerous sensors can be inscribed in a single optical fiber cable (Broadway et al., 2019).

7.2 Exoskeleton's angle assessment applications with intensity variation sensors

The applications of intensity variation-based sensors on angle assessment of wearable robots are discussed in this section. Despite the developments on FBG sensors (Leal-Junior et al., 2018b), interferometers (Leal-Junior et al., 2019a), and long period gratings (Barrera et al., 2018) for angle (or curvature assessment), intensity variation-based sensors present low cost and higher portability on the interrogation equipment that led to the development of practical applications in wearable devices. Such practical applications are divided into two case studies, one for an active orthosis for knee joint rehabilitation/assistance and the other is for a modular exoskeleton with multiple degrees of freedom. However, these sensors suffer from the influence of the modal distribution, which is mitigated by positioning the fiber with supports to avoid undesired movements outside the curvature region (and sensitive zone) of the sensor. In addition, the POF with 1 mm diameter is used in the intensity variation-based applications due to its lower cost (when compared with SOFs and POFs with smaller diameters) and higher strain limits and fracture toughness when compared with SOFs. As discussed in Chapters 5 and 6, the polymer presents a frequency-dependent response and generally shows a high hysteresis, especially when compared with glass. In order to mitigate these effects that can harm the sensor's performance in terms of accuracy, repeatability, and reliability, different compensation techniques have been proposed in the last few years. These techniques include data regressions on characterization tests under different angular velocities (Leal-Junior et al., 2018f) to mitigate the sensor's hysteresis and dynamic compensators for the sensitivity variation under different velocities due to the elastic modulus variation of the polymer under different time/frequency conditions (Leal-Junior et al., 2020).

As a compensation technique for sensitivity variations and hysteresis in POF sensors using the material features, the Prony series (discussed in Chapter 5) is applied on the sensors responses. In this case, the response is fitted on a sum of two exponentials due to the POF's composition, where the core/cladding material (PMMA) is different from the polyethylene used on the fiber jacket. Fig. 7.1 shows the experimental setup used on the sensor characterization and compensation technique validation, where a DC motor with position and velocity control performs the flexion-extension cycles in the plane indicated in Fig. 7.1. The reference angle is measured by a potentiometer attached to the rotation axis. As intensity variation sensors are sensitive to light source power deviations as well as temperature and humidity variations, a reference arm is employed to monitor the optical power variation in an optical fiber in the same environmental and light source conditions, but without bending or stress. The performed flexion and extension cycles range from 0° to 90° with constant and predefined angular velocities, five tests are performed with each angular velocity. The physical principle of the compensation technique is related to the differences in the polymer relaxation in loading (flexion) and unloading (extension) cycles, which lead to nonlinearities and hysteresis in the sensor response. One manner of modeling this response is through differential models that consider the polymer viscoelastic response as a combination of springs (for the elastic component) and dashpots (for the viscous component) (Lakes, 2009), where the general equation for the relaxation function is known as the Prony series (Lakes, 2009). Thus an experimental fit is made for flexion and another for extension to approximate the POF sensor response to a reference response, where there is linearity and low errors. The regression results in two compensated equations, one for flexion (Eq. (7.1)) and one for extension (Eq. (7.2)), where in both equations the compensated response is linear and presents low errors when compared to the reference, which also yields in low hysteresis in the flexion-extension responses. In addition, the compensated equations are obtained in flexion-extension cycles in a range of different angular velocities from 0.05 rad/s to 1.1 rad/s. Five cycles are performed at each angular velocity and the parameters of Eqs. ((7.1) and (7.2)) are the mean of the values obtained in each test. The employed angular velocity range is within the range of velocities that the robotic devices are able to provide. Furthermore, the time between steps is about 5 seconds, where the fiber is maintained under bending for 3 minutes before performing the flexion and extension cycles (Leal-Junior et al., 2018d),

$$\alpha_f = -0.0753 \cdot e^{(5.8314 \cdot (\frac{P}{P_0})_f)} + 161.6809 e^{(-2.0131 \cdot (\frac{P}{P_0})_f)} \tag{7.1}$$

$$\alpha_e = -0.099 \cdot e^{(0.4358 \cdot (\frac{P}{P_0})_e)} - 0.098.6809 e^{(-0.4360 \cdot (\frac{P}{P_0})_e)} \tag{7.2}$$

where α_f and α_e are the angles for flexion and extension, respectively. In addition, $(\frac{P}{P_0})_e$ and $(\frac{P}{P_0})_f$ are the normalized POF power output for extension and flexion cycles, respectively.

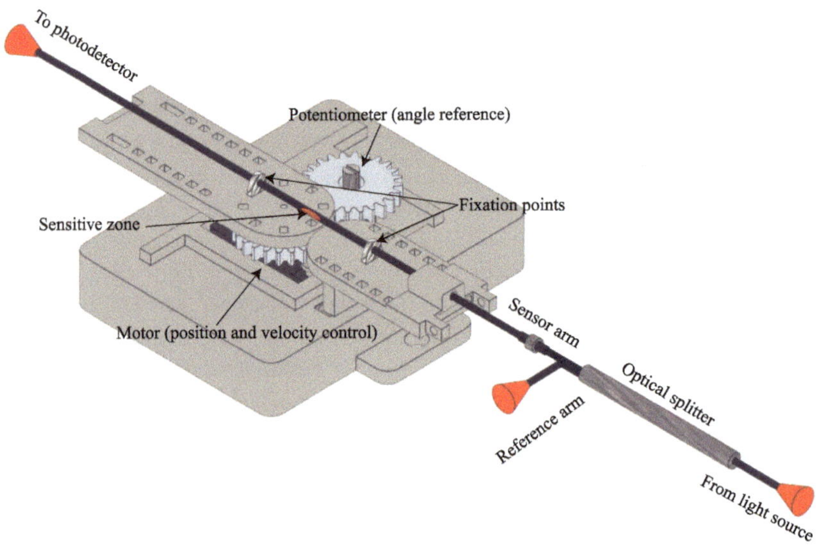

FIGURE 7.1 Experimental setup employed on the POF curvature sensor characterization and compensation technique development.

Fig. 7.2 shows the sensor's flexion and extension responses after the application of the compensation technique discussed, using Eqs. (7.1) and (7.2) for flexion and extension, respectively. It is possible to observe a hysteresis of 0.14%, which can be considered a negligible hysteresis for practical applications in angle assessment, since it results in errors below the sensor resolution. Comparing with the sensor response without the compensation technique application, there is, at least, a tenfold reduction of the hysteresis, which is 2% without the compensation technique. In order to analyze the improvements on the sensor's responses after the application of the proposed compensation technique, Fig. 7.2 shows the hysteresis and root means squared error (RMSE) of the sensor's responses with and without the compensation technique, where it is possible to observe a reduction of the hysteresis and/or the RMSE in all analyzed

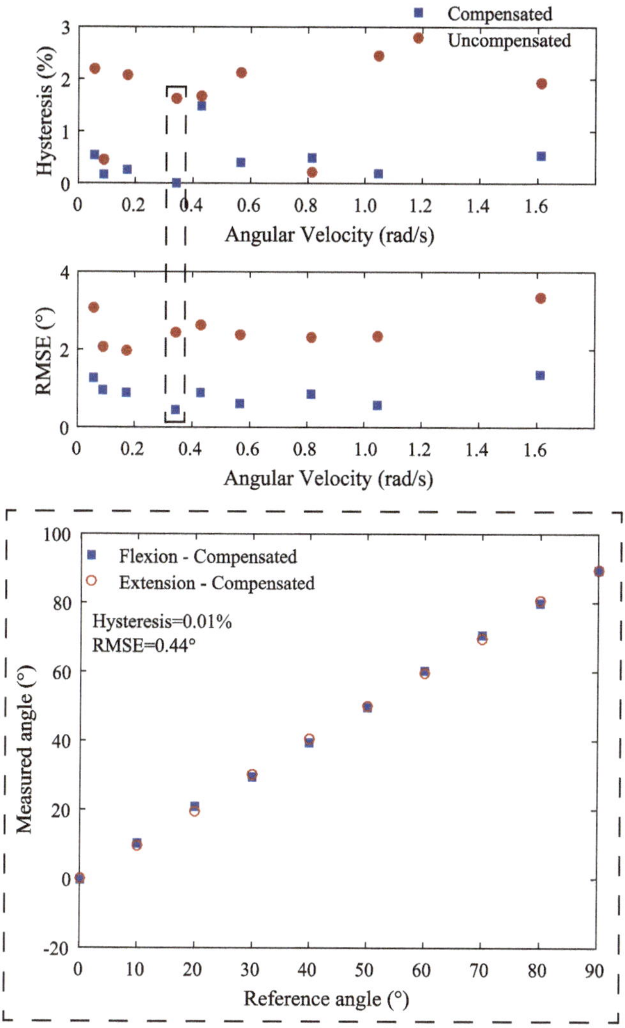

FIGURE 7.2 Comparison between compensated and uncompensated responses with respect to the hysteresis and RMSE. Figure inset shows a compensated response of the intensity variation-based angle sensor response for flexion and extension cycles.

cases. Nevertheless, it should be noted that since the compensation technique is based on an exponential regression, the proposed technique needs to be applied only on the angular velocity region in which the sensor was characterized. If different angular velocities are employed, an additional characterization is needed. However, the wearable robots operate with predefined angular velocities and the sensor characterization can be performed considering the angular velocity range of the robotic device. Thus the proposed compensation technique is a feasible

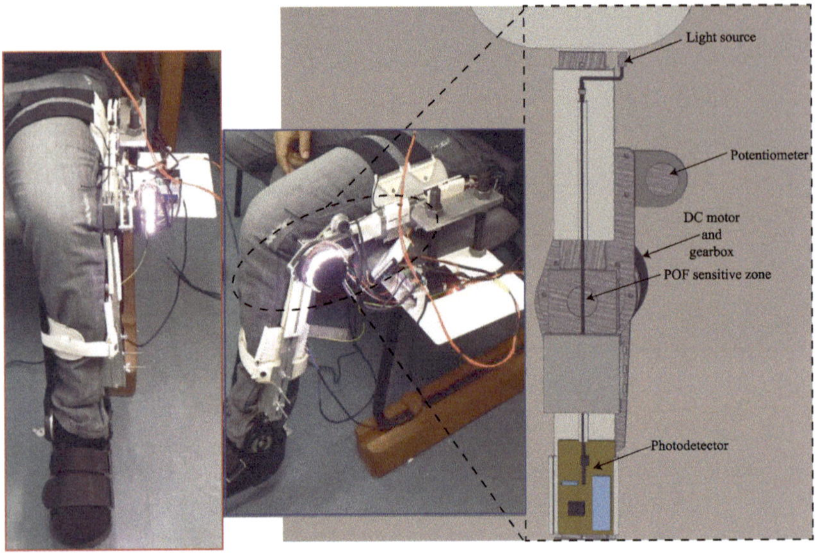

FIGURE 7.3 ALLOR with 3D printed supports for POF sensor positioning. Figure also shows the wearable robot positioned on the left knee of the user.

method for enhancing the performance of the optical fiber sensors in angle assessment applications, as discussed in the following case studies: one with a modular exoskeleton and the other with an active orthosis.

7.2.1 Case study: active lower limb orthosis for rehabilitation (ALLOR)

The ALLOR employed in this case study has an active degree of freedom on the knee, where the motor unit is a DC motor (Maxon Motors, Switzerland) and a harmonic drive gear reduction (Harmonic Drive AG, Germany). The active joint provides controlled angular movements on the range of 0 to 90°, where the angle is acquired with a potentiometer (10 kΩ and ±2% of linearity) attached to an additional structure on the ALLOR as shown in Fig. 7.3. The physical connection between the motor unit and potentiometer is performed using a toothed belt and a low pass filter is applied on the potentiometer response. The system is used for knee rehabilitation therapy, where the wearable robot is attached to a chair and the user performs flexion and extension movements with assisted (or resisted) torque of the ALLOR. In order to position the POF sensor on the ALLOR structure, without creating additional structure as the one of the potentiometer, 3D printed supports are fabricated and positioned on the robotic device as shown in Fig. 7.3 (Leal-Junior et al., 2017). Such supports also attached the light source and detectors on the structure without harm the ALLOR movement and the optical fiber bending region. Fig. 7.3 shows a photograph of the device with a subject ready to perform the tests. The test comprises of five flexion and extension movements on the range of 0 to 80° for higher comfort of the patient and the tests are performed in the different predefined angular velocities of the device, namely 0.05 rad/s, 0.10 rad/s, 0.16 rad/s, 0.21 rad/s, and 0.26 rad/s.

As shown in Fig. 7.2, the compensation technique leads to an improvement in the sensor's responses in terms of accuracy and hysteresis. The compensation technique is applied on the POF angle sensor responses at different angular velocities, which leads to the results shown in Fig. 7.4, where the comparison between potentiometer and POF angle sensors responses are presented at each angular velocity tested. The results show that the errors tend to be reduced as the angular velocity increases. Furthermore, the errors also vary with the angle measured. Thus Fig. 7.4(b) shows the deviations between the potentiometer and proposed POF sensor for each angle, where the solid line is the mean error for each angle. Regarding to the results shown in Fig. 7.4(b), the maximum error is around 2.4° at angles close to 47°. The region between 40° and 65° has the highest errors, which is a region at about the half of the strain (bending) cycle, where the hysteresis and viscoelastic features of the material responses are generally higher. Even in that region of such high hysteresis, the proposed compensation technique limited the errors to below 5% (when the whole cycle is considered). The mean error considering all five cycles at different angular velocities is around 1.2°, which can be regarded as low errors and shows the suitability and reliability of the proposed sensor as a new approach for the instrumentation of wearable robots applied on rehabilitation exercises.

7.2.2 Case study: modular exoskeleton

In the modular exoskeleton case study, the analysis of the sensor response in gait cycles is evaluated, which is a condition at which there is the simultaneous variation of the angle and angular velocity within the cycle. The exoskeleton presents active joints on the knee, ankle, and hip, where their connection is performed using seat-clamps (Dos Santos et al., 2017), which leads to higher modularity and enables the adjustment of each joint to the user. A linear bearing mechanism is used as hip support and provides the connection between the right and left leg. The actuators employed are direct current motors (EC 90 flat motor, Maxon Motors, Switzerland) with a harmonic drive (Harmonic Drive AG, Germany) that leads to a reduction ratio of 1:50. The angle assessments on each joint is performed through an encoder positioned inside the structure, whereas Fig. 7.5 shows the representation of the joint with the encoder attached and the case in which the proposed POF sensor is attached on the structure. The joint with optical fiber sensor has smaller dimensions and weights, which results in a lighter system more transparent to the user as it is less harmful for the user's natural movements.

Similar to the previous case study, the POF sensor is attached on the structure using 3D printed supports that also position the light source and photodetector

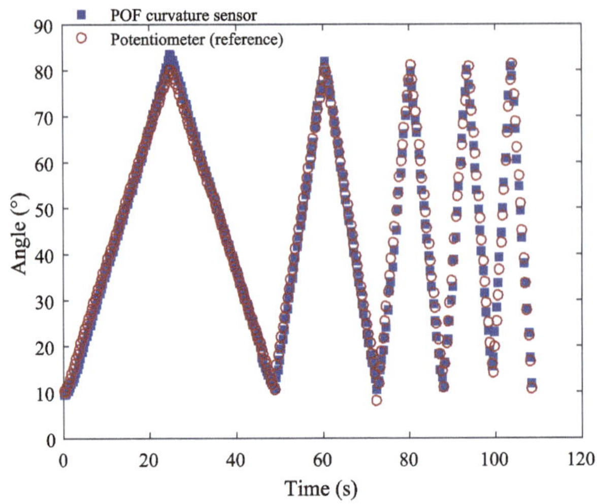

(a)

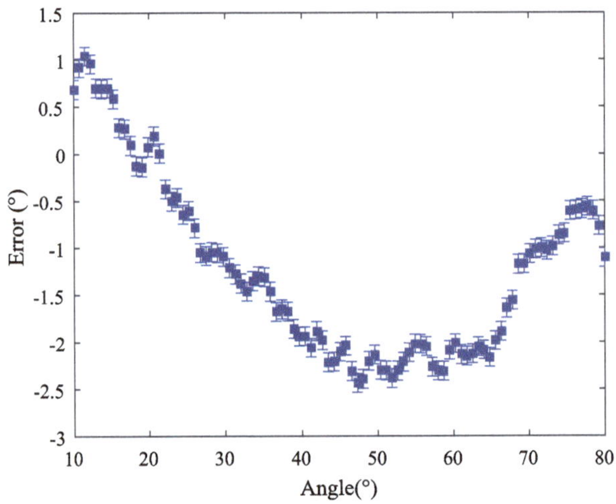

(b)

FIGURE 7.4 (a) Comparison between POF sensor and potentiometer responses for flexion and extension cycles with 5 different angular velocities, represented as regions 1 to 5 in the figure. (b) Mean error and standard deviation of the POF sensor at different angles for all tested angular velocities.

in order to obtain a compact structure. For the POF sensor validation and comparison with the encoder, the POF sensor structure is attached on the one with the encoder to enable the acquisition of both sensors responses simultaneously. Fig. 7.6 shows the structure used on the tests, where the first test is the application of sequential flexion and extension cycles on the range of 0° to 90°. In this case, there is the comparison of POF sensor and encoder responses as shown

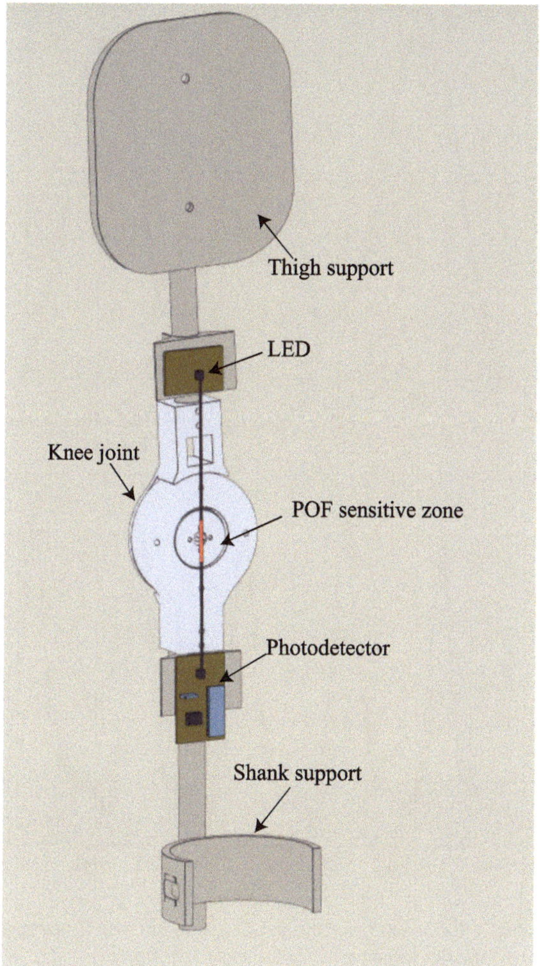

FIGURE 7.5 Representation of one knee joint with POF sensor system and an encoder for comparison purposes.

in Fig. 7.6 for the three performed cycles. It is worth noting that, once again, the compensation technique is applied on the POF sensor response in order to enhance its accuracy. In this first analysis, there is a high correlation between the encoder and POF sensor responses, where a mean deviation below 3.5° is found, considering all performed cycles. In the second test, the user wearing the exoskeleton, is asked to walk in a treadmill with constant angular velocity during 2 minutes, whereas the knee joint angles are acquired with the encoder and POF sensor, simultaneously. For brevity, the results of fiver sequential cycles are shown in Fig. 7.6, where it can be seen a good agreement between the encoder and POF sensor responses in which the mean error is about 3.8°. Therefore, the

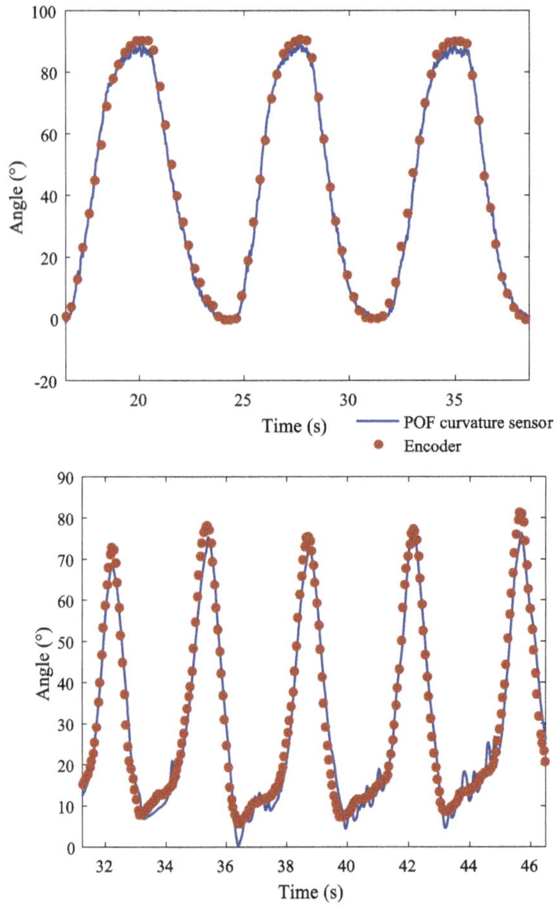

FIGURE 7.6 Results of flexion/extension cycles and gait cycles with the comparison between encoder and POF-based intensity variation sensor for angle assessment.

results indicate the feasibility of using POF-based intensity variation sensors on joint angle assessment in robotic devices, where a high accuracy can be obtained with a compact system that does not add significant weight to the wearable robot with the additional advantage of electromagnetic field immunity, which is important for wearable robots with constant activation of electric motors.

7.3 Human-robot interaction forces assessment with Fiber Bragg Gratings

The flexibility and small dimensions of optical fibers enable their embedment directly on the exoskeleton's structure. In this case, the ALLOR is used on the same configuration as discussed in Section 7.2. In order to obtain a transparent sensor embedded in the wearable robot structure, the 3D printing is used to create optical fiber-embedded structures. The embedment of optical fiber sensors in 3D printed structures was already discussed in the literature, where the temperature dependency (Leal-Junior et al., 2018a) and infill density (Leal-Junior et al., 2018e) influence on the optical fiber sensors responses were unveiled. The embedment of optical fibers in other polymer (such as the ones used in the 3D printing) results in a variation on the sensor's responses. Regarding to the temperature response, the sensor presents higher sensitivity since there is not only the thermooptic and optical fiber thermal expansion, but also the thermal expansion of the host material, resulting in an additional mechanism for optical signal variation as a function of the temperature. The infill density, i.e., the amount of material used on the 3D printing, e.g., an infill density of 10% indicates that the printed part only has 10% of material within, whereas the infill density of 100% indicates a massive 3D printed part without any region without material. Such parameter is also important on the temperature response, since it directly affects the sensitivity and linearity of the sensor embedded in 3D printed material (see Fig. 7.7), since higher infill density indicates higher influence of the host material thermal expansion on the sensor responses. Similarly, the force response is also influenced by the infill density. However, higher infill density leads to lower sensitivity, as shown in Fig. 7.7. The reason of such lower force sensitivity is the increase of the 3D printed part stiffness with higher infill densities, which, following the well-known Hooke's law, results in smaller strain on the optical fiber region, leading to smaller optical signal variation as a function of the applied force. Nevertheless, for that same reason, the structures with higher infill densities can withstand higher forces, which results in larger dynamic range for the force sensors. These effects are indicated in Fig. 7.7, where FBG sensors were embedded in 3D printed structures with different infill densities.

The development of optical fiber-embedded 3D printed structures resulted in the possibility of measuring the human-robot interaction forces directly on the exoskeleton's structures. Since FBG sensors are used in this application, the grating inscription, which, in this case, is performed in a perfluorinated POF, the gradient index multimode CYTOP fiber (Chromis Fiberoptics Inc). This optical fiber has a core diameter of 120 μm, a cladding thickness of 20 μm, and a polycarbonate overcladding. The grating inscription is performed using a femtosecond (fs) laser operating at 517 nm (repetition rate of 5 kHz and pulse energy of 80 nJ) with 220 fs pulse duration (HighQ laser femtoREGEN), where the inscription is made using the plane-by-plane direct write method, discussed in Chapter 6. After the inscription an annealing is performed in the optical fiber at a temperature of 70 °C for approximately 12 hours. As discussed in Chapter 5, the annealing treatment leads to a relaxation of the polymer molecular alignment, which leads to the reduction of internal stress created in the fiber manufacturing process (Stajanca et al., 2016). One of the key advantages of this thermal treatment in this application is the reduction of the POF Young's modulus, which leads to higher force sensitivity for the proposed sensor. Thereafter,

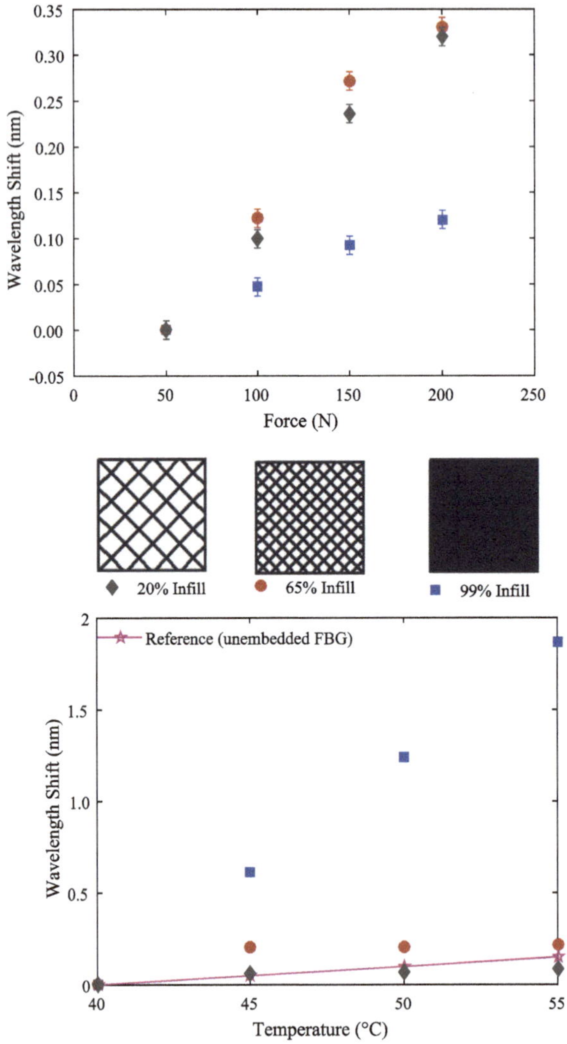

FIGURE 7.7 Influence of the infill density on force and temperature responses in FBG-based sensors.

the 3D-printed structures are manufactured with predefined infill densities. As depicted in Fig. 7.8, the 3D-printed structure used as exoskeleton leg support has two materials, one Acrylonitrile butadiene styrene (ABS) with 70% infill density to increase its stiffness, since it is attached to the exoskeleton and the other is the Thermoplastic polyurethane (TPU) with 50% density to increase the sensor sensitivity, since it is the region in which the optical fiber sensor is embedded. The ABS and TPU structures are attached to each other using a thermoplastic glue. Thus it provides a stability, flexibility, and comfort in the interface between the robotic device and the user. Fig. 7.8 shows the schematic representation of the proposed 3D-printed supports, as they are positioned in two regions of the user's leg to provide better stability on the flexion and extension movements; two supports are fabricated and an array with 2 FBGs is inscribed in order to incorporate all sensors in the same optical fiber cable. A photograph of the fabricated structures is also shown in Fig. 7.8.

For the characterization and validation tests of the proposed FBG sensor array for human-robot interaction forces assessment, the POF is connectorized to a single mode fiber (SMF) using the butt-coupling method (discussed in Chapter 4). The optical signal acquisition is performed by using the spectrometer I-MON 512 (Ibsen Photonics, Denmark) for a 10-kHz acquisition frequency. A Gaussian fit is used for the peak detection on the spectra, where the peaks are identified after setting a threshold. It is also worth to mention the possibility of changing the spectrometer integration time, which can act as a low pass filter for the spectra, reducing the side lobes of the spectra that makes the single peak detection easier.

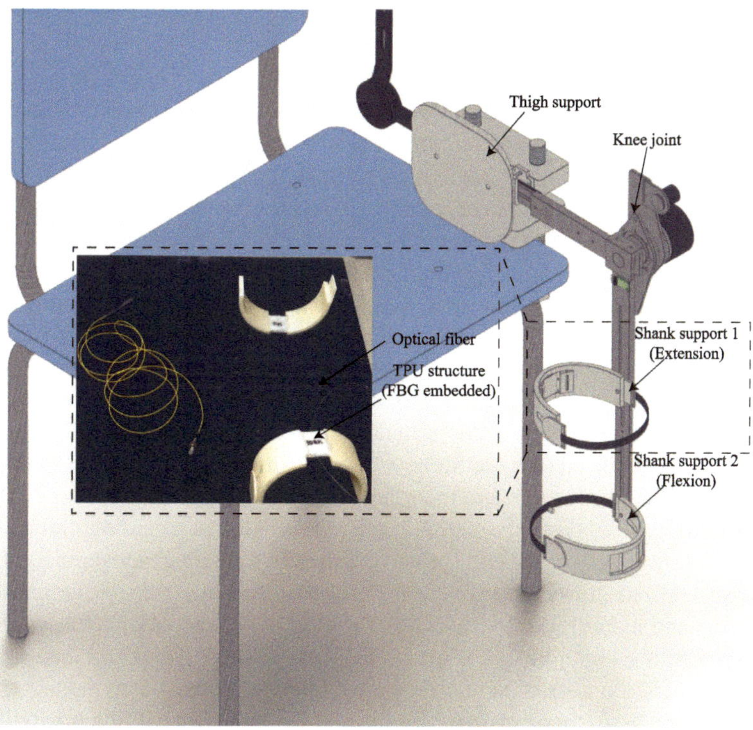

FIGURE 7.8 Representation of the proposed optical fiber-embedded 3D-printed support. Figure also shows a photograph of the device.

As previously discussed, the optical fiber embedment in 3D-printed structures can influence the sensors responses, as shown in Fig. 7.7, especially on the temperature responses. Thus for an accurate force sensing using the embedded devices, a compensation for the temperature effects is needed. The temperature compensation is performed using one of the techniques discussed in Chapter 6, where the temperature and force responses are characterized as shown in Fig. 7.9. The results obtained in the force characterization at a constant temperature of 25 °C for FBG 1 and FBG 2 shows sensitivities of 57.8 pm/N and 63.1 pm/N for FBG 1 and FBG 2, respectively. Then additional force characterization tests are performed at predefined temperatures for human thermal comfort, i.e., 30 °C, 35 °C, and 40 °C. In these tests, the force sensitivity variation as a function of the temperature is evaluated and, by considering the such variation on force sensitivity, it is possible to compensate the temperature effects on the sensor response, as shown in Fig. 7.9, where 7 tests were made with different forces and temperatures in which the force is varied by the application of different calibration weights (indicated as the reference in Fig. 7.9) on the sensor region, whereas the temperature variation is acquired by means of positioning the sensor on the top of a Peltier thermoelectric plate. Regarding to the tests shown in Fig. 7.9, tests 1 to 3 were made at room temperature (25 °C). The temperature was increased to 35 °C and tests 4 and 5 were performed, whereas the tests 6 and 7 were made at 45 °C. The comparison shows a root mean squared error (RMSE) of 0.50 N for FBG 1 and 2.29 N for FBG 2. The reason for the higher RMSE for FBG 2 is the lower sensor linearity, as presented in the characterization tests, which leads to higher errors in the force estimation when the linear regression is made. In the case of FBG 1, the errors are even lower, since a relative error of 1.0% was obtained, where such low error shows the possibility of achieving POF sensors with greater accuracy than the conventional technologies for robotic instrumentation.

For the validation of the proposed FBG array embedded in 3D-printed exoskeleton supports, the printed structures are positioned on the ALLOR as shown in Fig. 7.10, where the 3D-printed structures are positioned as shank supports of the ALLOR. In this case, the highest force is applied to shank support 1 when the flexion cycle is made, whereas in the extension cycle shank support 2 is subjected to the highest force. A thigh support is used to position and align the user's thigh and the robot structure, as indicated in Fig. 7.10. The ALLOR provides three types of rehabilitation exercises regarding to the degree of assistance: active assisted, passive, and active resistant movements. In the active assisted movement, the motor unit provides assistive torque to aid the flexion and extension movements of the user, whereas the opposite occurs in the active resistant mode. The passive operation mode, which is the one used in the validation tests is an intermediate operation mode, where the motor does not provide assistive or resistive torque and the impedance in the rehabilitation exercises is due to friction and back-drivability of the ALLOR. The validation tests are performed by means of positioning the user on the chair with the thigh

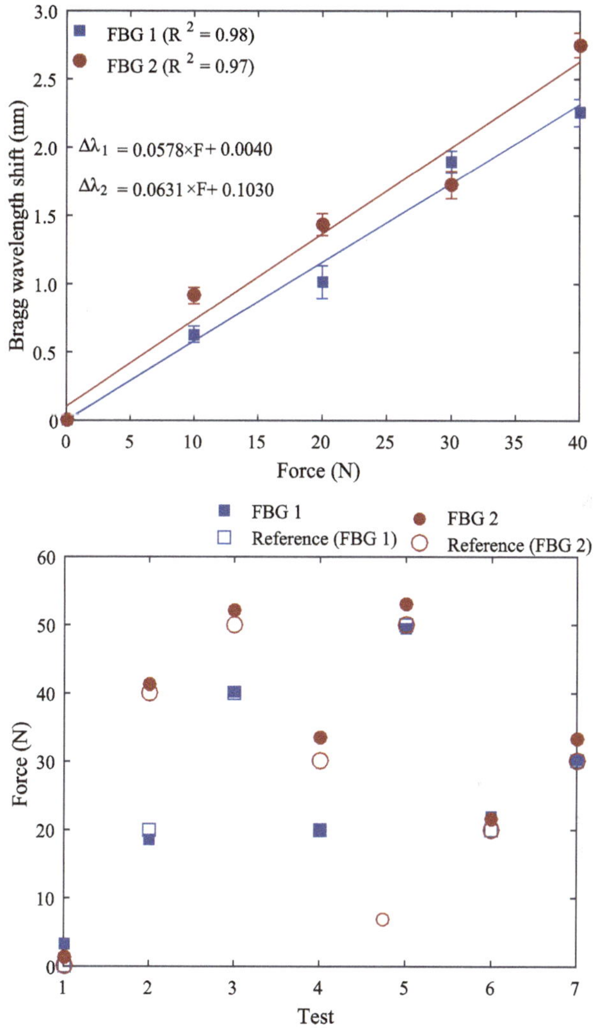

FIGURE 7.9 Force and temperature characterization of FBGs 1 and 2 as well as compensated force responses of the sensors under different temperature conditions.

and shank supports attached and the volunteer is asked to perform five flexion and extension cycles. Fig. 7.10 also shows estimated force using FBG 1 for extension movements and FBG 2 for flexion cycles. The extension cycles present similar measured forces in the range of 10 N and 15 N. Additionally, the human-robot interaction forces in flexion cycles have higher variation with a higher force application of the user as the cycles occur, which is related to variations of the user interaction with the exoskeleton along the tests as well as the adaptability on the wearable device. Thus the tests with the FBG-embedded shank supports attached to the exoskeleton have shown the feasibility of the proposed approach on the human-robot interaction forces assessment. For this reason, the 3D-printed FBG-embedded supports can be regarded as a viable option not only for the monitoring and validation of control techniques for the minimization of human-robot interaction forces (Duong et al., 2016), but also in the instrumentation of soft robotics wearable devices (Awad et al., 2017).

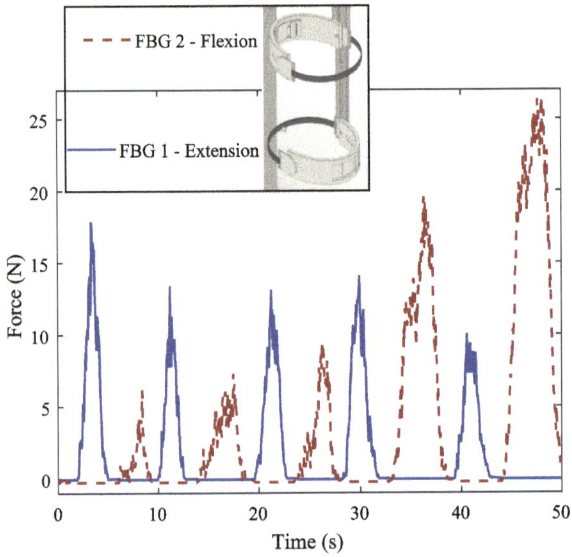

FIGURE 7.10 ALLOR for knee rehabilitation exercises positioned on the user and human-robot interaction forces measured with the proposed 3D-printed FBG-embedded sensors.

7.4 Interaction forces and microclimate assessment with intensity variation sensors

The microclimate assessment is an important feature for wearable devices, since the use of such wearable solutions leads to variations on the microclimate parameters (temperature and humidity) in the interface between the user and the wearable devices. Thus the assessment of such parameter is vital for wearable robots for rehabilitation exercises or functional assistance, since they are generally used for long periods. For this reason, an optical fiber-integrated solution for human-robot interaction forces assessment (as shown in Section 7.3) and microclimate measurements is proposed. In this case, intensity variation-based sensors are used due to their lower cost and higher portability, since the proposed sensor is a modular solution to be applied not only on the ALLOR, but also in other orthosis for knee movement assistance. The proposed sensor system is comprised of an optical fiber-embedded 3D-printed shank support for orthotic devices. In this case, there are four embedded intensity variation-based sensors, two sensors for human-robot interaction forces assessment in flexion and extension cycles, whereas the other two are used on the assessment of temperature and humidity. The sensors for temperature and humidity measurement are embedded in ABS structures made of ABS to isolate them from the stress or strain applied in the exoskeleton's shank support, which restricts the optical power variation to the change of environmental conditions. In contrast, the sensors for force assessment are embedded in TPU structures (similar to the ones presented in Section 7.3) to enable the strain transmission to the optical fiber under stress, resulting in an optical power variation proportional to the applied force.

Fig. 7.11 shows the schematic representation of the proposed 3D-printed smart device. The 3D-printed smart shank support is fabricated in different steps due to its multimaterial nature. The optical fiber used in this development a multimode polymethyl methacrylate (PMMA) POF with a core diameter of 980 μm and cladding thickness of 10 μm. The first step is the lateral section on the optical fibers, which is performed through abrasive removal of material and has 14 mm length and 0.6 mm depth due to its higher sensitivity and linearity (Leal-Junior et al., 2018c). Then an annealing thermal treatment is performed in the POFs to reduce the internal stress created in the fiber manufacturing process, which increases the transverse force sensitivity. The shank support structure (represented in gray on Fig. 7.11) is fabricated using ABS material with 100% infill. Similarly, the microclimate sensors 1 and 2 (represented in red on Fig. 7.11) are also made of ABS with 100% infill and the part has an opening to position the lateral section of the optical fiber close to the interface between user and wearable device, but without touching the user's leg. In addition, a torsion is made on each POF during the embedment in the ABS rigid structure. Such torsion improves the humidity responsivity of the PMMA due to the higher strain on the optical fiber as discussed in Zhang and Webb (2014). The force sensors 1 and 2 are embedded in flexible structures made of TPU for the transmission of applied force to the optical fiber, which deforms in conjunction with the flexible TPU structure, resulting in a curvature on the POF and optical signal attenuation due to the macrobending losses and refractive index variations due to the stress-optic effect (as discussed in Chapter 6). In order to obtain the interaction forces in flexion and extension cycles, force sensor 1 is positioned in the front of the 3D-printed support for the force assessment during leg extension movement and force sensor 2 is positioned on the elastic band positioned on the back of the support to measure the interaction force during flexion movements, as shown in Fig. 7.11. As the device has a modular configuration, a thermoplastic resin is used to attach the 3D-printed structures on the configuration presented in Fig. 7.11.

The characterization of the temperature, humidity, and force sensors was performed with respect to each desired parameter and the compensation techniques for the cross-sensitivities are based on the direct difference between the responses of each sensor, as discussed in Chapter 6. For the microclimate

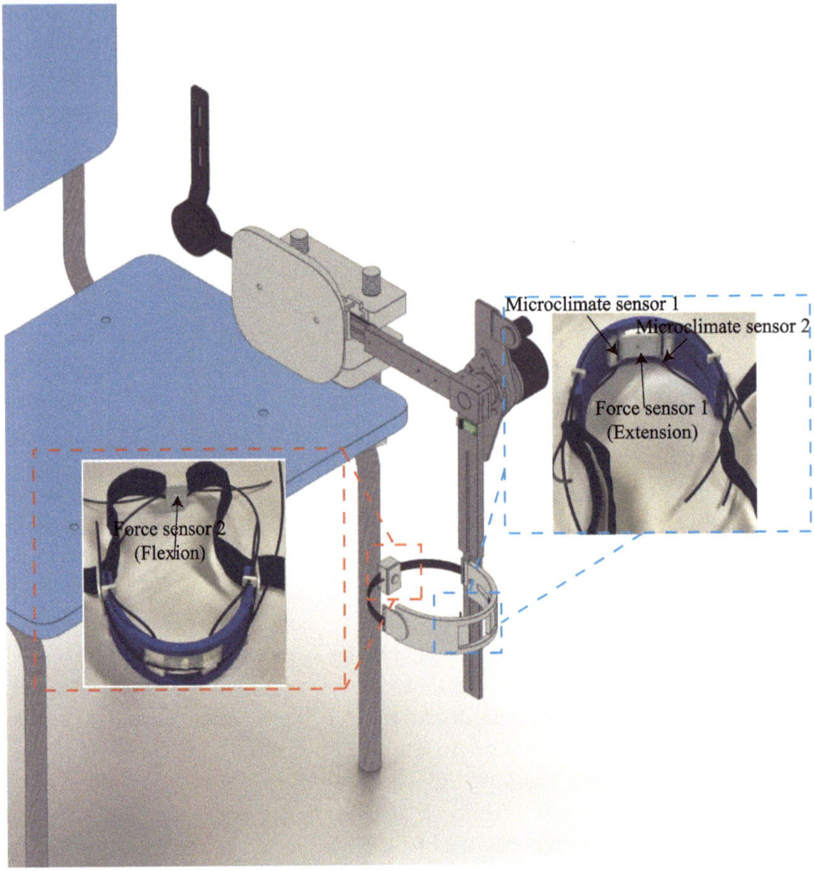

FIGURE 7.11 Schematic representation of the optical fiber-embedded instrumented shank support.

sensors 1 and 2, each sensor is characterized with respect to temperature and humidity, as shown in Fig. 7.12(a), where the humidity tests were performed in constant temperature conditions (around 25 °C). Similarly, the temperature tests were made without humidity variations. From the temperature and humidity characterizations, it is possible to obtain the sensitivities of microclimate sensors 1 and 2 with respect to these parameters, which is used in Eq. (7.3) to obtain the humidity and temperature using the combined responses of microclimate sensors 1 and 2:

$$\begin{bmatrix} P_{M1} \\ P_{M2} \end{bmatrix} = \begin{bmatrix} K_{1,H} & K_{1,T} \\ K_{2,H} & K_{2,T} \end{bmatrix} \cdot \begin{bmatrix} \Delta H \\ \Delta T \end{bmatrix} \tag{7.3}$$

Similar to the interaction force sensors presented in Section 7.3, Force sensors 1 and 2 also have cross-sensitivity to temperature. However, differently

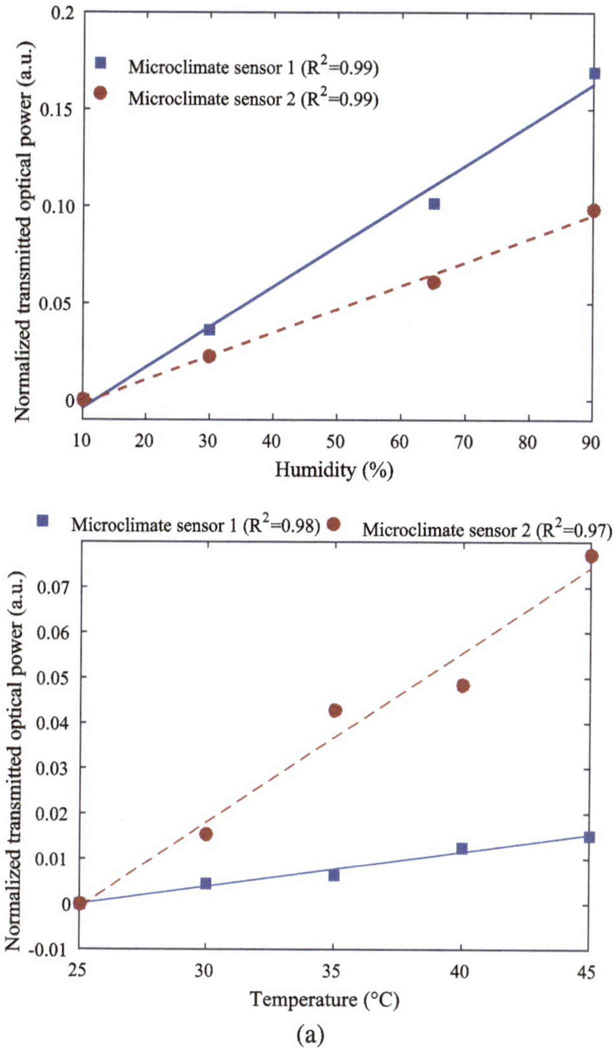

FIGURE 7.12 (a) Temperature and humidity responses of microclimate sensors 1 and 2. (b) Characterization of force sensors 1 and 2 under different temperature and humidity conditions.

from CYTOP fibers used on the system presented in Section 7.3, the PMMA POF also has cross-sensitivity to humidity, as discussed in Chapter 5. Thus both temperature and humidity can lead to elastic modulus variation on PMMA POFs, which affect the sensor's force sensitivity. For this reason, the force characterization of force sensors 1 and 2 is performed under different temperature and humidity conditions as shown in Fig. 7.12(b). In this case, the compensation for temperature and humidity effects is similar to the one presented in

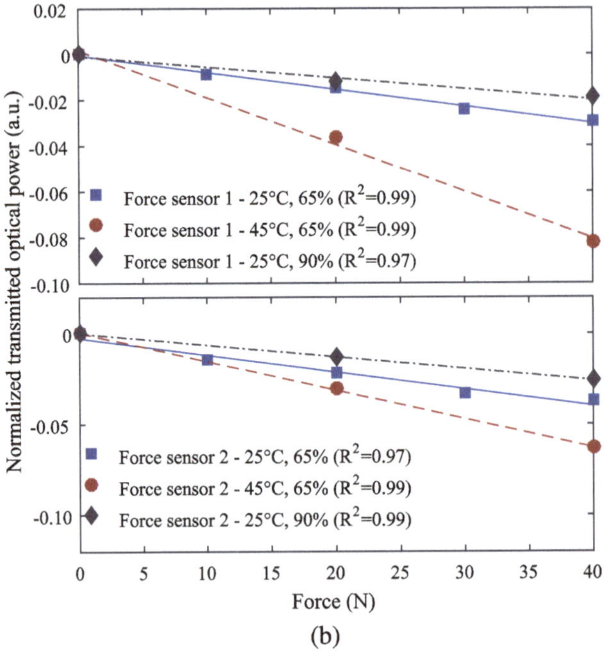

FIGURE 7.12 *(continued)*

Section 7.3. The only difference is the use of a polynomial regression of the force sensitivity as a function of two terms, i.e., temperature and humidity (instead of just temperature, as shown in Section 7.3). The force characterization results show similar trend for both sensors 1 and 2 in which the increase in the temperature leads to an increase of the sensor force sensitivity. In contrast, the increase of the humidity results in a reduction of the sensor sensitivity with respect to the applied force. The reason for this behavior is related to the changes on the POF Young's modulus under which environmental condition.

After the characterization and validation of the proposed 3D-printed instrumented support, the proposed smart device is used as a shank support for the ALLOR as presented in Fig. 7.13(a) and a passive orthosis for gait assistance shown in Fig. 7.13(b). With the lower limb exoskeleton, the tests made are the flexion and extension of the knee, whereas the tests with the passive orthosis are performed in a treadmill with constant linear velocity. The results of both applications obtained with the proposed 3D-printed instrumented support are shown in Fig. 7.13(a) and (b), where the 10 flexion/extension cycles were performed with the exoskeleton (see Fig. 7.10 (a)) and the gait cycles with the passive orthosis are shown for an interval of 30 seconds (see Fig. 7.13 (b)). Since Force sensor 1 is activated in the extension cycle and Force sensor 2 in the flexion, the response of Force sensor 2 was inverted in order to provide a better visualization of the results obtained in the exoskeleton application. It can be observed

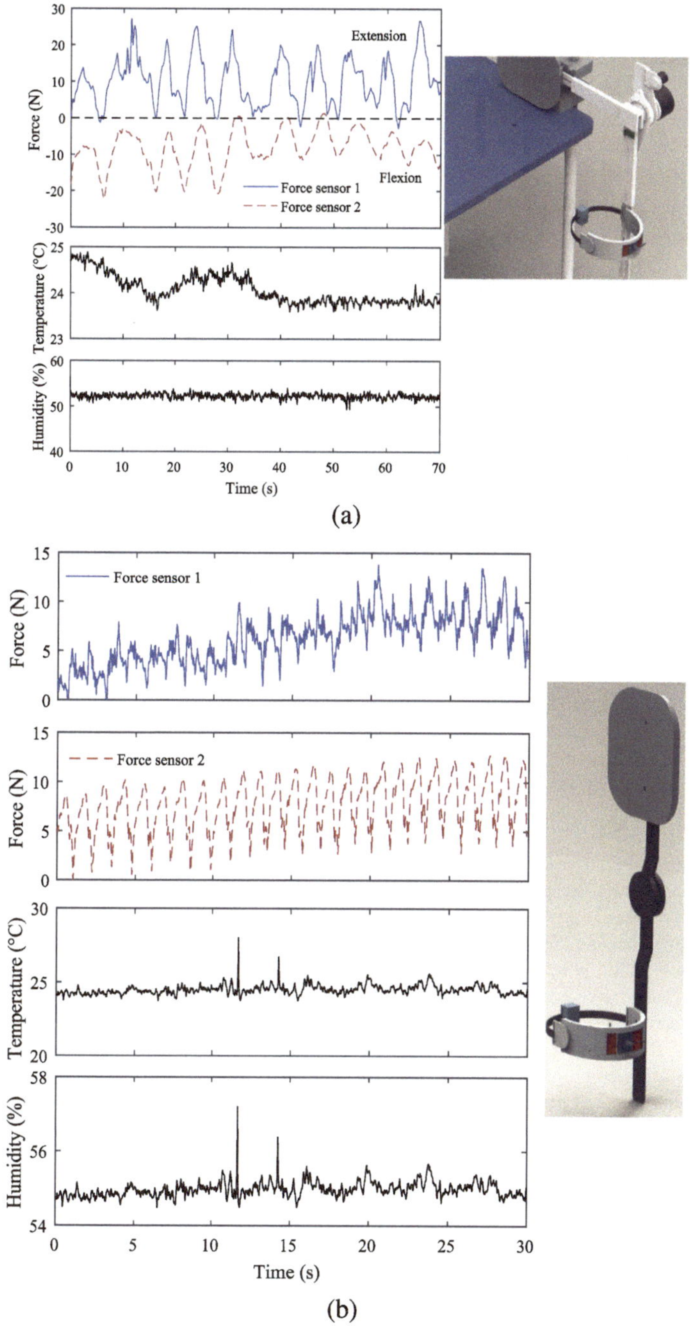

FIGURE 7.13 Human-robot interaction forces and microclimate assessment applied on the (a) AL-LOR and (b) passive orthosis for gait assistance.

in Fig. 7.13(a) that both sensors respond to the force input in their respective cycles, i.e., sensor 1 in extension and 2 in flexion. Thus the maximum force of each sensor occurs in distinct instants of the flexion/extension cycle, which is in agreement with what is expected in those cycles and the higher force variation on the extension cycle. In addition, the temperature and humidity responses remain stable during the whole test and within the range of environmental conditions of the room, where the tests were performed.

Regarding the gait tests shown in Fig. 7.13(b) with the passive orthosis, a higher force variation occurs in sensor 2, which is expected due to the dynamics of the gait cycle, where there is higher degree of flexion than extension (Kirtley, 2006). Once again, the temperature and humidity remained stable during the test. These results show not only the feasibility, but also the advantages of the proposed 3D-printed instrumented support, since it possible to measure the forces in the region where it is applied due to the high flexibility of POFs. In addition, the possibility of fabricating parts of a wearable device with embedded sensors, such as the one proposed, result in novel approaches on the wearable robots design. Moreover, the microclimate measurements performed also can prevent injuries on the user and increase their comfort during the use of different robotic devices.

References

Awad, L.N., Bae, J., O'Donnell, K., De Rossi, S.M.M., Hendron, K., Sloot, L.H., Kudzia, P., Allen, S., Holt, K.G., Ellis, T.D., Walsh, C.J., 2017. A soft robotic exosuit improves walking in patients after stroke. Science Translational Medicine 9, eaai9084. https://doi.org/10.1126/scitranslmed.aai9084.

Barrera, D., Madrigal, J., Sales, S., 2018. Long period gratings in multicore optical fibers for directional curvature sensor implementation. Journal of Lightwave Technology 36, 1063–1068. https://doi.org/10.1109/JLT.2017.2764951.

Broadway, C., Min, R., Leal-Junior, A., Marques, C., Caucheteur, C., 2019. Toward commercial polymer fiber Bragg grating sensors: review and applications. Journal of Lightwave Technology 37. https://doi.org/10.1109/JLT.2018.2885957.

Dos Santos, W.M., Nogueira, S.L., De Oliveira, G.C., Peña, G.G., Siqueira, A.A., 2017. Design and evaluation of a modular lower limb exoskeleton for rehabilitation. In: IEEE International Conference on Rehabilitation Robotics, pp. 447–451.

Duong, M.K., Cheng, H., Tran, H.T., Jing, Q., 2016. Minimizing human-exoskeleton interaction force using compensation for dynamic uncertainty error with adaptive rbf network. Journal of Intelligent and Robotic Systems: Theory and Applications 82, 413–433. https://doi.org/10.1007/s10846-015-0251-x. From Duplicate 1 (Minimizing human-exoskeleton interaction force using compensation for dynamic uncertainty error with adaptive rbf network – Duong Mien Ka, Cheng Hong, Tran Huu Toan, Jing Qiu).

El-Gohary, M., McNames, J., 2012. Shoulder and elbow joint angle tracking with inertial sensors. IEEE Transactions on Bio-Medical Engineering 59, 2635–2641. From Duplicate 2 (Shoulder and elbow joint angle tracking with inertial sensors – El-Gohary Mahmoud, McNames James).

Huo, W., Mohammed, S., Moreno, J.C., Amirat, Y., 2016. Lower limb wearable robots for assistance and rehabilitation: a state of the art. IEEE Systems Journal 10, 1068–1081. https://doi.org/10.1109/JSYST.2014.2351491. From Duplicate 2 (Lower limb wearable robots for assistance and rehabilitation: a state of the art – Huo Weiguang, Mohammed Samer, Moreno Juan C., Amirat Yacine).

Kirtley, C., 2006. Clinical Gait Analysis: Theory and Practice. Elsevier, Philadelphia.
Lakes, R., 2009. Viscoelastic Materials. Cambridge University Press, Cambridge.
Leal-Junior, A.G.A., Frizera-Neto, A., Pontes, M.M.J., Botelho, T.T.R., Leal Junior, A., Frizera Neto, A., Pontes, M.M.J., Botelho, T.T.R., Leal-Junior, A.G.A., Frizera-Neto, A., Pontes, M.M.J., Botelho, T.T.R., 2017. Hysteresis compensation technique applied to polymer optical fiber curvature sensor for lower limb exoskeletons. Measurement Science and Technology 28, 125103. https://doi.org/10.1088/1361-6501/aa946f.
Leal-Junior, A., Casas, J., Marques, C., Pontes, M., Frizera, A., 2018a. Application of additive layer manufacturing technique on the development of high sensitive fiber Bragg grating temperature sensors. Sensors 18, 4120. https://doi.org/10.3390/s18124120.
Leal-Junior, A., Theodosiou, A., Díaz, C., Marques, C., Pontes, M.M., Kalli, K., Frizera-Neto, A., 2018b. Polymer optical fiber Bragg gratings in cytop fibers for angle measurement with dynamic compensation. Polymers 10, 674. https://doi.org/10.3390/polym10060674.
Leal-Junior, A.G., Frizera, A., José Pontes, M., 2018c. Sensitive zone parameters and curvature radius evaluation for polymer optical fiber curvature sensors. Optics & Laser Technology 100, 272–281. https://doi.org/10.1016/j.optlastec.2017.10.006.
Leal-Junior, A.G., Frizera, A., Marques, C., Pontes, M.J., 2018d. Viscoelastic features based compensation technique for polymer optical fiber curvature sensors. Optics & Laser Technology 105, 35–40. https://doi.org/10.1016/j.optlastec.2018.02.035.
Leal-Junior, A.G., Marques, C., Ribeiro, M.R.N., Pontes, M.J., Frizera, A., 2018e. Fbg-embedded 3-d printed abs sensing pads: the impact of infill density on sensitivity and dynamic range in force sensors. IEEE Sensors Journal 18, 8381–8388. https://doi.org/10.1109/JSEN.2018.2866689.
Leal-Junior, A.G.A., Frizera, A., Pontes, M.M.J., 2018f. Dynamic compensation technique for pof curvature sensors. Journal of Lightwave Technology 36, 1112–1117. https://doi.org/10.1109/JLT.2017.2752361.
Leal-Junior, A.G.A., Frizera, A., Vargas-Valencia, L., Dos Santos, W.M.W., Bo, A.A.P.L., Siqueira, A.A.A.G., Pontes, M.M.J., 2018g. Polymer optical fiber sensors in wearable devices: toward novel instrumentation approaches for gait assistance devices. IEEE Sensors Journal 18, 7085–7092. https://doi.org/10.1109/JSEN.2018.2852363.
Leal-Junior, A.G., Avellar, L.M., Diaz, C.A., Frizera, A., Marques, C., Pontes, M.J., 2019a. Fabry-Perot curvature sensor with cavities based on uv-curable resins: design, analysis, and data integration approach. IEEE Sensors Journal 19, 9798–9805. https://doi.org/10.1109/JSEN.2019.2928515.
Leal-Junior, A.G., Diaz, C.A., Avellar, L.M., Pontes, M.J., Marques, C., Frizera, A., 2019b. Polymer optical fiber sensors in healthcare applications: a comprehensive review. Sensors 19, 3156. https://doi.org/10.3390/s19143156.
Leal-Junior, A.G., Theodosiou, A., Diaz, C.A.R., Avellar, L.M., Kalli, K., Marques, C., Frizera, A., 2020. Fpi-pofbg angular movement sensor inscribed in cytop fibers with dynamic angle compensator. IEEE Sensors Journal 20, 5962–5969. https://doi.org/10.1109/JSEN.2020.2974931.
Peters, K., 2011. Polymer optical fiber sensors—a review. Smart Materials and Structures 20, 013002. https://doi.org/10.1088/0964-1726/20/1/013002. From Duplicate 2 (Polymer optical fiber sensors—a review – Peters, Kara).
Stajanca, P., Cetinkaya, O., Schukar, M., Mergo, P., Webb, D.J., Krebber, K., 2016. Molecular alignment relaxation in polymer optical fibers for sensing applications. Optical Fiber Technology 28, 11–17. https://doi.org/10.1016/j.yofte.2015.12.006.
Zhang, W., Webb, D.J., 2014. Humidity responsivity of poly(methyl methacrylate)-based optical fiber Bragg grating sensors. Optics Letters 39, 3026. https://doi.org/10.1364/OL.39.003026.

Chapter 8

Smart structures and textiles for gait analysis☆

In addition to their use in wearable robots instrumentation, the optical fiber sensors can be directly applied on wearable system (including clothing accessories) to perform the assessment of kinematics, kinetics, and spatiotemporal parameters on the human movement. Thus it is possible to obtain a complete movement analysis tool using only optical fiber sensors. In order to discuss some of these wearable sensors applications using optical fiber sensors, this chapter divides the sensors applications into three groups. Kinematic parameters assessment, discussed in Section 8.1, where modular systems using intensity variation sensors and a portable fiber Bragg grating (FBG) sensor system are discussed. Then the use of optical fiber sensors on plantar pressure and ground reaction force (GRF) measurement is presented in Section 8.2, where once again the developments on FBG and multiplexed intensity variation sensors are discussed. Finally, in Section 8.3 the spatiotemporal parameters are discussed and measured using an optical fiber-embedded smart carpet through the multiplexed intensity variation sensor system (as discussed in Section 8.3).

8.1 Optical fiber sensors for kinematic parameters assessment

8.1.1 Intensity variation-based sensors for joint angle assessment

The intensity variation-based sensors (discussed in Chapter 6) are applied on the joint angle assessment in the sagittal plane. A polymer optical fiber (POF) made of polymethyl methacrylate (PMMA) with 980 μm core diameter and 10 μm cladding thickness is used in this application (Leal-Junior et al., 2018a). A lateral section is performed on the optical fiber to create a sensitive zone; the section is performed through abrasive removal of material with controlled shape, length and depth. Thereafter, an annealing treatment is performed to the optical fiber for an internal stress relaxation, which leads to hysteresis reduction and sensitivity increase (Woyessa et al., 2016). In addition to the optical fiber, the sensor system has two modules: (i) an emitter module, supplied by a 9 V battery, comprised a of light emitting diode (LED) IF-E97 (Industrial Fiber Optics,

☆ This chapter is carried out with the participation of Leticia Munhoz de Avellar.

https://doi.org/10.1016/B978-0-32-385952-3.00018-4

USA) with central wavelength at 660 nm. (ii) A receiver module with a photodiode IF-D91 (Industrial Fiber Optics, USA) and transimpedance amplifier, which also stores the data in a SD card with the additional possibility of providing a Bluetooth connection for real time data processing and analysis. All modules are encapsulated in 3D-printed structures. Fig. 8.1 shows the schematic representation of the proposed system (and its modules) as well as the sensor responses to angle variations compared with a potentiometer shown in Fig. 8.1. The compensation technique discussed in Section 7.1 is also applied in this case, resulting a root mean squared error (RMSE), between the POF curvature sensor and the potentiometer, below 1° with negligible hysteresis.

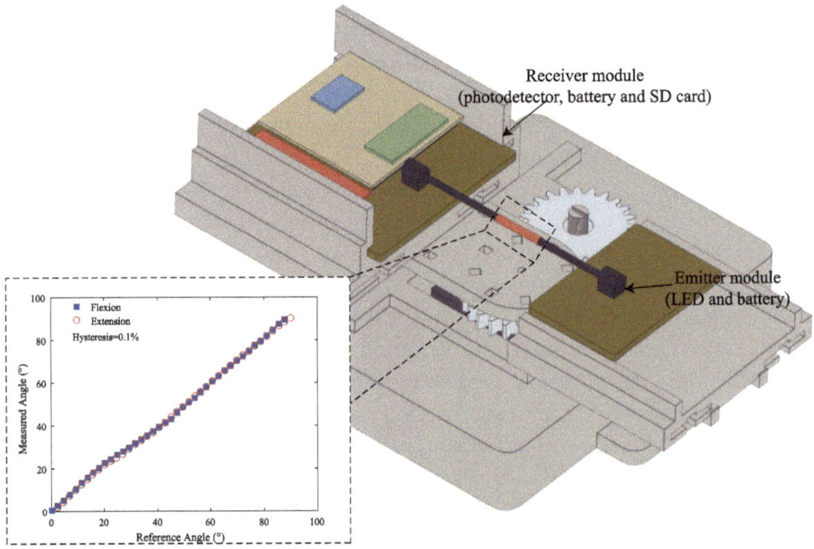

FIGURE 8.1 Schematic representation of the sensor system positioned on the setup for angle assessment and its angle response.

The designed wearable sensor for joint angle assessment is applied on joint angles measurement in the elbow and knee joints as shown in Fig. 8.2, where the POF curvature sensor is used on the assessment of the flexion and extension cycles of the elbow as well as in the knee angles during the gait. The modular nature of the POF-based wearable sensor enables its application on different human joints and even on soft robotic devices (Manti et al., 2016). In addition, its compactness and small dimensions do not inhibit the natural pattern of the movement when it is positioned on the user's joint. The wearable sensor installation is simple, since it only needs the positioning of the modules on the joint at which the angle is assessed using elastic bands to avoid slippage of the sensor units (see Fig. 8.2).

The wearable POF sensor was the elbow of a female subject and sequential flexion and extension cycles were made. The results obtained are presented in

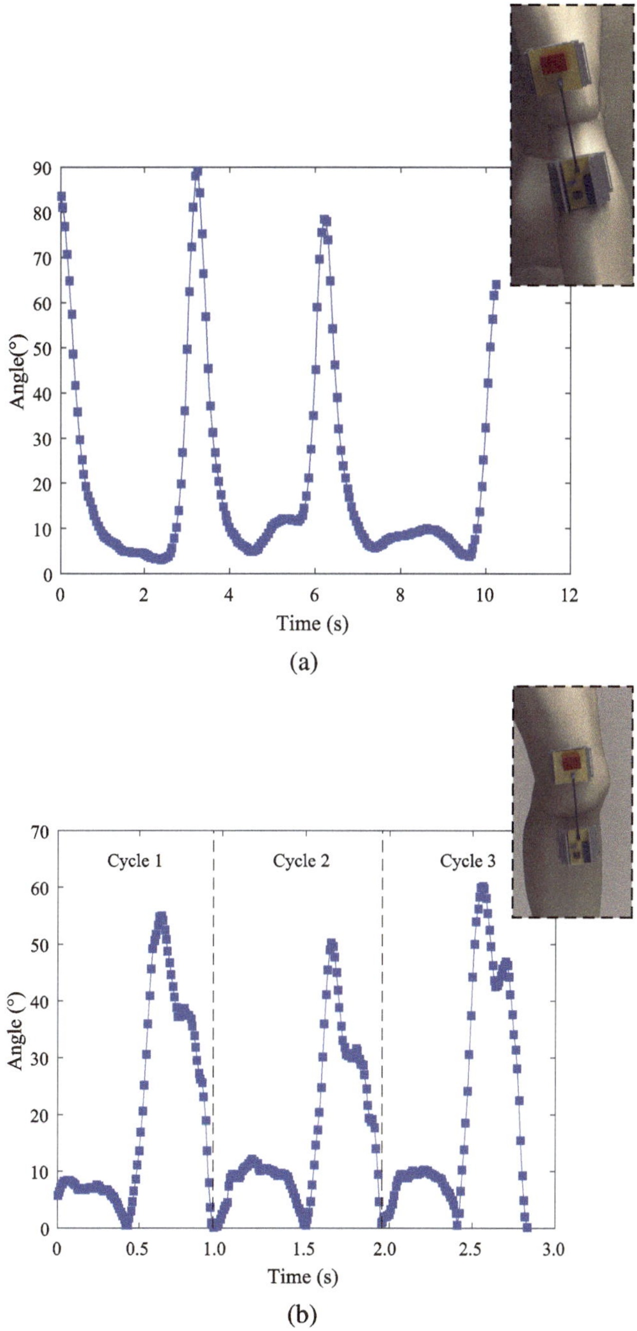

FIGURE 8.2 POF curvature sensor, its modules positioned and the angle measured with the wearable device on (a) knee joint and (b) elbow joint.

Fig. 8.2(a) and this first angle evaluation was made to evaluate the sensor easiness of positioning and the comfort for the patient. The test begins with the user's elbow at around 90°, shown in Fig. 8.2(a), with the sequential flexion and extension cycles detected by the sensor. The sensor resolution, considering an analog-to-digital converter (ADC) of 8 bits is about 1.3°, which is suitable for measuring knee and elbow angles due to the angular range of these joints. However, if higher resolution is desired, it is possible to simply change the ADC to one with higher resolution such as the ones with 16 bits, where resolutions in the order of 0.5° can be obtained. In order to assess the sensor repeatability on the joint angle analysis, the sensor was positioned on the knee of the subject and sequential gait cycles were made as depicted in Fig. 8.2(b). The results indicate that knee angles present a well-defined pattern on the sagittal plane during gait. The tests presented in Fig. 8.2(b) show a high repeatability of the angle measured by the proposed wearable sensor, where a mean angle deviation of 6.60° is found between cycles, which is related to the human movement since it presents angle variations between cycles (Kirtley, 2006).

8.1.2 Fiber Bragg gratings sensors with tunable filter interrogation for joint angle assessment

The portability is a critical requirement in wearable sensor systems, as the majority of commercially available optical interrogators for FBG sensors are bulky and nonportable, interrogation techniques and/or customized systems are needed to enable the application of FBG sensors in wearable applications outside a clinical environment (Diaz et al., 2019). To that extent, an interrogation technique based on tunable Fabry–Perot interferometer (FPI) used as an optical filter is employed for the development of a wearable FBG-embedded textile for joint angle assessment. The proposed interrogation method comprises of one optical stage that includes tunable filter, light source, photodetector, and optical circulator, whereas an electronic stage with the microcontrollers with an on-chip digital-to-analog converter (DAC) of with 12 bits resolution and analog-to-digital converter (ADC) with 16 bits resolution and small single-board computer (Raspberry pi 3) is used on the signal acquisition with the capacity of wireless transmission. The operation principle is based on the FPI tuning using the microcontrollers to obtain a fast- and high-resolution sweep on the optical spectrum. Then a convolution between the filter and the FBG sensor is acquired by the photodetectors at each position of the filter during its sweep, which enables a spectral reconstruction of the grating spectrum with a portable and lower cost system (when compared with commercially available ones). Fig. 8.3 shows a schematic representation of the proposed optical interrogator and acquisition unit. The FPI has a free spectral range (FSR) of 47 nm and is connected to the FBG using an optical circulator; the photodector with a transimpedance amplifier is responsible for the optical power acquisition. The light source is superluminescent light-emitting diode (SLED) centered at 1550 nm with 70 nm

3 dB spectral width and 12 mW maximum optical power. The FBG used in this application was inscribed via the phase mask method (described in Chapter 6) in photosensitive silica optical fiber using an Nd:YAG laser centered at 266 nm with 8 ns of pulse duration and 1 kHz of repetition rate. The optical fiber also has an acrylate protection to increase its fracture toughness and strain limits, especially for bending on the grating region.

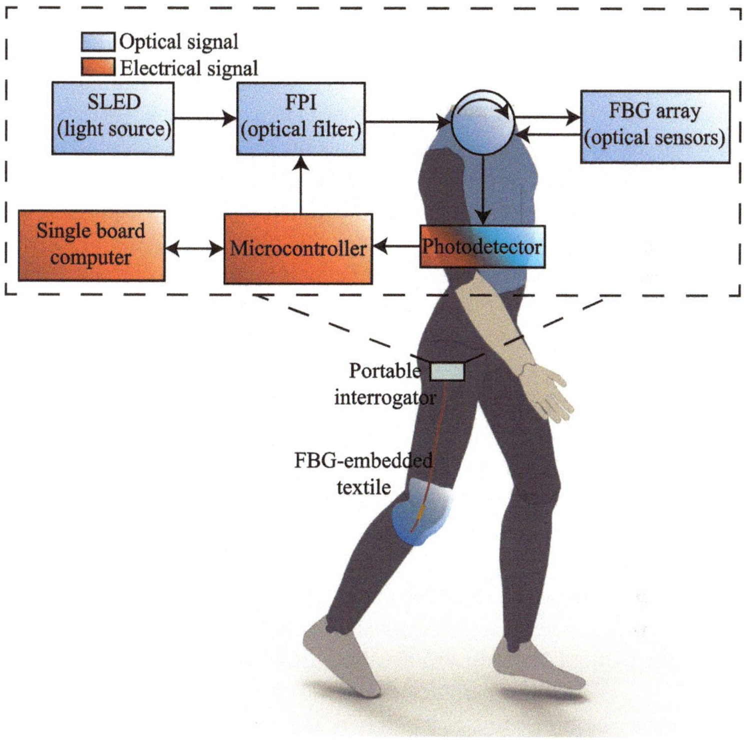

FIGURE 8.3 Schematic representation of the FBG-embedded smart textile on a subject for knee angle and displacement assessment. Figure insets show the portable interrogator schematics.

The smart textile for knee angle assessment positioned on a subject is schematically represented in Fig. 8.3, where the figure insets show the FBG spectrum as well as the schematics of the portable interrogator. It is worth to mention that knee is a polycentric joint, and which has a nonconstant center of rotation. One approach for knee modeling is through a four-bar linkage model. This polycentric design allows knee stability throughout the gait cycle since the stability criteria varies during gait, as heel-strike and toe-off phases need different stiffness for the knee joint. Thus it is important to measure two dimensional displacements in the knee joint since the center of rotation of this joint is posterior to the ground reaction vector during heel-strike and sequential loading phases, whereas the center of rotation is anterior to the ground reaction vector during the late stance of the gait cycle (Kirtley, 2006). Conventional methods for two-dimensional knee displacement methods include magnetic resonance imaging and computer modeling, which have enabled advances in knee kinematics, enabling the analysis of kinematics in different planes of motion and incorporation of conjoint rotation (Smith et al., 2003).

To that extent, FBG sensors can be employed in plane displacements analysis by the analysis of the whole grating spectra (instead of only the wavelength shift, as commonly employed). The different mechanical loading, i.e., axial strain, bending and torsion, result in different variation in the grating spectra regarding the reflectivity (or optical power), wavelength shift and full width half-maximum (FWHM). If the optical fiber is under bending, there is not only the wavelength shift, but also the variation on the grating reflectivity due to intensity variation effects. Similarly, when the FBG is submitted to torsion conditions, there is a linear variation of the axial strain on the grating region that induces a chirp on the FBG (Leal Junior et al., 2018c). Thus the FBG presents an increase of the FWHM as it shows a behavior similar to chirped FBGs discussed in Chapter 6. Such effects can be used to decouple the axial strain and bending on the optical fiber during the gait cycle with the optical fiber sensor attached on the user's knee. In this case, the reflectivity and the wavelength shift data are acquired by the portable interrogator to decouple both parameters, where it is possible to obtain the vertical and horizontal components of the two-dimensional displacement using the knee angle at the sagittal plane and the axial strain on the same plane. In order to obtain such decoupled response, the sensor is characterized as a function of axial strain (without bending) and only bending at a single plane. Thus the FBG-embedded knee brace is positioned on a goniometer (with fixed center of rotation) for the angle characterization. Then the sensor is positioned on a linear translation setup in order to obtain the axial displacement without bending or curvature. Such characterizations resulted in linear regressions for both parameters, as shown in Fig. 8.4, where the sensitivities of each parameter to strain and bending are presented. The strain range tested was from 0 to 2500 $\mu\epsilon$ on steps of 500 $\mu\epsilon$, whereas the angle was assessed on the range between 0 and 100° on 25° steps, both parameters are within the knee range of motion. The results show higher variations on reflectivity and wavelength shift when the optical fiber is subjected to bending and axial strain, respectively, with only minor variations of the opposite spectrum feature at each case, i.e., when axial strain is applied, there are only minor variations on the reflectivity, whereas the wavelength shift shows the highest variation and vice versa.

As each parameter presented a linear response as a function of the applied strain and bending, it is possible to apply the coefficients obtained on the linear regression on Eq. (8.1) to decouple the axial strain from the bending angle, where the matrix with the sensitivities is needs to be well conditioned, i.e., its determinant should be different than zero, otherwise the decoupling technique is not applicable since the matrix would not be invertible if its determinant is

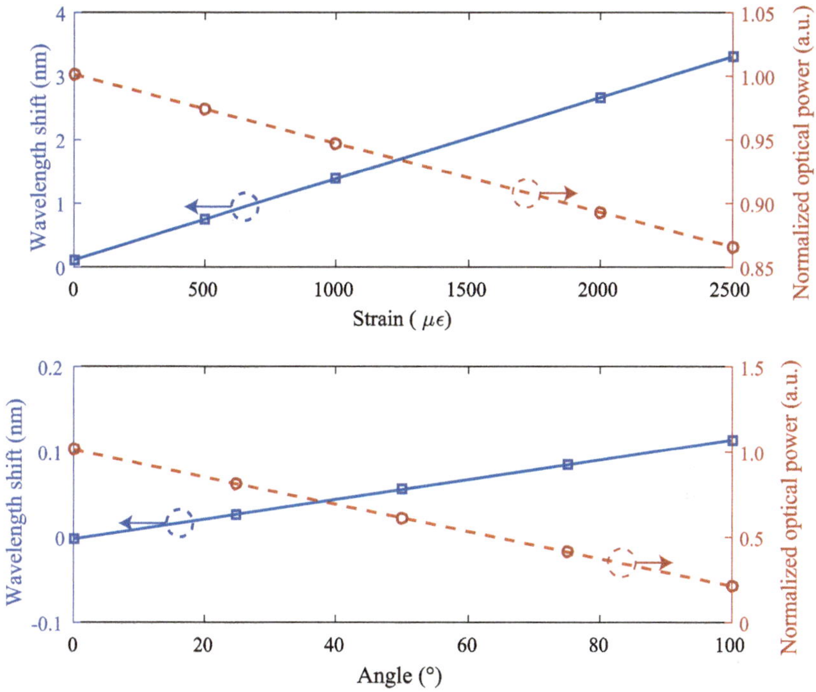

FIGURE 8.4 Characterization of reflectivity and wavelength shift on the FBG sensor for knee joint assessment as a function of strain and angle variation.

zero,

$$\begin{bmatrix} \epsilon \\ \alpha \end{bmatrix} = \begin{bmatrix} s_{\lambda,\epsilon} & s_{r,\epsilon} \\ s_{\lambda,\alpha} & s_{r,\alpha} \end{bmatrix} \cdot \begin{bmatrix} \Delta\lambda_B \\ \Delta r \end{bmatrix}, \tag{8.1}$$

where ϵ is the axial strain, α is the bending angle, $\Delta\lambda_B$ is the wavelength shift, Δr is the variation on the grating reflectivity. Moreover, $s_{a,b}$ are the coefficients obtained on the linear regression, where a is the spectral characteristic analyzed (λ for the wavelength and r for the reflectivity) and b is the parameter analyzed, i.e., axial strain and bending angle in this case.

The gait analysis with the presented FBG-embedded knee assessment system is performed in a healthy subject without history of knee surgical interventions. The subject was asked to perform three steps in a straight line with self-defined linear velocity. The reflectivity and wavelength shift of the FBG are acquired by the portable interrogator and the decoupling of strain and bending is performed using Eq. (8.1) with the coefficients obtained from the slopes in Fig. 8.4. The results obtained in the tests are shown in Fig. 8.5, which include the knee angles as a function of time as well as the vertical and horizontal positions of the knee joint. The angle variation showed in Fig. 8.5 presents similar trend as

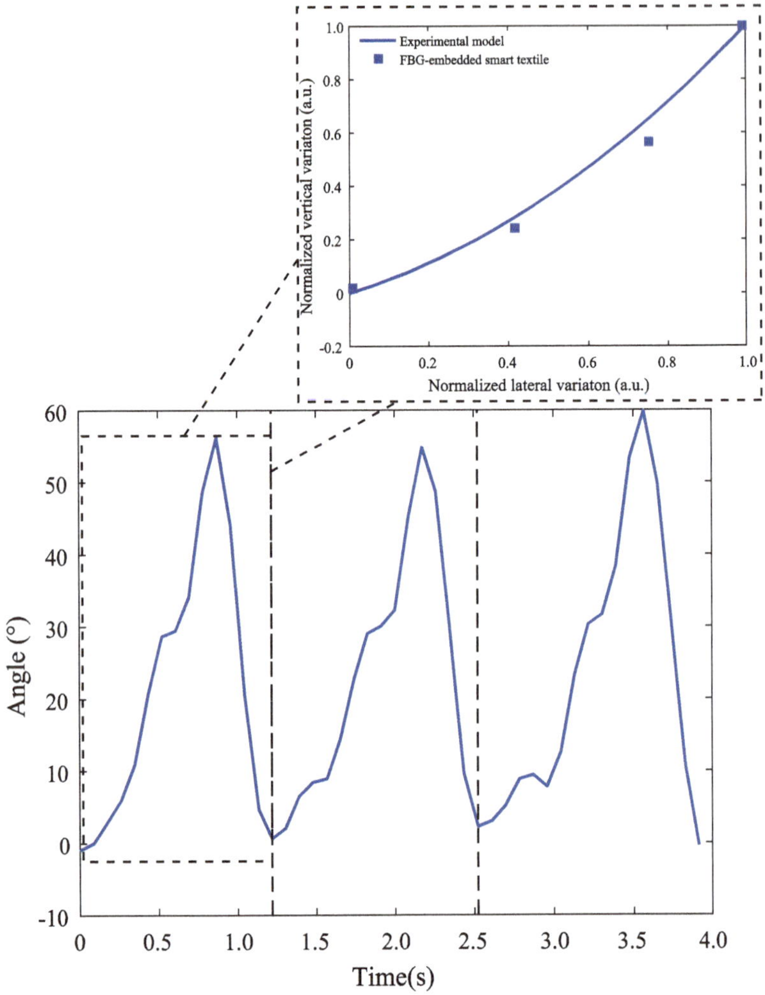

FIGURE 8.5 Bending angle as a function of the knee joint measured by the FBG-embedded smart textile. Figure inset shows the horizontal and vertical displacements on the FBG for knee flexion cycles compared with the model presented in Walker et al. (1985).

the ones obtained with gold standard measurement systems (such as marker-based camera systems) and it is within the range of angle of the knee joint in healthy volunteers in the gait cycle. Furthermore, the differences between different cycles are related to the natural variability of the gait and is within the standard deviation of the gait in an individual (Kirtley, 2006). The vertical and horizontal displacements are obtained from the measured angles and strain using trigonometric relations between the measured parameters and the initial curvature radius (12.5 cm). The results in Fig. 8.5 inset are compared with the ones obtained in a widely adopted model for knee polycentric joints, where the model was obtained in Walker et al. (1985) by the direct measure of the knee joint center of rotation in ex-vivo and in-vivo subjects. Comparing both responses, it is possible to observe a good agreement and determination coefficient higher than 0.9 between them, which indicate that the measured horizontal and lateral displacements have high correlation with the model commonly adopted for the knee center of rotation analysis. For this reason, the proposed sensor system, including the portable interrogator, can be used as a reliable tool for joint kinematics assessment with the advantages of portability, compactness and it is nonintrusive tool.

8.2 Instrumented insole for plantar pressure distribution and ground reaction forces evaluation

8.2.1 Fiber Bragg grating insoles

The key areas of the foot for plantar pressure monitoring are presented in Fig. 8.6(a), which include the heel, midfoot, metatarsal, and toe areas. These areas are divided into 15 foot anatomical areas that supports the weight and are responsible for the body balance as presented in Shu et al. (2010). In addition, there are general guidelines in instrumented insole for dynamic measurements of plantar pressure, such as wireless connection, low power consumption, portability, low cost and sensors distributed along foot pressure points (Abdul Razak et al., 2012), indicated in Fig. 8.6(a). For the FBG-embedded cork insole, 5 sensors are employed in a gradient index multimode CYTOP fiber (Chromis Fiberoptics Inc) with core diameter of 120 μm, a 20 μm cladding layer, and an additional polyester and polycarbonate over-cladding structure to protect the fiber, resulting in a total diameter of 490 μm. The FBGs are inscribed via the plane-by-plane direct inscription technique (discussed in Chapter 6) using a femtosecond (fs) laser system (HighQ laser femtoREGEN) operating at 517 nm with 220 fs pulses duration and pulse energy of 80 nJ to modify the material with the advantageous feature related to the ability of inscribing the FBGs without the necessity of removing the overcladding region, which is a limitation of commonly employed phase mask technique (Theodosiou et al., 2017). Moreover, the repetition rate is set at 5 kHz with the gratings' period is close to 2.2 μm resulting in the inscription of fourth-order FBGs with minor variations between them to obtain an array of 5 FBGs. This approach combines the advantage of high strain limits of polymer optical fibers (POFs) with the multiplexing capabilities of FBG sensors, since all 5 sensors are multiplexed in a single optical fiber cable.

As the length of the gratings is around 1.2 mm and such small dimensions enable their embedment in an insole structure for distributed measurement of different plantar pressure points. For the integration of the sensor array, a groove with 2.5 mm depth and 2 mm width was caved in a cork insole, where the optical fiber is positioned along the insole. An opening of 5 mm depth and a diameter of

5 mm is manufactured in the cork structure and the FBG regions are positioned on the openings. Thus 5 openings are made along the insole, in the regions shown in Fig. 8.6(b), which is then filled with an epoxy resin (Liquid LensTM) to provide stability and enable the deflection on the FBG region when the plantar pressure is applied due to the high flexibility of the material. When a force is applied on the epoxy structure, it undergoes an axial strain. Such strain provokes an embedded-FBG related wavelength shift proportional to the applied force on it. To compensate temperature changes, an FBG temperature sensor was incorporated in the insole, in order to ensure an effective thermal isolation provided by the cork and to guarantee that the FBG plantar pressure sensors are not affected by the body temperature or external temperature changes. On the other hand, the cork material is employed due to its thermal isolation properties as well as small Poisson ration, which leads to lower crosstalk between the sensors in the FBG array.

The FBG-embedded insole is connected to an interrogation system composed of a battery, a miniaturized broadband optical ASE module (B&A Technology Co., As4500), an optical circulator (Thorlabs, 6015-3), and an optical spectrometer (Ibsen, I-MON 512E-USB) with an acquisition rate of 960 Hz and a wavelength resolution of 5 pm. Furthermore, the CYTOP is connected to a silica single mode fiber (SMF) using a UV-curing optical adhesive via the butt-coupling method (discussed in Chapter 4). The pressure response of each FBG is characterized prior to the application on the plantar pressure assessment, which is performed by positioning each of the FBGs on the array in a universal testing machine in the compression tests configuration (Shimadzu® AGS-5kND), where pressures ranging from 10 to 1500 kPa with 100 kPa steps using a probe with a 10 mm diameter. The wavelength shift of each sensor as a function of the applied pressure is presented in Fig. 8.6(c), where it is possible to observe similar sensitivities of the 5 FBGs, which also show determination coefficient higher than 0.9, indicating a high correlation between the wavelength shift and applied pressure.

The pressure induced in the sensing elements during a normal gait movement is analyzed on the gait tests with a female subject weighting 55 kg. The response of each pressure-sensing element during a gait cycle was recorded and presented in Fig. 8.7(a), where it is possible to verify that the similar trend is obtained in all four cycles, suggesting the human limits the repeatability of the sensors' response. Furthermore, it is possible to notice the relationship between the activation time for each sensor and the stance phase of the gait cycle. The FBG 5, located at the heel region, shows an increase of the pressure at the beginning of the cycle (IC phase), and FBG 4, 3, 2, and 1 are sequentially activated when the TO phase approaches. In Fig. 8.7(a), the dashed lines of the last cycle represent the activation of each FBG. The slight differences of each FBG sensor for each passage is a result of the differences of foot positioning on the platform during the tests. Moreover, it is possible to use the analysis of the pressures recorded during the stance phase to infer and monitor pressures of in-

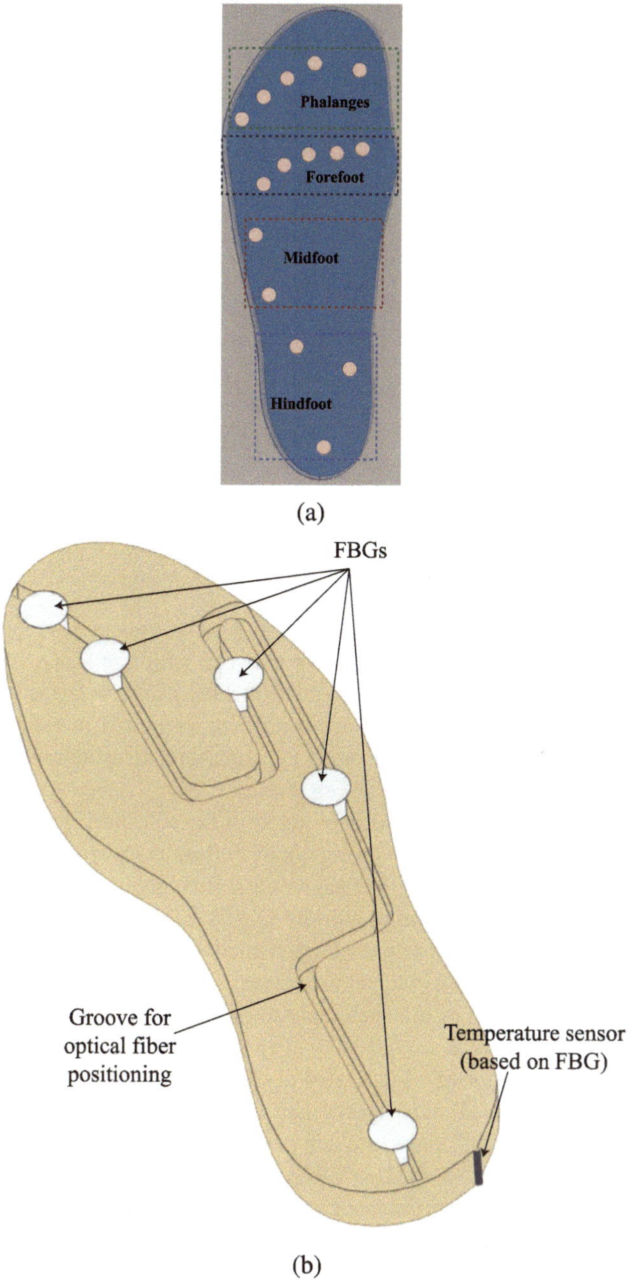

FIGURE 8.6 (a) Schematic representation of foot anatomical areas. (b) FBG array embedded in the cork insole. (c) Responses of each FBG as a function of the applied pressure.

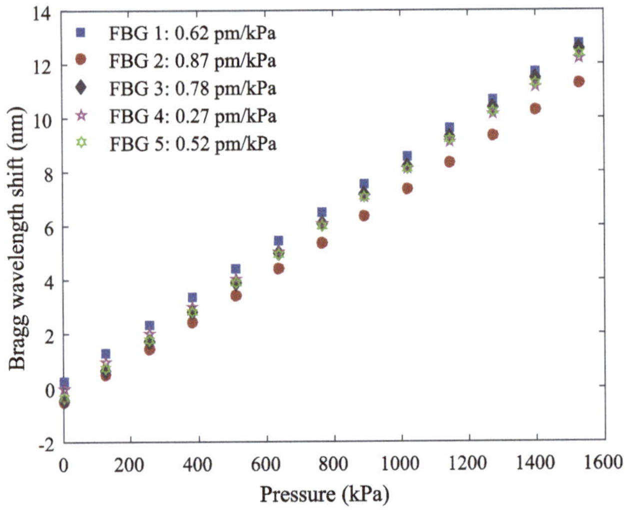

(c)

FIGURE 8.6 *(continued)*

dividuals. In order to perform a dynamic and autonomous pressure monitoring during gait, the insole which has a sensor network with five-FBG, was adapted to a shoe. The foot plantar pressure fluctuation during gait is also induced in the instrumented cork sole. Such pressure oscillations resulted in reflected Bragg wavelength shifts, which are converted into pressure measurements using the characterization results presented in Fig. 8.6(c).

The plantar pressure distribution induced in the sole during the gait movement was applied to the 5 FBGs during 4 gait cycles, as shown in Fig. 8.7(b). In addition, it is possible to notice the data repeatability, given the similar response of the sensors over the gait cycles showed in Fig. 8.7(b). Also, it is possible to detect the sensors activation sequence (maximum amplitude registered), which corresponds to the expected in a gait movement. The first activated sensor is the FBG 5, at the beginning of the stance phase of the gait cycle, with the contact between the heel and the floor. After this, FBG 4 (located at the middle-foot and beginning of the metatarsal) is activated at the start of the foot-flat stage at the single support moment. Following the gait movement, FBG 2 and FBG 3, located at the metatarsal positions, are activated at the forefoot contact during the middle of the stance phase. Finally, FBG 1, located at the toe area, is activated at the toe-off moment, before the foot leaves the ground, marking the end of the stance phase and the beginning of the swing phase of the gait cycle. The typical curve of the plantar pressure during the gait movement can be obtained by the sum of the 5 sensors' feedback. The resulting curve represents the vertical ground reaction forces on the stance phase of the gait. On the swing phase, the foot leaves the ground leading the response of each FBG near to zero. Fig. 8.7

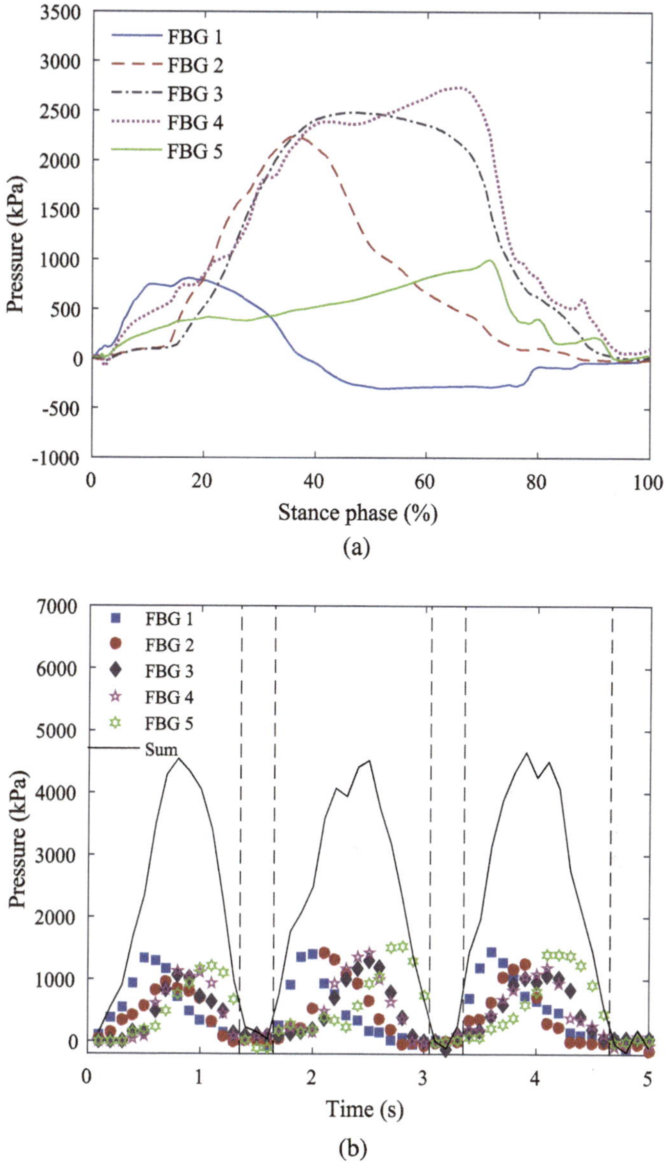

FIGURE 8.7 (a) Responses of each FBG during a gait cycle. (b) Pressure obtained during the four steps for the FBG-embedded insole, where the sum of the responses obtained in each FBG is presented.

presents the ground reaction forces resulted by the sum of the five FBGs showing a RMSE of 160 kPa between the four cycles, which represents an error of 5% when compared to the average of the sum of the five FBGs, over four cycles.

The description of the dynamic measurements was made to show the ability of the proposed system to detect the subdivisions of the stance phase discussed in the introduction of this paper, which can aid on the detection of gait-related pathologies, such as spine anomalies or foot ulcerations in patients with diabetes. In addition, the gait phase detection can be applied to control wearable robots for gait assistance.

Finally, any limitations for the maximum allowable velocity of the system measurement capability are related to the material response and acquisition frequency of the interrogator. Since the sampling frequency of 200 Hz is sufficient for acquiring the gait activities (Kirtley, 2006), the proposed system is able to cover the velocities employed on typical human gait. Moreover, based on simple stress-strain relations, the proposed sensor system can, in principle, measure the plantar pressure of subjects with body mass higher than 200 kg. Additionally, the sensitivity of each POFBG and the 0.5 pm resolution of the employed interrogator, enable the detection of weights lower than 1 kg, which leads to a high dynamic range of the proposed insole. It is also worth noting that the presented device is also capable of performing additional analysis, such as the center of mass displacement as depicted in (Vilarinho et al., 2017). In addition, the FBG array can be embedded in different insoles using a myriad of methods and configurations (Eguchi et al., 2017) not only for the gait analysis, but also for pronation and supination assessment in foot positioning as well as the prevention of pressure ulcers (Suresh et al., 2015). Additional sensors can also be embedded in the insole with orthogonal positioning in order to evaluate not only the normal components of ground reaction forces (derived from the plantar pressure), but also the shear stress during the gait, which is an important parameter for the investigation of gait biomechanics and its abnormalities (Tavares et al., 2018).

8.2.2 Multiplexed intensity variation-based sensors for smart insoles

The advances in additive layer manufacturing enable the development and widespread of low cost and highly customizable fabrication methods, such as the 3D printing, which has experienced a large growth in the last decade. The possibility of providing a fast and low cost fabrication of a device with customized dimensions has a major impact on the wearable devices development, since it is possible to design custom devices for each user. Aiming at these advantages, the insole structure consists of a 3D-printed using the Sethi3D S3 (Sethi, Brazil) with two different materials: TPU for the insole base and PLA for the top of the insole, which envelops the fiber and sensors. In the printing, the layer height and material infill density were set to 0.2 mm and 80%, respectively. The optical fiber employed was the commercially available PMMA POF (HFBR-EUS100Z, Broadcom Limited). This fiber has a PMMA core with 980 μm diameter, a cladding of fluorinated polymer with 10 μm thickness and a numerical aperture of 0.47, in addition to an overcladding made of polyethylene for mechanical protection of the POF. The light sources used were LEDs

(ASMT-BR20-AS000 Broadcom Limited) with a central wavelength of 628 nm and 650 mcd luminous intensity, whereas the photodetectors employed were photodiodes IF-D91 (Industrial Fiber Optics, USA). The signal acquisition and the LED control are performed by the microcontroller FRDM-KL25Z board (Freescale, Austin, TX, USA) and transmitted using the Bluetooth module HC-05 (Logoele Electronics Technology, China).

The presented insole was designed according to general guidelines of plantar pressure assessment systems discussed in Section 8.1. In order to accord to those design requirements, the sensor consists of commercial PMMA POF (HFBR-EUS100Z, Broadcom Limited) with 15 lateral sections. When the pressure is applied on each lateral section, there is an optical power variation, which is acquired by two photodetectors, one at each end of the fiber. In addition, light emitting diodes (LEDs) are laterally coupled to each lateral section. In this way, there is a side coupling of the light at each section of the fiber. Thus two photodetectors are employed, in which one works as a reference to the other, resulting in a self-referencing system to mitigate errors due to the light source power deviations, which is a common issue in intensity variation-based sensors (Leal-Junior et al., 2018b). The location of each sensor was based on (Shu et al., 2010), which presented the 15 foot anatomical areas that supports the weight and are responsible for the body balance. These locations are shown in Fig. 8.6(a), where the foot is divided into hindfoot (sensors 1 to 3), midfoot (sensors 4 and 5), forefoot (sensors 6 to 10) and phalanges (sensors 11 to 15) (Abboud, 2002). In this case, a multiplexed intensity variation-based system is obtained, where all 15 sensors are positioned in a single POF cable using the multiplexing technique discussed in Chapter 6 in which there is a side-coupling between the light source and the optical fiber with a sequential activation of the light sources with on–off keying, resulting in a system where the responses of all 15 sensors are decoupled (Leal-Junior et al., 2019).

The sensors sensing region is limited by the fiber diameter of 1 mm. Moreover, with all the 15 sensors positioned at each anatomical area of the foot, the proposed insole complies with the requirement of sensor positioning. When the lateral sections are concluded, the POF is positioned on a 3D-printed insole made of thermoplastic polyurethane (TPU), which is a flexible material suitable for insole production as shown in Fig. 8.8. Thereafter, the thermoplastic resin is applied on the POF in regions close to the lateral section to fix the sensor in two positions (before and after the lateral section) in order to keep the fiber deformation restricted to the sensing region, which inhibit the crosstalk between sensors. This is especially important in the forefoot and phalange regions, where the sensors are adjacents. Then a 3D-printed structure made of polylatic acid (PLA) with 0.1 mm thickness is positioned on the top of the TPU structure and is glued on the TPU structure with a thermoplastic resin. In this structure, the LEDs are aligned to each lateral section and located in selected areas, the LED's housing, as also depicted in Fig. 8.8. Furthermore, the housing of each LED also has region with contact to the fiber in a way that when a pressure is

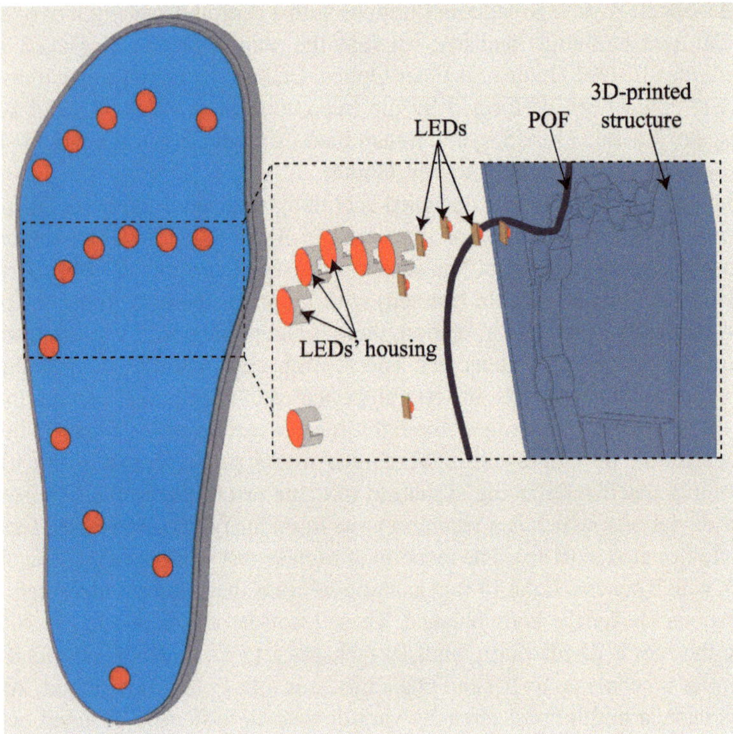

FIGURE 8.8 Schematic representation of the POF-embedded insole assembly.

applied to this region, there is a bending of the fiber and an increase of the stress on such region, which also results in power variation. The sensor assembly illustrated in Fig. 8.8 avoid the stray light from the ambient, since the groove on the TPU structure in conjunction with the LED housing cover the light coupling from the ambient. Then the PLA structure is positioned on the top of the insole, which isolates the fiber from ambient light. Thus there are three different attenuation mechanisms on each of the 15 sensors: the macrobending principle when the pressure is applied, the stress-optic effect due to the stress applied on the fiber that leads to a refractive index variation and the light coupling between the LED and the POF's lateral sections, as discussed in Chapter 6. All lateral sections were manually fabricated with the same tool, but minor deviations on the surface roughness, section depth and length can occur, leading to different sensitivities, since the sensor sensitivity is dependent on the sensor's radius of curvature. After the assembly, the whole 3D-printed multiplexed POF insole has a total weight of about 150 g, which does not influence the gait natural pattern, according to (Abdul Razak et al., 2012). Regarding the power consumption, the proposed instrumented insole consumes about 70 mA when all the components are active with the signals transmitted via Bluetooth. Thus the system can be powered by a 9 V alkaline battery with about 565 mAh, resulting in an autonomy higher than 8 hours. The more efficient batteries can also be employed on the system, increasing the system's autonomy. Therefore such low power consumption is also in accordance with the aforementioned design requirements.

The system consists of 15 LEDs coupled to 15 lateral sections on the fiber with controlled length and depth, creating 15 sensitive zones which increase the sensitivity when a pressure is applied. The multiplexing technique is based on the sequential activation of each LED with a predefined frequency of 30 Hz and activation sequence (from LED 1 to 15), where only one LED is activated at a time. The activation frequency and sequence is controlled by a microcontroller, which is also responsible for the signal acquisition when each LED is activated, as also indicated in Fig. 8.8. As discussed in Chapter 6, the optical powers acquired when each LED is activated are elements of the acquisition matrix in which the number of rows is related to the acquired samples over time and the columns are related to the number of light sources. In this case, the acquisition matrix has 15 columns, which are represented by the terms $R_{LED1-15}$. The signals acquired by the microcontroller are transmitted to the local processing unit via Bluetooth, which is performed all the signal processing. Hence, the proposed insole works via wireless connection, enabling the online monitoring by a computer or even a smartphone device.

Similar to the FBG-embedded cork insole, the pressure (or force) responses of each sensor embedded in the 3D-printed insole needs to be characterized prior to its application on gait analysis. To that extent, controlled pressures are applied on each sensor region in the range of 0 to 50 N in 10 N steps, which result in a pressure range from 0 to around 1700 kPa, considering the area at which the force is applied. For each sensor the measurements were repeated and Fig. 8.9 shows the linear regressions of sensors 1 to 15 relating the applied pressure and force to the normalized power of the POF sensors. The responses of the 15 sensors are presented in Fig. 8.9, where it is possible to observe differences in their sensitivities. Such differences are related to deviations on the lateral section parameters (e.g., depth and length) that directly affect their sensitivities (Leal-Junior et al., 2018b). Different sensors positions lead to different thermoplastic resins locations when compared to the sensitive zones. Thus the curvature radius of each sensor has differences when compared to the other ones, e.g., sensor 8 is close to sensor 9 and 7, which leads to an application of the thermoplastic resin close to its sensitive zone, and hence, a lower curvature radius when compared with a sensor distant to the other ones, such as sensor 2, which has a thermoplastic resin farther from the sensitive zone. It is also worth to mention that all sensors showed a linear response, where the correlation coefficient ($R^2 2$) between the sensor responses and the linear regression is higher than 0.99 in all cases, which is regarded as high correlation.

The static tests were performed with 4 subjects standing on the instrumented insole for about 100 seconds while the data of each sensor is recorded. The validation tests consisted of the comparison between the proposed insole and the

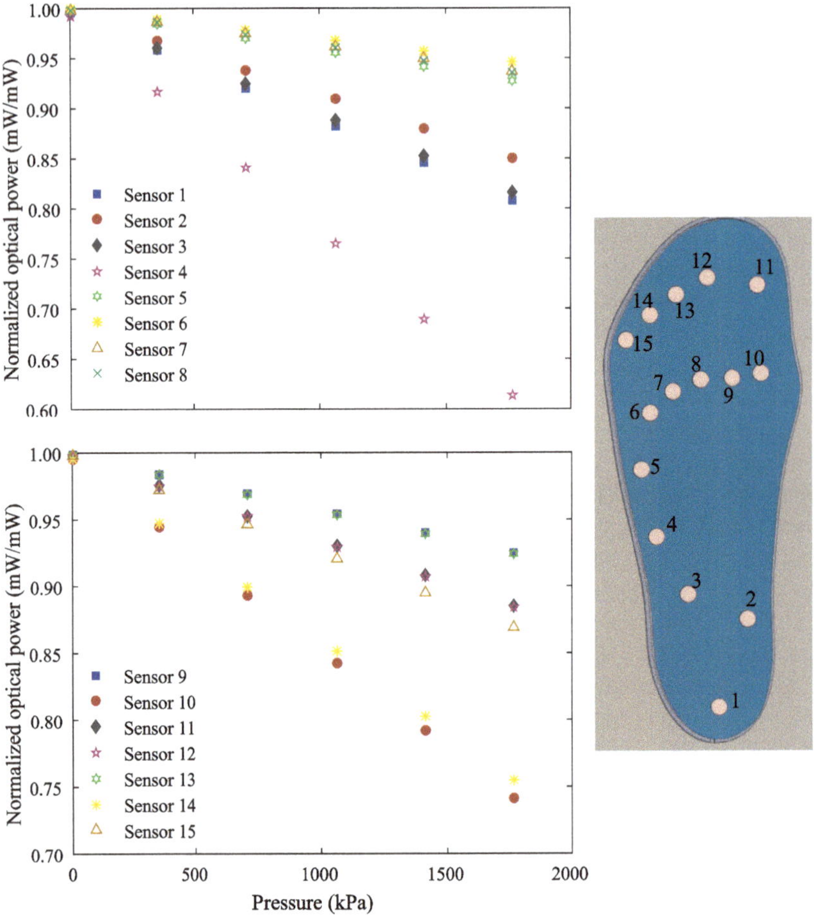

FIGURE 8.9 Linear regression of sensors 1 to 15 as a function of force and pressure.

force platform (EMG system, Brazil), where five subjects were asked to perform displacements on their center of pressure (CoP) for about 10 seconds, the test was repeated 3 times. Finally, on the dynamic gait tests the subjects walked in a room for 15 m (in a straight line). All the subjects were healthy adults with ages between 18 to 36 years without history of gait related pathologies. In addition, the shoe sizes of all participants were between Euro 39 (US 7 for male and 8.5 for female) and Euro 44 (US 10.5 for male and 12 for female), which are within the spatial resolution of the sensors. Participants with higher or lower foot sizes needed to be excluded from the study, since their feet will not cope with the spatial resolution of all sensors. Thus 10 male and 10 female participants were chosen with the body masses ranging from 46 kg to 97 kg.

In the first test, four subjects were asked to stand still on the instrumented insole for about 2 minutes. A pressure map of each participant was obtained. In this way, the weight estimation of each person was accomplished by the sum of all the pressure points, since the sensing area is known. The weight and height of each participant is shown in Fig. 8.10 in addition to a weight estimated by the instrumented insole. The sensors responses for subject 1 as well as the pressure map obtained on the sensor responses for this participant are also presented in Fig. 8.10. These results show the capability of the instrumented insole of estimating the body mass, but also leads to an important discussion about the number of sensors needed in an instrumented insole. Some works report dozens of pressure sensors, which brings important issues on the characterization of each sensor (Abdul Razak et al., 2012). Thus the reported ability of the proposed instrumented insole of estimating the body mass by the sum of the force responses between 15 sensors indicates that if the sensors are correctly positioned on the 15 anatomical areas of the foot (Shu et al., 2010), only 15 sensors are needed for a reliable estimation of the plantar pressure distribution and GRF. However, an important issue should be noted, since all the subjects considered in the study have normal foot and the 15 anatomical areas defined in (Shu et al., 2010) also considers a normal foot. Thus if the studies involve subjects that have foot abnormalities (such as flat foot) different sensor positioning should be considered to obtain reliable measurements of plantar pressure and GRF, which can be easily accomplished with the methodology proposed in this work, where the insole design and sensor positioning can be optimized for each user or application.

The pressure maps obtained for the subjects are similar to the one expected for subjects with foot without abnormalities (Kirtley, 2006) since they are similar to the ones obtained in reference works on the literature. In addition, the instrumented insole was capable of estimating the weight of each participant by means of summing the response of all 15 sensors. A root mean squared error (RMSE) between the participants' weight and the ones estimated by the insole is about 1.9 kg, which represents a relative error of only 2.7%, considering the mean weight of the participants. Furthermore, it shows the ability of the sensor of measuring the plantar pressure on a high range of weights, which in this case were from close to 50 kg to almost 100 kg. Then the presented 3D-printed insole was validated in a commercial force platform, which can measure the GRF and the two-dimensional center of pressure (CoP) displacements. In this case, the insole was positioned on the top of the force platform and five subjects were asked to position their foot on the insole, each test takes about 6 seconds and was repeated 3 times. The subjects were asked to perform on the sagittal and frontal planes as depicted in the arrows of Fig. 8.11(a). Fig. 8.11(a) presents the comparison between the force platform and instrumented insole on the measurement of the GRF and variations of the CoP on the x and y directions for subject 3 on test 2. The correlation coefficients presented in Fig. 8.11(a) for all 5 subjects show a high correlation between the force platform and instrumented

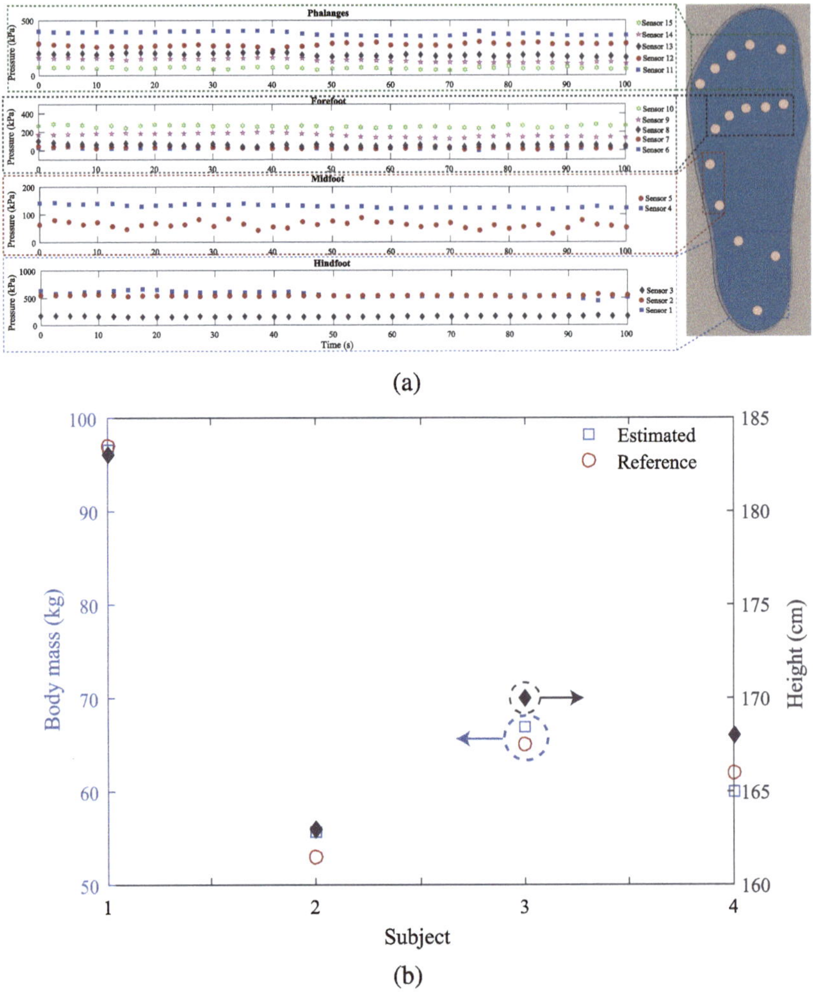

FIGURE 8.10 (a) Plantar pressure map of subject 1 using the 3D-printed insole. (b) Weight estimation of four subjects through the responses of all 15 sensors.

insole responses, since the lowest one found is for the GRF estimation in subject 5 (R^2=0.878). The reason for this higher error is related to errors in the force estimation, which can be due to sensor nonlinearities or differences on the foot of this specific subject that have led to variations in the sensors' responses. Nevertheless, the correlation coefficient is higher than 0.85, which can be considered a high correlation

The force and CoP responses of subject 3 are shown in Fig. 8.11(a) inset, where there is a good agreement between the proposed 3D printed insole and the commercial force platform. It is important to mention that the CoP measurements were normalized, since the CoP center for the platform (CoP(0,0)) is in the center of the platform, and it can presents different CoPs of the insole if the CoP center of the insole is not precisely positioned on the center of the platform. Thus in order to reduce this influence on the comparison, both estimated CoPs (from the platform and insole) were normalized. Since force platforms can be regarded as the gold standard for GRF and CoP monitoring, this high correlation between the proposed insole and the force platform indicates the reliability of the proposed device on the measuring of both parameters with the additional advantages of higher portability and lower cost than force platforms. Furthermore, the proposed insole with the wireless connection used can be an important asset on remote health monitoring applications (Majumder et al., 2017). Thus the GRF and plantar pressure distribution of each person can be monitored at home, performing their daily activities and the data can be sent to a clinician through a local gateway, where the data can also be processed offline and sent to a clinician through the cloud.

The last set of tests is the application of the proposed 3D-printed POF insole on gait analysis. In this analysis, 20 participants (10 males and 10 females) were asked to walk in a straight line for about 15 m and the plantar pressure as well as the GRF were monitored using the insole. Fig. 8.11(b) shows the response of all 20 subjects, where the GRF is normalized with respect to the stance phase percentage and to the weight of each participant. The GRF results show the possibility of identifying gait events on the stance phase. Among many subdivisions reported for the gait cycle in the literature, we used the five phases subdivision depicted in Taborri et al. (2016). Heel strike (HS) is when the heel makes contact with the ground and flat foot (FF) is when the foot is parallel to the ground. Furthermore, there is also the heel off (HO) phase and toe off (TO), where the former is when the heel loses contact with the ground and the latter is in the vicinity of the swing phase when the toe is not in contact with the ground. In this analysis, we also considered a maximum weight acceptance (MA) phase when there is a local peak of the GRF, which resulted in the well-defined M-shape of the GRF during the gait (Kirtley, 2006). The plantar pressure distribution on each gait phase is also shown, where the presented values are the mean of the pressure response of each cycle.

The normalized responses considering all 20 subjects present a considerable standard deviation. However, this deviation is similar to the ones previously reported in the literature with similar number of participant (Karatsidis et al., 2017). Thus it is possible to assume that these deviations are due to the gait inherent variability between different participants. In addition, it is interesting to note that a higher variability occurs in the HS region, which can confirm the variability of the gait cycle in this region. Moreover, the region with highest variation in Fig. 8.11(b) is HO region, which is also related to the gait velocity variation of each participant. Furthermore, the plantar pressure distribution at each gait phase is in accordance with the well-known plantar pressure distribution reported in the literature with gold standard measuring devices (Kirtley,

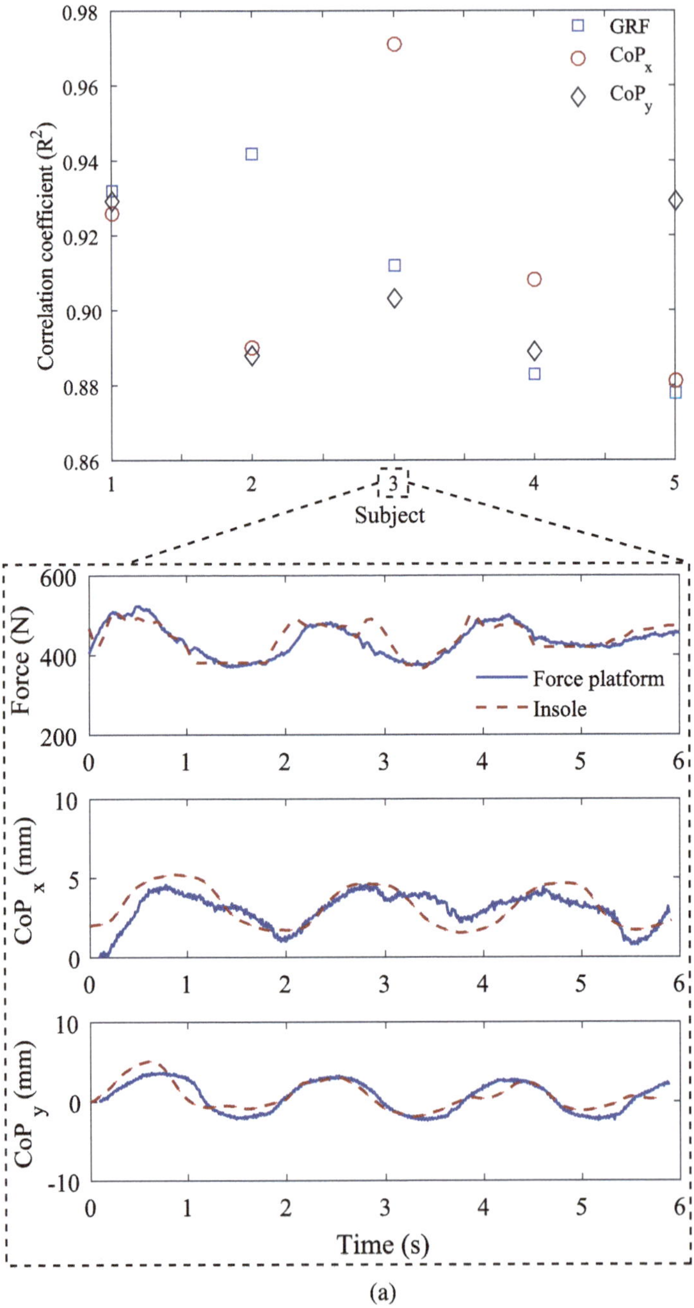

FIGURE 8.11 (a) POF-embedded 3D-printed insole validation in a force platform for the estimation of GRF, CoPx and CoPy. (b) Mean (solid line) and standard deviation (shaded curve) of the GRF during gait of 20 participants the plantar pressure map on each gait phase is also presented.

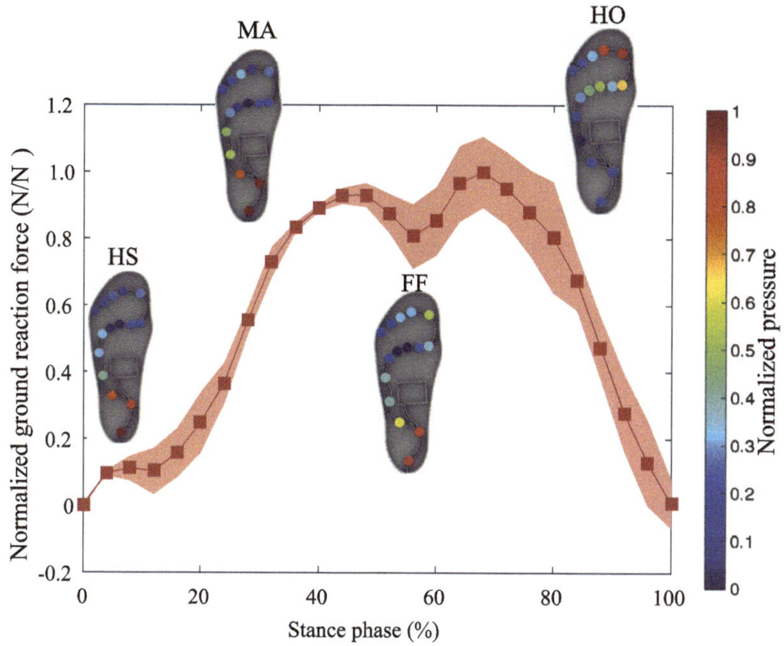

(b)

FIGURE 8.11 *(continued)*

2006). In these cases, a well-defined M-shape for the GRF was obtained and it was possible to identify 5 different gait events on the stance phase. These identified events can aid on the control of gait assistance devices and can also aid on the detection of gait related pathologies. Similarly, the plantar pressure distribution at each gait event also are similar to the pattern widely reported in the literature for healthy individuals during the gait (Kirtley, 2006), which is another important evidence of the accuracy and reliability of the proposed 3D-printed instrumented insole. This insole can also overcome the issue of low dynamic range reported in commercially available and in some previously reported insoles (Abdul Razak et al., 2012), since it was possible to measure the plantar pressure and GRF of subjects from 46 kg to 97 kg of body mass, which indicates the possibility of using this device for clinical evaluations of subjects with a large span of body masses (more than 51 kg). This feature, in addition to the validation on force platform and the weight estimating ability of the instrumented insole, indicates the reliability of the proposed 3D-printed POF insole for measuring plantar pressure in different scenarios (static and dynamic tests). Thus it is possible the large-scale fabrication of the proposed insole, in addition to the customization of the device for each individual, optimizing the number and positioning of the sensors for each user considering the foot size and anatomy. These advantages are well aligned with the state-of-the-art requirements for the development of assistive devices in which the device is optimized for each user, the so-called human-in-the-loop design (Walsh, 2018).

8.3 Spatiotemporal parameters estimation using integrated optical fiber sensors

The embedment of the multiplexed POF intensity variation-based system in textiles have been explored. In one nonwearable approach, the POF was embedded in a carpet, resulting in the so-called POF smart carpet (Avellar et al., 2019). In this case, the carpet has a length of 2 meters and width of 60 cm with two polyethylene layers, where the POF and the flexible LED strip are embedded in between those layers. The smart carpet system proposed in Avellar et al. (2019) also has the previously discussed advantages of low cost, portability and transparency to the user, especially when compared to commercial force platforms commonly used for gait analysis. The system has 20 measurement points distributed on the carpet as two rows of 10 sensors as shown in Fig. 8.12. In addition, the POF smart carpet is able of detecting the spatial-temporal gait parameters such as the stride/step length, cadence and time of double support (i.e., the time at which both feet are touching the ground during the gait). The system can also estimate the GRF pattern during gait that can also indicate the times of stance and swing phases. Fig. 8.12 also shows the results of the spatial characterization test performed on the smart carpet, in which the system was characterized by placing calibrated weights at predefined positions on the carpet and the system showed the capability of detecting the different positions of the applied forces. Then, gait tests were performed and a volunteer walked on the whole extent of the carpet and the step length and GRF pattern are shown, in addition to the double support time. It is noticeable that the GRF pattern can be used to detect gait phases such as the stance and swing phases and their times during each phase (Kirtley, 2006). Therefore the embedment of the multiplexed POF system in a textile resulted in a portable system for gait analysis that can be employed outside a clinical environment due to its portability and capability of remote monitoring. These features enable its use on home activity monitoring and remote monitoring of patients; such technologies are well aligned with the requirements of Healthcare 4.0. Comparing with commercially available technologies such as piezoelectric ones presented in zebris Medical GmbH (2020), the POF-smart carpet does not employ additional cameras (or vision systems) as the one presented in zebris Medical GmbH (2020), which is a high cost component with the necessity of constant calibrations and careful positioning. In addition, POF-smart carpet has higher scalability, since the commercial systems using electronic technologies are generally limited to a few steps with maximum lengths below 10 meters, whereas POFs can transmit optical signals to several meters (about 100 meters for optical signals in the visible spectrum). The optical signal transmission capability of POF leads to the possibility of developing

POF-embedded smart carpet with longer lengths and even integrating it in the area of a whole room.

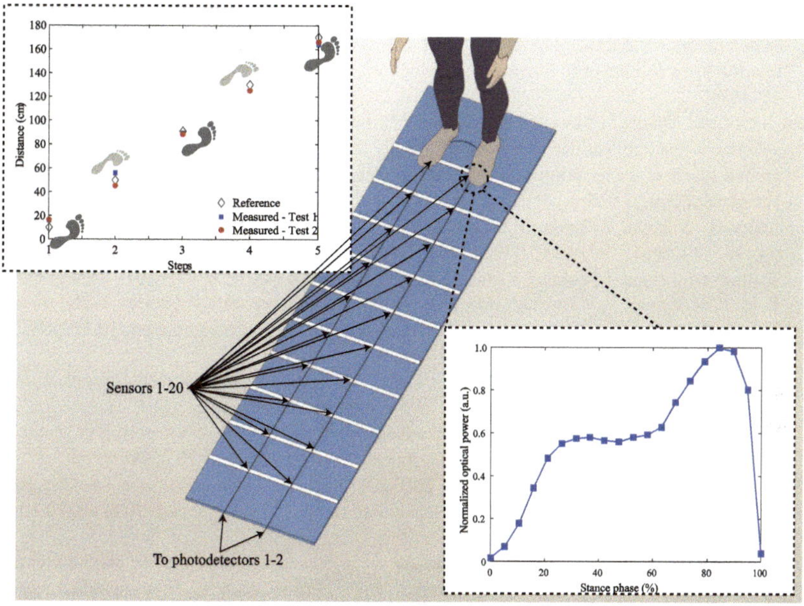

FIGURE 8.12 Experimental setup and results for POF smart carpet for step length estimation and GRF.

References

Abboud, R.J., 2002. (i) relevant foot biomechanics. pp. 165–179. https://doi.org/10.1006/cuor.2002.0268.

Abdul Razak, A.H., Zayegh, A., Begg, R.K., Wahab, Y., 2012. Foot plantar pressure measurement system: a review. Sensors 12, 9884–9912. https://doi.org/10.3390/s120709884.

Avellar, L.M., Leal-Junior, A.G., Diaz, C.A.R., Marques, C., Frizera, A., 2019. Pof smart carpet: a multiplexed polymer optical fiber-embedded smart carpet for gait analysis. Sensors 19, 3356. https://doi.org/10.3390/s19153356.

Diaz, C.A.R., Leal-Junior, A.G., Avellar, L.M., Antunes, P.F.C., Pontes, M.J., Marques, C.A., Frizera, A., Ribeiro, M.R.N., 2019. Perrogator: a portable energy-efficient interrogator for dynamic monitoring of wavelength-based sensors in wearable applications. Sensors 19, 2962. https://doi.org/10.3390/s19132962.

Eguchi, R., Yorozu, A., Fukumoto, T., Takahashi, M., 2017. Ground reaction force estimation using insole plantar pressure measurement system from single-leg standing. In: IEEE International Conference on Multisensor Fusion and Integration for Intelligent Systems, pp. 109–113.

Karatsidis, A., Bellusci, G., Schepers, H.M., de Zee, M., Andersen, M.S., Veltink, P.H., 2017. Estimation of ground reaction forces and moments during gait using only inertial motion capture. Sensors (Switzerland) 17, 1–22. https://doi.org/10.3390/s17010075.

Kirtley, C., 2006. Clinical Gait Analysis: Theory and Practice. Elsevier, Philadelphia.

Leal-Junior, A.G., Frizera, A., Avellar, L.M., Pontes, M.J., 2018a. Design considerations, analysis, and application of a low-cost, fully portable, wearable polymer optical fiber curvature sensor. Applied Optics 57, 6927. https://doi.org/10.1364/AO.57.006927.

Leal-Junior, A.G., Frizera, A., José Pontes, M., 2018b. Sensitive zone parameters and curvature radius evaluation for polymer optical fiber curvature sensors. Optics & Laser Technology 100, 272–281. https://doi.org/10.1016/j.optlastec.2017.10.006.

Leal Junior, A.G., Theodosiou, A., Diaz, C., Marques, C., Pontes, M.J., Kalli, K., Frizera, A., 2018c. Simultaneous measurement of axial strain, bending and torsion with a single fiber Bragg grating in cytop fiber. Journal of Lightwave Technology 37, 971–980. https://doi.org/10.1109/JLT.2018.2884538.

Leal-Junior, A., Díaz, C., Marques, C., Pontes, M., Frizera, A., 2019. 3d-printed pof insole: development and applications of a low-cost, highly customizable device for plantar pressure and ground reaction forces monitoring. Optics and Laser Technology 116. https://doi.org/10.1016/j.optlastec.2019.03.035.

Majumder, S., Mondal, T., Deen, M., 2017. Wearable sensors for remote health monitoring. Sensors 17, 130. https://doi.org/10.3390/s17010130. from Duplicate 2 (Wearable sensors for remote health monitoring – Majumder Sumit, Mondal Tapas, Deen M.).

Manti, M., Cacucciolo, V., Cianchetti, M., 2016. Stiffening in soft robotics: a review of the state of the art. IEEE Robotics and Automation Magazine 23, 93–106. https://doi.org/10.1109/MRA.2016.2582718.

Shu, L., Hua, T., Wang, Y., Li Qiao, Q., Feng, D.D., Tao, X., 2010. In-shoe plantar pressure measurement and analysis system based on fabric pressure sensing array. IEEE Transactions on Information Technology in Biomedicine: a Publication of the IEEE Engineering in Medicine and Biology Society 14, 767–775. https://doi.org/10.1109/TITB.2009.2038904.

Smith, P.N., Refshauge, K.M., Scarvell, J.M., 2003. Development of the concepts of knee kinematics. Archives of Physical Medicine and Rehabilitation 84, 1895–1902. https://doi.org/10.1016/S0003-9993(03)00281-8.

Suresh, R., Bhalla, S., Hao, J., Singh, C., 2015. Development of a high resolution plantar pressure monitoring pad based on fiber Bragg grating (fbg). Sensors 23, 785–794. https://doi.org/10.3233/THC-151038.

Taborri, J., Palermo, E., Rossi, S., Cappa, P., 2016. Gait partitioning methods: a systematic review. Sensors 16, 66. https://doi.org/10.3390/s16010066. From Duplicate 1 (Gait partitioning methods: a systematic review – Taborri Juri, Palermo Eduardo, Rossi Stefano, Cappa Paolo).

Tavares, C., Domingues, M., Frizera-Neto, A., Leite, T., Leitão, C., Alberto, N., Marques, C., Radwan, A., Rocon, E., André, P., Antunes, P., 2018. Gait shear and plantar pressure monitoring: a non-invasive ofs based solution for e-health architectures. Sensors 18, 1334. https://doi.org/10.3390/s18051334.

Theodosiou, A., Lacraz, A., Stassis, A., Koutsides, C., Komodromos, M., Kalli, K., 2017. Plane-by-plane femtosecond laser inscription method for single-peak Bragg gratings in multimode cytop polymer optical fiber. Journal of Lightwave Technology 35, 5404–5410. https://doi.org/10.1109/JLT.2017.2776862.

Vilarinho, D., Theodosiou, A., Leitão, C., Leal-Junior, A.A., Domingues, M., Kalli, K., André, P., Antunes, P., Marques, C., de Fátima Domingues, M., Kalli, K., André, P., Antunes, P., Marques, C., 2017. Pofbg-embedded cork insole for plantar pressure monitoring. Sensors 17, 2924. https://doi.org/10.3390/s17122924.

Walker, P.S., Kurosawa, H., Rovick, J.S., Zimmerman, R.A., Holbrook, J., Clarke, T., Carbone, P., Soltesz, K., Chesonis, A., Walker, P.S., 1985. External knee joint design based on normal motionc. Journal of Rehabilitation Research 22, 9–22.

Walsh, C., 2018. Human-in-the-loop development of soft wearable robots. https://doi.org/10.1038/s41578-018-0011-1. From Duplicate 2 (Human-in-the-loop development of soft wearable robots – Walsh Conor).

Woyessa, G., Nielsen, K., Stefani, A., Markos, C., Bang, O., 2016. Temperature insensitive hysteresis free highly sensitive polymer optical fiber Bragg grating humidity sensor. Optics Express 24, 1206. https://doi.org/10.1364/OE.24.001206.

zebris Medical GmbH, 2020. The plantar pressure distribution measurement systems fdm / pdm. https://www.zebris.de/en/medical/stand-analysis-roll-analysis-and-gait-analysis-for-the-practice.

Chapter 9

Soft robotics and compliant actuators instrumentation☆

9.1 Series elastic actuators instrumentation

Concerning human safety, a low output impedance is desirable in exoskeleton actuators to reduce the risk of an accident with the patient (dos Santos et al., 2015). To perform the desired movements of the physiotherapy section, the actuator's bandwidth must be close to that of human movement. These requirements can be achieved by the series elastic actuator (SEA), which consists of an elastic element placed between the load, namely the human limb, and the actuator, as proposed in Pratt and Williamson (1995). This elastic element, i.e., spring, can significantly reduce the system's output impedance, while also reducing the system's bandwidth (Robinson et al., 1999). As humans move in low frequencies (Kirtley, 2006), exoskeleton's actuators can be successfully composed of SEAs (Keemink et al., 2018).

A current trend for the design of robots and actuators is to reduce their stiffness (Gul et al., 2018), as seen in the so-called soft robotics, where flexible structures and actuators are used. Next generation robots have been developed using soft robotics (Manti et al., 2016). Biomedical applications can especially benefit from stiffness reduction, as further requirements of biocompatibility and biomimicry can be achieved by the soft robotic devices (Cianchetti et al., 2018). One of the first technologies to break the "stiffer is better" paradigm that endured for many decades are series elastic actuators (SEAs) (Pratt and Williamson, 1995), and since their first report in 1995, many robotic devices have been designed using this technology, such as robots, prostheses (Rouse et al., 2013), orthosis (dos Santos et al., 2015), and exoskeletons (Ragonesi et al., 2011). Even after 20 years, SEAs are still considered a core technology on actuator's design of flexible/compliant robots (Calanca et al., 2014).

The spring placed between the motor unit and the load in a SEA increases the compliance between the wearable robot for gait rehabilitation and assistance and the user (dos Santos et al., 2015). Therefore the actuator has a low output impedance, which reduces the risk of accidents. Furthermore, despite the bandwidth reduction induced by the spring, SEAs can be used in wearable applications since human movement also occurs at low frequencies (Robinson et al.,

☆ This chapter is carried out with the participation of Wagner Coimbra de Moraes Junior.

https://doi.org/10.1016/B978-0-32-385952-3.00019-6

1999). Thus the use of wearable robots comprising SEAs for gait rehabilitation in robot-assisted therapy present advantages over conventional therapies such as their higher repeatability, the quantitative feedback of the patient recovery and possibility of treatment customization (Kwakkel et al., 2008).

As stated by Hooke's law, the force (or torque) of the spring is proportional to the displacement when it operates in its linear region, where its proportionality is related to the spring stiffness. Hence, the force (or torque) exerted by the actuator can be estimated from the spring displacement, simplifying actuator instrumentation (Pratt and Williamson, 1995) for robust force control and impedance controllers (Calanca et al., 2014). Encoders and potentiometers are conventionally employed in the SEAs instrumentations (Leal Junior et al., 2016). Nonetheless, these sensors have disadvantages. Besides being affected by electromagnetic interference, they are also sensitive to backlash on the motor unit as they are not directly positioned on the spring, which generally needs compensation, as reported in dos Santos et al. (2015). The requirements for precise alignment of these sensors to the spring axle may also result in a bulky system (Leal-Junior et al., 2018a).

9.1.1 Torque measurement with intensity variation sensors

A SEA composed of a customized spring and a DC motor with an encoder is shown in Fig. 9.1. An angular contact bearing ensures the alignment and freedom of movement of the motor axis, which has its movement transmitted to the spring by a worm gear. The spring deflection is estimated by the difference between the motor encoder and another encoder positioned on the actuator output. The design and construction of this SEA are described in detail in dos Santos et al. (2015). A simpler experimental setup, focused only on the spring is proposed in order to analyze the spring deflection and torque. As shown in Fig. 9.1, a lever is linked to the spring to provide angular deflection and position the spring. The spring with its base is then attached to a wooden support with two holes which are at 4° and 10° from the lever position with no deflection. Below this wooden support, the encoder E5 series (US digital, USA) is positioned on the spring axles, and serves as reference for the dynamic tests of the POF sensor. This simple experimental setup facilitates the sensor validation on both static and dynamic tests. The polymer optical fiber (POF) sensor is positioned perpendicular to the face of the spring, in order not to interfere in the spring extension movement due to fiber strain and prevent slip on the compression movement, which would decrease the sensor sensitivity. Thus the fiber supports used allowed fiber positioning without hindering the spring movement or sensor sensitivity.

Prior to the application in dynamic measurements, POF intensity variation-based sensors need a calibration phase on static or quasistatic conditions. In this setup, the calibration curve is obtained by measuring the sensor response at each of the angles of the support, namely 0°, 4°, and 10°. The lever, which is directly connected to the spring and turns with its same angular displacement,

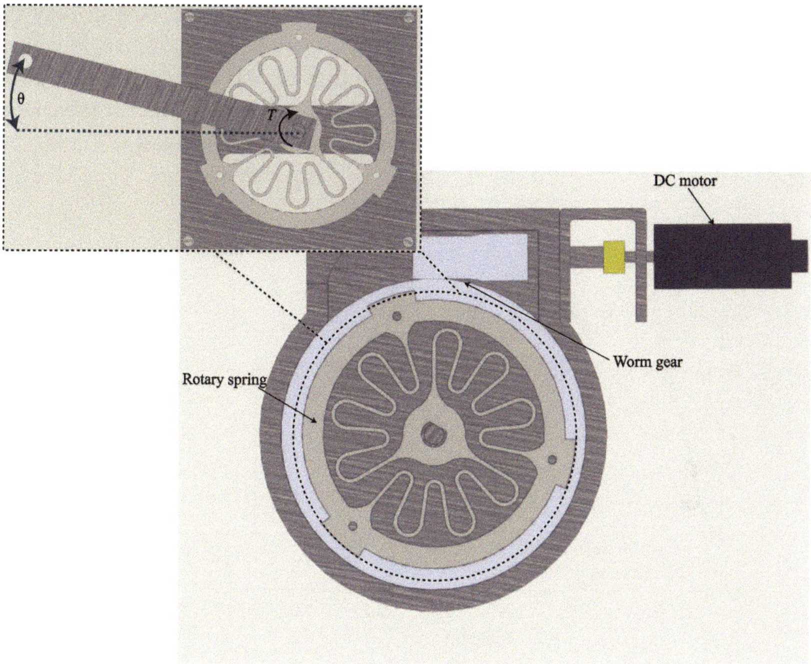

FIGURE 9.1 Schematic representation of the rotary series elastic actuator.

is held for about 150 seconds at each angle. The sensor system is composed of a laser with 3 mW at 650 nm connected to one end of the fiber and a photodiode with a transimpedance amplifier circuit connected to the other end, with a acquisition frequency of 200 Hz. Temperature and humidity was maintained constant along all tests, and another POF sensor, connected to the same light source and without any strain applied, is positioned close to the spring to monitor and compensate these parameters, since variations on them can affect the POF sensor and induce measurement errors. It is also worth to mention that the stress relaxation is analyzed and compensated using the Maxwell's model discussed in Chapter 6. The compensation responses for 0°, 4°, and 10° were 1.92 V, 2.01 V, and 2.14 V, respectively. A linear regression was made to relate the spring angles to the voltage responses, as shown in Fig. 9.2, along with the sensor's calibration curve and correlation coefficient. To evaluate the dynamic response of the sensor, sequential compression cycles are performed, in which the lever is loaded and unloaded for about 3.5 seconds. The results for the POF sensor, continuous line, and the encoder measurements, showing points instead of continuous line for better visualization of the POF response, are shown in Fig. 9.2. The root mean squared error (RMSE) between the POF sensor and the encoder is calculated, and an RMSE of about 0.57° was obtained for this test. However, although the encoder response is in steps of 0.3°, the POF sensor has a resolution of 0.1° when considering the 8-bit ADC. Hence, RMSE may be lower if a higher resolution reference system is employed, since an angle within the POF resolution but below encoder resolution can be calculated as an error of the POF sensor in the present setup. Nonetheless, the POF sensor presented errors lower than 4% when considering the entire range of the test.

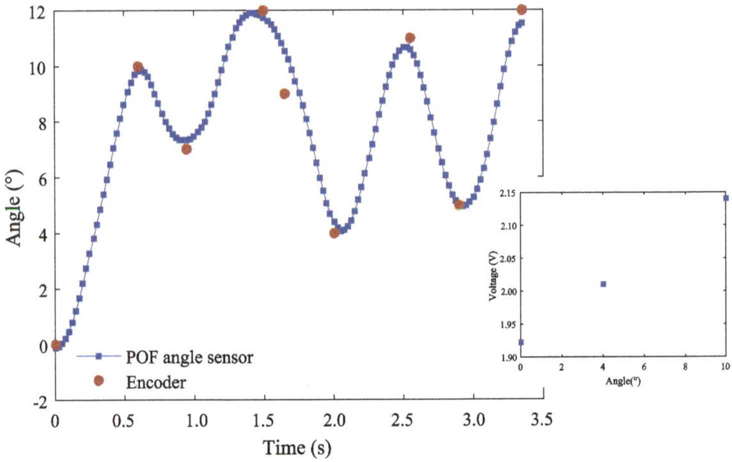

FIGURE 9.2 POF intensity variation-based sensor response as a function of the applied angle (or spring deflection).

In spite of the good accuracy of the POF curvature sensor, its response generally has a saturation trend in angles higher than 90°. In this case, the initial position is already on a 180° bend and the sensor has a linear behavior. The response of the sensor is generated by the radiation losses due to the curvature and the attenuation generated by the refractive index variation that the stress-optic effect causes. In this setup, a high stress on the fiber is present, since the fiber is at 180° and the distance between the supports is about 9 mm. All the attenuation of the optical fiber of the POF torque sensor is thereby assumed to be caused by the variation of the refractive index by stress-optic effect. Moreover, all the torque generated by the spring deflection is directly transmitted to the fiber.

The output power varies almost linearly with the refractive index changes, as shown in Fig. 9.3. Maxwell's model, as discussed in Chapter 6, can be used to isolate the time-varying component of the sensor's viscoelastic response from the static one. Since all output signal attenuation is primarily assumed to be caused by the stress-optical effect, which changes the core refractive index as discussed in Chapter 6, Eq. (9.1) can be applied for the stress response. Considering pure bending stress, the static stress (σ_0) can be calculated as

$$\Delta n = \frac{n_c^3 \cdot q_{11} \cdot \sigma(t)}{2}, \tag{9.1}$$

$$\sigma_0 = \frac{T \cdot x}{I}, \tag{9.2}$$

where T is the torque, x is the perpendicular distance between the bending axis, and the neutral line and I is the moment of inertia around the neutral axis. The torque equation for the POF torque sensor, Eq. (9.3), is obtained by rewriting Eq. (9.1) in terms of stress instead of optical power, as assumed by the stress-optical effect predominance, and by substituting Eq. (9.2) in Eq. (9.1) using the compensated response of the sensor:

$$T = \frac{\Delta n_c \cdot 2 \cdot I}{n_c^3 \cdot q_{11} \cdot x}. \tag{9.3}$$

The moment of inertia of the POF only depends on its total diameter due to its circular cross-sectional area. Considering the core, cladding, and outer coating, the total diameter of the fiber is 2.2 mm. The distance between the bending axis and the neutral line is about 3.5 mm and considered constant. The refractive index variation can be estimated by the output power through linear regressions obtained in the characterization tests, as shown in Eq. (9.4) for this case

$$n_{est} = 1.571 \cdot \frac{P}{P_1} + 0.04474. \tag{9.4}$$

The POF torque sensor measurements are compared with the expected torque for each angle of the static characterization test. Since the spring has a linear behavior on both extension and compression (dos Santos et al., 2015), the torque on it is proportional to its deflection angle. Hence, the torque on spring is evaluated by multiplying the spring deflection angle by the spring constant. The spring constant for compression of 92 Nm/rad of this spring gives a torque of 6.42 Nm for the 4° deflection angle, whereas the POF sensor estimated an 6.78 Nm torque for the same angle. For the 10° deflection angle, the torque on the spring is 16.05 Nm and the estimated one is 16.57 Nm. As indicated by the differences on torques, the error reduces as the spring deflection angle increases, with an error of 5.6% for 4° and 3.2% for the 10°.

In dynamic measurements, the POF torque sensor is validated by comparison with the torque estimations from the angle measurements of the encoder, as shown in Fig. 9.3. The RMSE between the POF sensor and the encoder estimations is 0.33 Nm. The RMSE for the torque and the angle are compared by calculating the percentage contribution of each RMSE over the total range of the test, i.e., the ratio between the RMSE and the test range. The assumptions made on POF torque sensor equations, such as constant moment of inertia and constant perpendicular distance between the bending axis and the neutral line, can induce errors on torque estimation, as indicated by the larger error contribution of 5.7% for the torque sensor when compared to the 4.7% one for the curvature sensor. Thus the error of the POF torque sensor may be further reduced by including the variations in these parameters for different spring angles. Fig. 9.3

also shows the comparison between the torque measured by the proposed technique for the POF torque sensor and the one estimated by applying the spring constant to the POF curvature sensor, which can be an alternative technique to determine the torque on a torsional spring. The POF torque sensor presented a resolution of about 0.15 Nm, which is better than the one provided by the encoder of about 0.5 Nm. The proposed POF sensor may be considered of high precision, since it presented errors lower than 5% for the torque measurements.

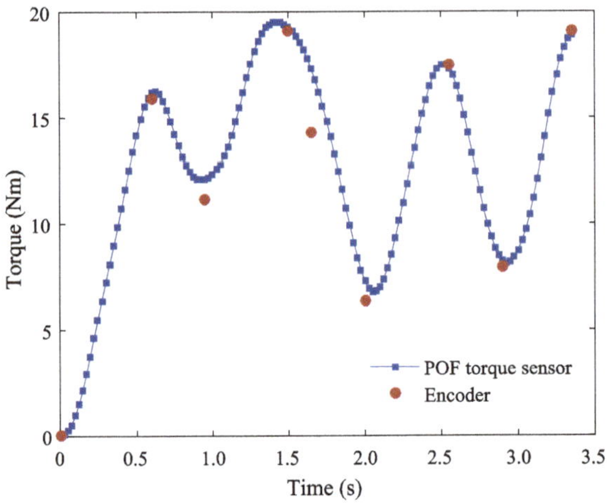

FIGURE 9.3 Torque measured by the presented POF intensity variation-based sensor compared with the one estimated from the angle obtained by the encoder (considering the spring stiffness previously measured).

9.1.2 Torque measurement with intensity variation sensors

Eight FBGs inscribed on two POFs, four on each fiber, were employed on torque estimation of the torsional spring, as shown in Fig. 9.4. 3D-printed supports were used to place them on the spring, and they were then fixed by a thermoplastic glue. The spring, along with its base, was attached to a fixed support to limit its flexion and extension movements to only one plane. The spring was displaced by a lever connected to the front of its output shaft, and its deflection angle was measured by an encoder E5 series (US digital, USA) positioned on the back of the shaft. The angular displacement measured by this encoder was used for comparison with the FBG sensors, through torque estimation using the stiffness constants of 92 Nm/rad for flexion and 96 Nm/rad for extension, as characterized in dos Santos et al. (2015), since the spring is considered linear for torque less than 30 Nm.

The FBGs, with a physical length of 1.4 mm, were inscribed at 1550 nm wavelength region in a commercial multimode CYTOP fiber (Chromis Fiberoptics Inc.) with graded index. The fiber, which has a total diameter of 490 μm, comprises a 120 μm core diameter, a 20 μm cladding thickness, and a polycarbonate overcladding. The direct-write, plane-by-plane inscription method (Lacraz et al., 2016) technique was used to inscribe the FBGs on the fiber. A fs laser system (HighQ laser femtoREGEN) centered at 517 nm with a 220 fs pulse duration provided pulses with repetition rate of 5 kHz and energy of about 80 nJ, while an air-bearing translation stage with nanometer resolution moved the POF during inscription. A ×50 objective lens focused the beam on the POF, which modifies the refractive index at the center of its core. An annealing is performed on the arrays (as described in Leal-Junior et al. (2018b)). As discussed in Chapter 4, a smooth variation of the core diameter is desired, and thereby a UV-curing glue (Loctite AA 3936, USA) was used to butt-couple the FBG array to a multimode silica fiber, which is then fusion spliced to a single mode silica pigtail.

The spring deflection at each of the FBG sensor locations was evaluated by means of a FEM analysis (Ansys Workbench 15.0), in which the outer cylinder is fixed and a 20 Nm flexion torque is applied to the inner cylinder corresponding to the output shaft. As expected by Hooke's law, the simulation indicates that the strain at all points analyzed varies proportionally to the applied torque. The direction is given by the strain signal, in order that a negative signal indicates the counter-clockwise direction when the clockwise direction (extension cycle) is considered as the positive direction.

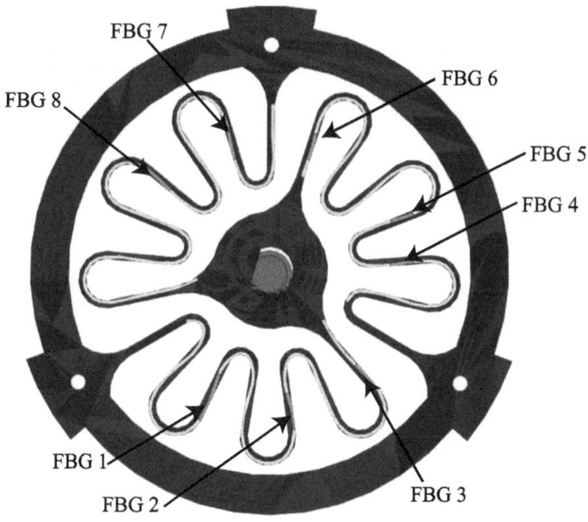

FIGURE 9.4 Schematic representation of the spring of the FBG positions.

The different FBG responses were fused using a Kalman filtering (Leal-Junior et al., 2019b) for a reliable measurement of the flexion and extension of the spring. The filter was referenced by the known displacements applied to the springs. In this filter, the state of a system is estimated by the measurements of each sensor and a reference to compose an estimation of the system as a whole, which has a lower error than that of each separate sensor (Sasiadek and Hartana, 2000). For each sensor, a weight in the estimation of the state is given according to the sensor uncertainty, in order that the lower the covariance of the sensor, the higher the weight of its result in the final estimation (Feng et al., 2014). The Kalman filter is suitable for sensor fusion due to its simplicity and ease of implementation and optimization (Feng et al., 2014), and a detailed description of the Kalman filter is given by (Auger et al., 2013). Prediction is made through Eqs. (9.5) and Eqs. (9.6) and updated through Eqs. (9.7), (9.8), and (9.9), in state space representation:

$$\hat{x}_{k|k-1} = A \cdot \hat{x}_{k-1|k-1}, \tag{9.5}$$

$$P_{k|k-1} = A \cdot P_{k-1|k-1} \cdot A', \tag{9.6}$$

$$G = P_{k|k-1} \cdot C' \cdot (C \cdot P_{k|k-1} \cdot C' + R), \tag{9.7}$$

$$P_{k|k} = (I - G \cdot C) \cdot P_{k|k-1}, \tag{9.8}$$

$$\hat{x}_{k|k} = \hat{x}_{k|k-1} + G \cdot (z_k - C \cdot \hat{x}_{k|k-1}), \tag{9.9}$$

where A and C are the transition matrices of state (at time k) and observability, respectively, which have unitary values in this case, since, presumably, the sensor does not alter the system dynamics. G is the Kalman gain, P is the covariance matrix, z_k is the measured signal, R is the sensors covariance, I is the identity matrix and is the estimated state at time k given observations up to and including at time k. Thus the response of each FBG results in the z_k matrix used on the state prediction in conjunction with the transition matrix A (unitary) and the sensor covariance matrix P. Then the prediction is updated by means of estimating the Kalman gain (G) from Eq. (9.7) in which the covariance (P) of each sensor obtained in the calibration procedure is used to provide such estimation. Thereafter, the covariance matrix and the estimated state are updated, resulting in a more accurate estimation of the torque.

It is also worth noting that if a transverse force is applied on the spring, the FBGs close to the region of force application detect the increase in strain due to stress concentrated on this region. Thus by analyzing the response of all of the sensors and comparing it with the case where only flexion and extension torques are applied to the spring, the wavelength shift of FBGs can be used to differentiate flexion and extension from transverse mechanical disturbances in a simple and straightforward manner.

As expected by the simulation in Fig. 9.5(a), each FBG measured a different strain corresponding to the deformation at each region of the spring. A torque close to that used on the simulation (20 Nm) was exerted on the spring by rotating and maintaining the lever by 0.2 rad in the clockwise direction (extension), which corresponds to a torque of about 18.4 Nm considering the 92 Nm/rad spring stiffness (dos Santos et al., 2015). The experimental wavelength shift obtained for each FBG at a constant torque is presented in Fig. 9.5(a), along with the equivalent simulated strain for each region. Since the FBGs wavelength shift increases with increasing strain, the behavior presented by the FBGs are similar to that predicted by the numerical analysis, as evidenced qualitatively by Fig. 9.5(a). This comparison is only performed to check if the strain pattern measured by the FBGs follows that expected by the simulation, since material anisotropy and numerical errors diverge the simulated strains from the measured ones, as verified in dos Santos et al. (2015). Thus as indicated by the analysis of Fig. 9.5(a), it is possible to reconstruct the shape of the spring under flexion and extension by means of placing a colormap with the measured wavelength shift of each FBG at each region. The response of each of the 8 FBGs are presented in Fig. 9.5(b), where the positive signal corresponds to flexion and extension torques, respectively. The regions of FBGs 3 and 6 show the largest variation whereas the ones of FBGs 1, 2, and 8 show the lowest one. These results were expected by the simulation and confirmed by the experimental tests, as seen in Figs. 9.5(a) and (b), and are strictly related to the FBGs position. Each FBG response has a linear relation to the applied torque depending on its position, in order that linear regressions for each FBG was performed, as shown in Fig. 9.5(b). The RMSE between the estimated torque estimated by each FBG and the applied one was calculated for three measurements. Besides its ease of installation, the proposed POFBG sensors present sensitivities up to 8 times higher than silica FBGs (Sanchez et al., 2018), which make them more suitable for this particular application.

The determination coefficients (R^2 comparing with a linear regression) shown in Fig. 9.5(b) are caused by nonlinearities on each FBG response, whereas the coefficients a and b are directly influenced by the position of each FBG, as already anticipated by the numeral analysis. The response of the FBG 3 presented the lowest R^2 for the linear regression, hence the highest RMSE on torque estimation, due to an asymmetry between flexion and extension cycles that can be related to this specific region on the spring. The other FBGs presented a more linear response, with determination coefficients greater than 0.99. The mean RMSE of all 8 FBGs is 1.39±0.73 Nm, with a relative error of about 4.2% regarding the whole torque range of about 32 Nm. Despite having low errors individually, the information of all of the 8 FBGs can be merged by a sensor fusion algorithm for a more accurate torque estimation, as all the sensors measure the flexion/extension torque on the rotary spring. Thus the torque is estimated by the application of the Kalman filter on the sensor responses, and compared with the applied one in Fig. 9.6, which also shows the parameters of the filter. The parameters of the filter are estimated recursively by comparing the sensor response after the sensor fusion with the reference torque in order to obtain the lowest possible error for the sensors. In this case, the covariance of each sensor (R1-8) is obtained based on the error of each sensor, where the lower errors resulted in lower covariance values as depicted in Fig. 9.6.

As shown in Fig. 9.6, the fusion of the 8 FBG sensors estimated the torque with a seven-fold reduction of the RMSE considering the applied torque, which

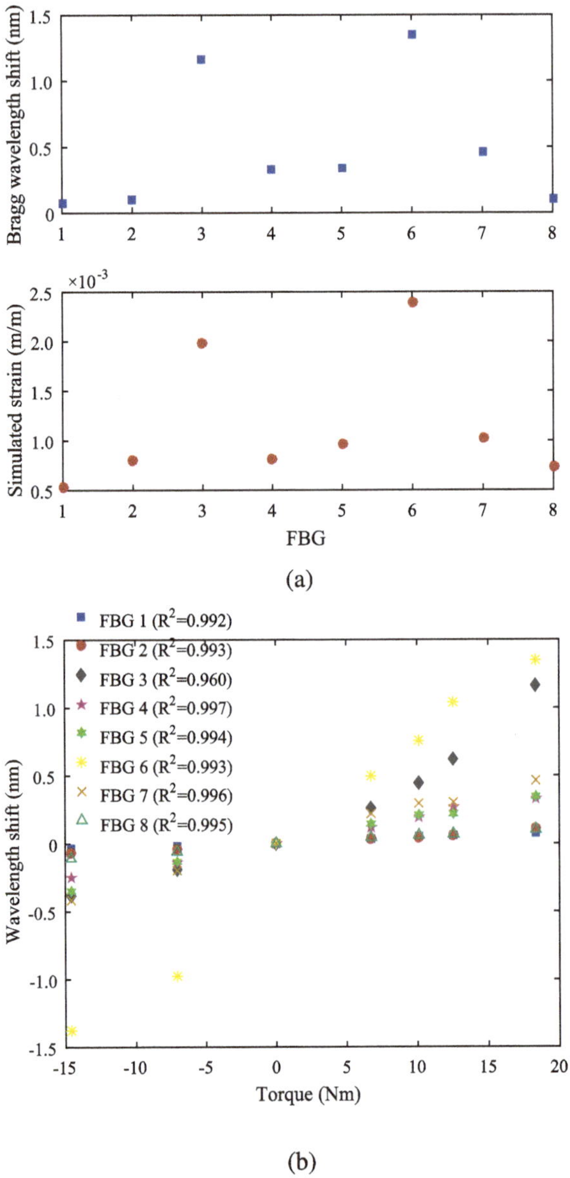

FIGURE 9.5 (a) Comparison between the wavelength shift and the simulated strain on the spring in the regions of sensors. (b) Wavelength shift of each FBG as a function of the applied torque on the spring.

is related to the increase of linearity of the sensor fusion ($R^2 = 0.999$). Thus the application of the Kalman filter contributed to a significant improvement in

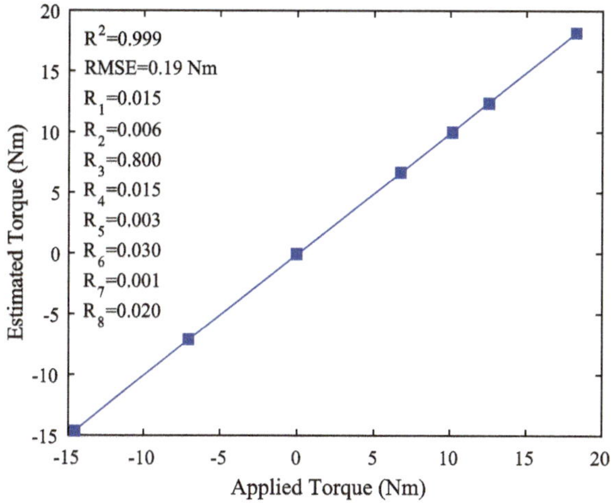

FIGURE 9.6 Torque measurement using the sensor fusion via Kalman filter applied on the 8 FBGs. The RMSE and covariance matrix values are also presented.

the accuracy of the measurement, with a relative error of only 0.59% regarding the whole torque range. As verified by these results, a FBG array with sensor fusion techniques can be used in this spring as a highly accurate and reliable sensor system. In addition, the quasidistributed sensor system has another advantageous feature, which is the ability to distinguish transverse forces from the flexion/extension torques on the spring. External mechanical disturbances, such as the transverse forces, can be estimated and possibly rejected from the strain measured on each region of the spring when compared to a reference condition. An experimental test is performed in order to verify this assumption, in which arbitrary loads are applied in 5 points of the spring, along with 3 different flexion/extension torques. The wavelength shift of each FBG can be presented by a set of colormaps in a spring shape reconstruction tool for forces 1 and 3. The FBG responses at the 3 different flexion/extension torques and at the 5 transverse force conditions are also shown at the top and bottom, respectively.

The wavelength shift, which measures the strain, increases the closer the FBG is to the point of application of transverse force, as shown in Fig. 9.7. As expected by the simulation and experimental analyses, FBGs 3 and 6 presented a higher shift due to their position in the spring. The strain distribution, i.e., wavelength shift pattern, of the FBG sensors for only flexion and extension torques can be used as a reference for transverse force detection, since each sensor presents a linear response that follows a well-defined pattern in this case. Hence, the strain distribution for a given loading can be compared with a reference strain condition in order to identify external disturbances on the spring in a simple and straightforward manner.

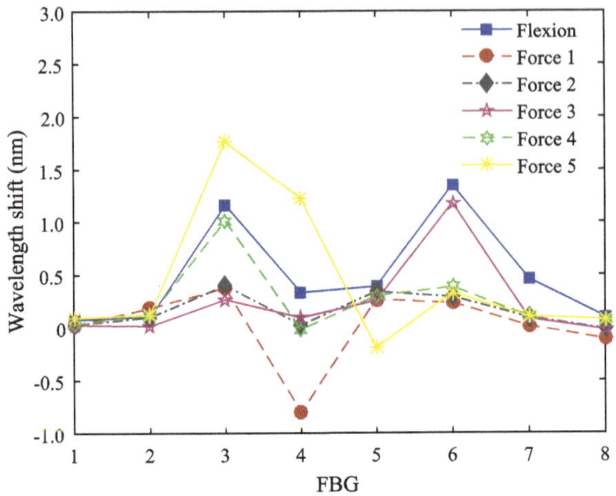

FIGURE 9.7 Wavelength shift obtained from each FBG at transverse force conditions as well as under flexion and extension.

9.2 Tendon-driven actuators instrumentation

The developments in robotics, along with innovations on its signal processing, control and instrumentation, have broadened the application of robotic devices (Shukla and Karki, 2016; Zhang et al., 2019). Among the diversity of actuation technologies for soft robots, as summarized in Chapter 2, tendon-driven actuators have a remote actuation with a low flexural stiffness when pulled (Manti et al., 2016), in order that they are suitable in many applications of continuously operated soft robots (Bundhoo et al., 2009; Kastor et al., 2020; Natale and Pirozzi, 2008).

Tendon-driven actuators can be better controlled by means of measuring the tendon displacement (or strain) for assessment of the system dynamics and the transmitted movement to the robot's joint, or limb in a tendon-driven wearable robot (Wen et al., 2017). However, the large deformation of the tendon, as well as the difficulty in sensor assembly, hinders the evaluation of its displacement and dynamics, which are thereby considered a black box model in the modeling and control or estimated through indirect measurements (Casas et al., 2019). While imaging methods and stress-strain tests with universal testing machines (Louis-Ugbo et al., 2004) are some of the methods proposed for tendon displacement evaluation, they can nonetheless provide real time measurements in unstructured environments or during the robot operation.

Optical fiber sensors are an emerging sensing technology which have been used for assessment of mechanical parameters in many fields, industrial (Diaz et al., 2019), healthcare (Leal-Junior et al., 2019a), and robotics applications (Xiong et al., 2018). These sensors were able to assess the strain in Achilles tendons and knee ligaments in ex-vivo experiments (Behrmann et al., 2012), (Ren et al., 2007) ex-vivo tests. In such applications, the Young's modulus of the embedded sensor should be close to that of the tendon, in order not to influence the tendon's mechanical parameters. The Young's moduli of silica and polymer optical fibers (POF), which are of tens of GPa and a few GPa, respectively, are nonetheless one order of magnitude higher than that of tendons, which are of hundreds of MPa.

9.2.1 Artificial tendon instrumentation with highly flexible optical fibers

Experimental tests were conducted in order to assess the measurement of optical fibers sensors glued to a filament of thermoplastic polyurethane (TPU) that emulates a tendon, as well as their mechanical properties. The TPU filament, which has a diameter of 2.85 mm, is used as a tendon as it is flexible and has a low Young's modulus of tens of MPa. As shown in Fig. 9.8, the tests were performed using a tensile test machine (Biopdi, Brazil), which has a displacement sensor and a load cell to measure position and force, respectively. In order to estimate the Young's modulus of the TPU and the LPS-POF following the ISO 527:2012 standard, tensile tests were conducted by means of applying the same strain rate on each and calculating the modulus from the slope of the stress-strain curve in the linear region following the ISO 527:2012 standard.

The stiffness of combinations of each optical fiber with the tendon, i.e., TPU/LPS-POF, TPU/CYTOP, TPU/silica, were analyzed in order to ascertain which combination is closer to the tendon alone. An UV-curing optical adhesive NOA 88 (Norland, USA) is used to attach the fibers to the tendon. After the material characterization, an strain sensing test was conducted, in which the ends of the fiber are connected to a laser source (5 mW at 660 nm) and a phototransistor (PD) IF-D92 (Industrial Fiber Optics, USA), as shown in Fig. 9.8. The compensation for light source power and environmental variations, such as temperature and humidity variations, was performed by normalizing the sensor response with respect to the reference signal of another fiber, connected to the same light source by means of an optical coupler (LC) IF-562 (Industrial Fiber Optics, USA) and another phototransistor. A microcontroller FRDM-KL25Z (NXP, Netherlands) with a 16-bits analog-to-digital converter acquired the signals of both phototransistors. These signals refer to optical power variations on the fibers as a result of refractive index changes due to the photoelastic effect of the strain on the optical fibers. Two configurations of sensors are tested as shown in Fig. 9.8. The fiber is first attached parallel to the tendon through two gluing points in Configuration 1, and then twisted around the tendon longitudinal axis with two turns in Configuration 2. While the first configuration allows the strain applied on the tendon to be measured by the fiber stretching, the second configuration is expected to present a higher sensitivity as a function of the force exerted on the tendon due to the stress-optic effect.

The Young's moduli of the materials characterized by the tensile tests, i.e., TPU and LPS-POF, and the ones of the silica and CYTOP fibers, as characterized in the literature, are presented in Table 9.1, along with their diameters.

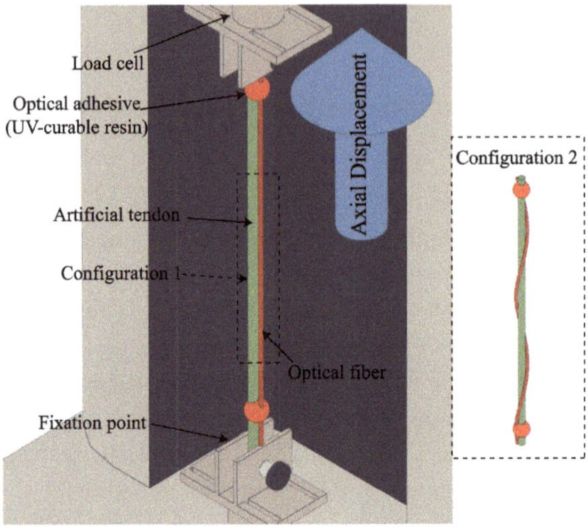

FIGURE 9.8 Schematic representation of the experimental setup used on the sensors analyses and sensors' configurations.

TABLE 9.1 Young's modulus and diameter of the materials used.

Material	Young's modulus	Diameter
TPU	72.45 MPa	2.85 mm
LPS-POF	19.67 MPa	0.90 mm
Silica Fiber (without coating)	69.00 GPa	0.125 mm
CYTOP	1.50 GPa	0.50 mm

Fig. 9.9 shows the stress-strain curves of TPU and LPS-POF as well as the force-displacement curves of all materials configurations, i.e., TPU/Silica, TPU/CYTOP, TPU/LPS-POF. Despite that both materials can withstand higher strains (Leal-Junior et al., 2020; Meckel et al., 1987) the mechanical characterization tests for TPU and LPS-POF were conducted in the 16% strain range. The Young's modulus of the TPU, which is used to emulate the tendon, is of about 72 MPa, while that of the LPS-POF, of about 19 MPa, is orders of magnitude lower than the ones of commercial fibers, such as silica and CYTOP in this case. The high flexibility of these materials is further evidenced by their low Young's moduli. Moreover, the linear behavior of the LPS-POF in the 16% strain range indicates its suitability for strain measurement on tendons.

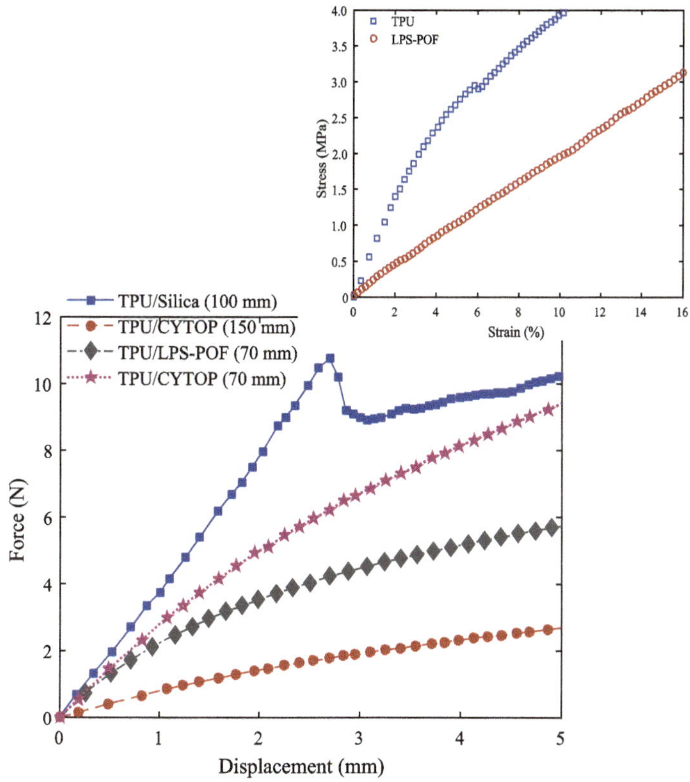

FIGURE 9.9 Force-displacement characterization curves for different materials combinations. Figure inset shows stress-strain curves of TPU and LPS-POFTPU and LPS-POF.

Thereafter, the mechanical properties of the TPU combined with each of the optical fibers were evaluated by means of a force-displacement test, since the Young's moduli of the combinations could not be obtained due to their non-uniform cross-sectional area, as opposed to circular or rectangular cross-sections. Therefore, the stiffness (k), which is the ratio of the force to displacement, was experimentally evaluated. Furthermore, the material properties and geometry can also be used in the definition of the stiffness, which for an axial loading can be defined as

$$k = \frac{A \cdot E}{L}, \tag{9.10}$$

where A is the cross-sectional area of the materials, E is the Young's modulus of the material combination, and L is the length.

Eq. (9.10) evidences the relation of the stiffness to the Young's modulus, cross-sectional area and length of the material. The length of the material, which must be defined prior to the application, has a huge impact on the stiffness of the

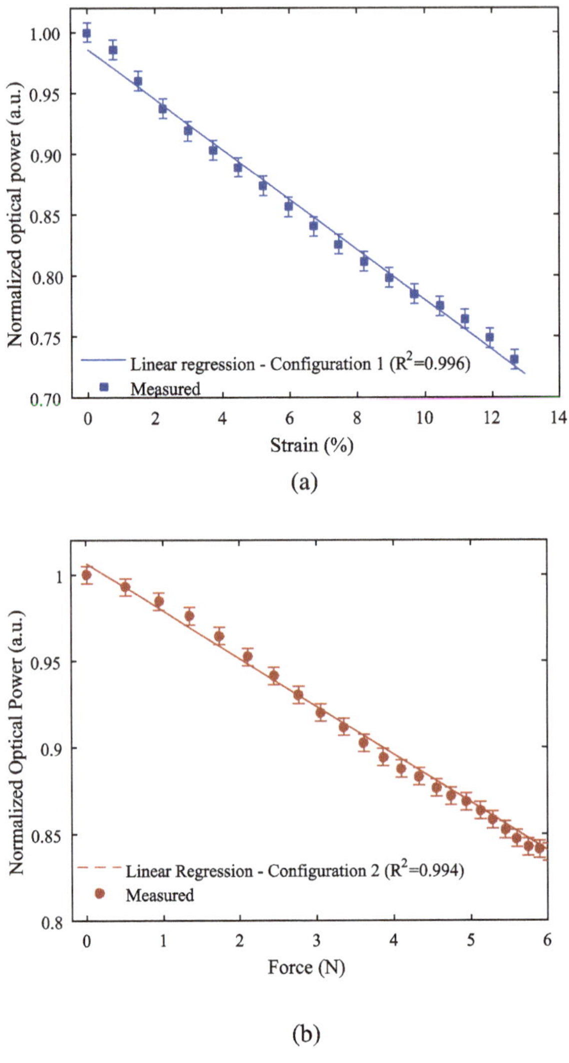

FIGURE 9.10 Force and strain responses of the presented intensity variation-based sensors in (a) configuration 1 and (b) configuration 2.

material, which is indicated by the results in Fig. 9.9. Despite the second highest Young's modulus of the CYTOP among the tested optical fibers, its combination with the TPU with a longer length, of 150 mm, presented the lowest stiffness. In contrast, the TPU/CYTOP combination with 70 mm length provided a higher stiffness, which surpassed that of the TPU/LPS-POF with the same length. The TPU/Silica fiber combination broke at about 2.7 mm displacement due to the brittle nature of the silica fiber. Hence, these results indicate that the material stiffness can be largely tuned by varying the lengths of material in the combination. The minimum influence of the tendon stiffness can therefore be achieved by knowing the maximum and minimum lengths of the sensing region. While the TPU presented a stiffness of 1790.2 N/m in the tests, the evaluated stiffness for the TPU/LPS-POF combination of 1783.5 N/m is close to that of the TPU. The LPS-POF does not significantly alter the material stiffness in regard to the combination of the optical fiber with the tendon and was used on the tendon displacement assessment application.

The results of the first and second configurations, for strain and force sensing respectively, are presented in Figs. 9.10(a) and 9.10(b). As intended by the first configuration, which measures the displacement in the tendon, the determination coefficient (R^2) of 0.996 of the linear regression ascertains the strong correlation of the sensor with the applied strain. Similarly, in the second configuration, the linear correlation between the sensor response and the measured force has a R^2 of 0.994. Therefore the proposed sensor in configuration 2 is able to estimate the force on the tendon, or the stress when the tendon cross-sectional area is known, whereas the sensor in configuration 1 can measure the strain on the tendon.

The sensor in configuration 1 comprises a fiber parallel to the tendon. In this configuration, both elements suffer the same displacement, and the fiber thereby measures the strain applied to the tendon. In configuration 2, in conjunction with the axial displacement, the fiber is also subjected to bending and torsion by the applied force due to its positioning around the tendon, and hence, the stress-optic effect can be used for a direct force measurement, as confirmed by the experiments. Therefore stress or strain can be measured in the tendon depending on the sensor-mounting configuration. Control and design of tendon-driven actuators can benefit from these sensors, which can be used individually or together, resulting in a complete dynamic characterization of the tendon responses.

References

Auger, F., Hilairet, M., Guerrero, J.M., Monmasson, E., Orlowska-Kowalska, T., Katsura, S., 2013. Industrial applications of the Kalman filter: a review. IEEE Transactions on Industrial Electronics 60, 5458–5471. https://doi.org/10.1109/TIE.2012.2236994.

Behrmann, G.P., Hidler, J., Mirotznik, M.S., 2012. Fiber optic micro sensor for the measurement of tendon forces. BioMedical Engineering Online 11, 1. https://doi.org/10.1186/1475-925X-11-77.

Bundhoo, V., Haslam, E., Birch, B., Park, E.J., 2009. A shape memory alloy-based tendon-driven actuation system for biomimetic artificial fingers, part I: design and evaluation. Robotica 27, 131–146. https://doi.org/10.1017/S026357470800458X. From Duplicate 2 (A shape memory alloy-based tendon-driven actuation system for biomimetic artificial fingers, part I: Design and evaluation – Bundhoo Vishalini, Haslam Edmund, Birch Benjamin, Park Edward J.).

Calanca, A., Capisani, L., Fiorini, P., 2014. Robust force control of series elastic actuators. Actuators 3, 182–204. https://doi.org/10.3390/act3030182.

Casas, J., Leal-Junior, A., Díaz, C.R., Frizera, A., Múnera, M., Cifuentes, C.A., 2019. Large-range polymer optical-fiber strain-gauge sensor for elastic tendons in wearable assistive robots. Materials 12. https://doi.org/10.3390/ma12091443.

Cianchetti, M., Laschi, C., Menciassi, A., Dario, P., 2018. Biomedical applications of soft robotics. Nature Reviews Materials 3, 143–153. https://doi.org/10.1038/s41578-018-0022-y.

Diaz, C.A.R., Leal-Junior, A., Marques, C., Frizera, A., Pontes, M.J., Antunes, P.F.C., Andre, P.S.B., Ribeiro, M.R.N., 2019. Optical fiber sensing for sub-millimeter liquid-level monitoring: a review. IEEE Sensors Journal 19, 7179–7191. https://doi.org/10.1109/JSEN.2019.2915031.

Feng, B., Fu, M., Ma, H., Xia, Y., Wang, B., 2014. Kalman filter with recursive covariance estimation—sequentially estimating process noise covariance. IEEE Transactions on Industrial Electronics 61, 6253–6263. https://doi.org/10.1109/TIE.2014.2301756.

Gul, J.Z., Sajid, M., Rehman, M.M., Siddiqui, G.U., Shah, I., Kim, K.H., Lee, J.W., Choi, K.H., 2018. 3d printing for soft robotics – a review. Science and Technology of Advanced Materials 19, 243–262. https://doi.org/10.1080/14686996.2018.1431862.

Kastor, N., Mukherjee, R., Cohen, E., Vikas, V., Trimmer, B.A., White, R.D., 2020. Design and manufacturing of tendon-driven soft foam robots. Robotica 38, 88–105. https://doi.org/10.1017/S0263574719000481.

Keemink, A.Q., van der Kooij, H., Stienen, A.H., 2018. Admittance control for physical human–robot interaction. International Journal of Robotics Research 37, 1421–1444. https://doi.org/10.1177/0278364918768950.

Kirtley, C., 2006. Clinical Gait Analysis: Theory and Practice. Elsevier, Philadelphia.

Kwakkel, G., Kollen, B.J., Krebs, H.I., 2008. Effects of robot-assisted therapy on upper limb recovery after stroke: a systematic review. Neurorehabilitation and Neural Repair 22, 111–121. https://doi.org/10.1177/1545968307305457. From Duplicate 2 (Effects of robot-assisted therapy on upper limb recovery after stroke: a systematic review – Kwakkel Gert, Kollen Boudewijn J., Krebs Hermano I.).

Lacraz, A., Theodosiou, A., Kalli, K., 2016. Femtosecond laser inscribed Bragg grating arrays in long lengths of polymer optical fibres; a route to practical sensing with pof. Electronics Letters 52, 1626–1627. https://doi.org/10.1049/el.2016.2238.

Leal Junior, A.G., de Andrade, R.M., Filho, A.B., 2016. Series elastic actuator: design, analysis and comparison. In: Recent Advances in Robotic Systems. InTech.

Leal-Junior, A.G.A., Frizera, A., Marques, C., Sanchez, M.R.A.M., Dos Santos, W.M.W., Siqueira, A.A.G., Segatto, V.M.M., Pontes, M.M.J., 2018a. Polymer optical fiber for angle and torque measurements of a series elastic actuator's spring. Journal of Lightwave Technology 36, 1698–1705. https://doi.org/10.1109/JLT.2017.2789192.

Leal-Junior, A.G.A., Theodosiou, A., Marques, C., Pontes, M.M.J., Kalli, K., Frizera, A., 2018b. Thermal treatments and compensation techniques for the improved response of fbg sensors in pofs. Journal of Lightwave Technology 36, 3611–3617. https://doi.org/10.1109/JLT.2018.2842461.

Leal-Junior, A.G., Diaz, C.A., Avellar, L.M., Pontes, M.J., Marques, C., Frizera, A., 2019a. Polymer optical fiber sensors in healthcare applications: a comprehensive review. Sensors 19, 3156. https://doi.org/10.3390/s19143156.

Leal-Junior, A.G., Theodosiou, A., Min, R., Casas, J., Diaz, C.R., dos Santos, W.M., Pontes, M.J., Siqueira, A.A.G., Marques, C., Kalli, K., Frizera, A., 2019b. Quasi-distributed torque and displacement sensing on a series elastic actuator's spring using fbg arrays inscribed in cytop fibers. IEEE Sensors Journal 1748, 4054–4061. https://doi.org/10.1109/JSEN.2019.2898722.

Leal-Junior, A.G., Coimbra, W., Marques, C., Frizera, A., 2020. Highly stretchable polymer optical fiber for mechanical sensing in artificial tendons: towards novel sensors for soft robotics. Actuators 9 (4), 125. https://doi.org/10.3390/act9040125.

Louis-Ugbo, J., Leeson, B., Hutton, W.C., 2004. Tensile properties of fresh human calcaneal (achilles) tendons. Clinical Anatomy 17, 30–35. https://doi.org/10.1002/ca.10126.

Manti, M., Cacucciolo, V., Cianchetti, M., 2016. Stiffening in soft robotics: a review of the state of the art. IEEE Robotics and Automation Magazine 23, 93–106. https://doi.org/10.1109/MRA.2016.2582718.

Meckel, W., Goyert, W., Wieder, W., 1987. Thermoplastic polyurethane Elastomers. Thermoplast. Elastomers a Compr. Rev.

Natale, C., Pirozzi, S., 2008. Minimally invasive torque sensor for tendon-driven robotic hands. In: 2008 IEEE/RSJ International Conference on Intelligent Robots and Systems, IROS, pp. 65–70.

Pratt, G., Williamson, M., 1995. Series elastic actuators. Proceedings 1995 IEEE/RSJ international conference on intelligent robots and systems. Human Robot Interaction and Cooperative Robots 1, 399–406. https://doi.org/10.1109/IROS.1995.525827.

Ragonesi, D., Agrawal, S., Sample, W., Rahman, T., 2011. Series elastic actuator control of a powered exoskeleton. In: Proceedings of the Annual International Conference of the IEEE Engineering in Medicine and Biology Society, EMBS, pp. 3515–3518.

Ren, L., Song, G., Conditt, M., Noble, P.C., Li, H., 2007. Fiber Bragg grating displacement sensor for movement measurement of tendons and ligaments. Applied Optics 46, 6867–6871. https://doi.org/10.1364/AO.46.006867.

Robinson, D., Pratt, J., Paluska, D., Pratt, G., 1999. Series elastic actuator development for a biomimetic walking robot. In: 1999 IEEE/ASME International Conference on Advanced Intelligent Mechatronics (Cat. No.99TH8399), pp. 561–568.

Rouse, E.J., Mooney, L.M., Martinez-Villalpando, E.C., Herr, H.M., 2013. Clutchable series-elastic actuator: design of a robotic knee prosthesis for minimum energy consumption. IEEE International Conference on Rehabilitation Robot 2013, 6650383. https://doi.org/10.1109/ICORR.2013.6650383.

Sanchez, M., Leal-Junior, A., Segatto, V.M., Marques, C., Dos Santos, W.M., Siqueira, A., Frizera, A., 2018. Fiber Bragg grating-based sensor for torque and angle measurement in a series elastic actuator's spring. Applied Optics 57, 7883. https://doi.org/10.1364/AO.57.007883.

dos Santos, W.M., Caurin, G.A.P., Siqueira, A.A.G., 2015. Design and control of an active knee orthosis driven by a rotary series elastic actuator. Control Engineering Practice, 1–12. https://doi.org/10.1016/j.conengprac.2015.09.008. From Duplicate 1 (Design and control of an active knee orthosis driven by a rotary Series Elastic Actuator – dos Santos, Wilian M; Caurin, Glauco A P; Siqueira, Adriano A G).

Sasiadek, J.Z., Hartana, P., 2000. Sensor data fusion using Kalman filter. In: Information fusion. FUSION 2000. Proceedings of the Third International Conference on, WED5/19–WED5/25, vol. 2, 2000.

Shukla, A., Karki, H., 2016. Application of robotics in offshore oil and gas industry-a review part ii. Robotics and Autonomous Systems 75, 508–524. https://doi.org/10.1016/j.robot.2015.09.013.

Wen, L., Li, Y., Cong, M., Lang, H., Du, Y., 2017. Design and optimization of a tendon-driven robotic hand. In: Proceedings of the IEEE International Conference on Industrial Technology, pp. 767–772.

Xiong, L., Jiang, G., Guo, Y., Liu, H., 2018. A three-dimensional fiber Bragg grating force sensor for robot. IEEE Sensors Journal 18, 3632–3639. https://doi.org/10.1109/JSEN.2018.2812820.

Zhang, Z., Wang, X., Liu, J., Dai, C., Sun, Y., 2019. Robotic micromanipulation: fundamentals and applications. Annual Review of Control, Robotics, and Autonomous Systems 2, 181–203. https://doi.org/10.1146/annurev-control-053018-023755.

Part IV

Case studies and additional applications

Chapter 10

Wearable multifunctional smart textiles

10.1 Optical fiber embedded-textiles for physiological parameters monitoring

Important applications of smart optical fiber sensors in healthcare are physiological parameters monitoring in which many sensors were proposed throughout the years for different parameters. For example, optical fiber sensors are developed for the blood pressure estimation, as it is an indicator of many cardiovascular diseases. For this reason, many configurations were presented throughout the years, which include fiber Bragg gratings (FBGs) embedded in a portable pen-like structure (Leitao et al., 2015) as well as low cost intensity variation-based approaches, where a reflective surface is positioned on the user's carotid artery's surface (Leitao et al., 2017). In this case, the intensity variation-based sensor uses the light coupling principle, since the light reflected from the reflective surface is acquired by a photodetector and the power variation is correlated with the carotid distension that also leads on variations of the reflective surface position. To enable the higher integration and transparency between sensor system and the user, optical fiber-embedded textiles have been proposed taking advantage of the compactness, high flexibility, electromagnetic field immunity, and multiplexing capabilities.

The integration of polymer optical fibers (POFs) in textiles, creating the so-called photonics textiles, was also investigated to create a sensing matrix for oximetry measurements (Rothmaier et al., 2008). To that extent, an intensity variation-based configuration is used, illuminating the textiles with two different wavelengths, namely near infrared (at around 850 nm) and visible (at 650 nm), where the sensor is based on the absorption differences of the oxyhemoglobin, deoxyhemoglobin at each wavelength, which enable to obtain the SpO2 (percentage of hemoglobin with bound oxygen measured in the pulse). This smart textile can also be used to estimate the heartbeat rate using the frequency analysis of the sensor responses. A detailed application of oximetry sensors using optical fibers is discussed in Chapter 11. As the majority of textiles applications are developed on the heart and breath rates monitoring as well as the body temperature assessment, this section focuses on the recent developments of optical fiber-embedded smart textiles for the assessment of these parameters.

https://doi.org/10.1016/B978-0-32-385952-3.00021-4

10.1.1 Breath and heart rates monitoring

The development of new healthcare technologies and their integration devices commonly used in the daily life resulted in health condition assessment outside clinical environments (Korhonen et al., 2003). Nowadays, it is possible to monitor different physiological and physical parameters of patients at their homes, which is especially desirable for the elderly population and people with chronic disabilities/diseases (Majumder et al., 2017). Among many important physiological parameters, abnormalities on the heart rate (HR) and breathing rate (BR) are important indicators of some cardiovascular diseases (Bohm et al., 2015), fatigue (Nishyama et al., 2011), and apnea and respiratory abnormalities (Strau et al., 2014). To address these important parameters, several sensors based in different approaches such as piezo-electric films, dry textile electrodes, flexible capacitive electrodes (among others) have been proposed throughout the years, which such technologies are summarized in the following review works (Majumder et al., 2017; Nag et al., 2017). In general, these sensors are sensitive to electromagnetic interferences, which inhibit their application in magnetic resonance imaging (MRI). In order to overcome this limitation, optical fiber sensors that have been proposed may be used and present the advantage of immunity to the electromagnetic field. In addition, optical fiber sensors also are compact, lightweight, present chemical stability, and multiplexing capabilities, as discussed in Chapter 4.

The optical fiber advantages motivated the development of different sensor systems for breath rate (BR) and heart rate (HR) monitoring, especially using fiber Bragg gratings (FBGs) due to its precision, resolution, and multiplexing capabilities. In general, these developments include a wearable sensor embedded in a textile for BR (De Jonckheere et al., 2009) and for HR (Witt et al., 2011), where there is also the possibility of distributing the FBGs along the textile to measure different points in the thoracic region. However, the cost and bulkiness of the optical interrogation equipment can be regarded as important drawbacks that limit the use of such sensors in the clinical environment, especially during magnetic resonance imaging (MRI) due to the electromagnetic field immunity of FBGs.

The necessity of specialized (and generally costly) equipment for FBGs inscription is a drawback of using FBGs in these applications, since the BR and HR measurements can be performed with a single sensor, which reduces the necessity multiplexing capabilities for the sensor system. To that extent, a Fabry–Perot interferometer (FPI) can be used its simplicity on fabrication and higher sensitivity when compared with FBGs. Although FPIs also need an optical spectrum analyzer or optical interrogator for its signal acquisition, the development of portable interrogators (such as the one presented in Chapter 7) enables the possibility of using such sensors outside the clinical environment and even for remote health monitoring. In this case, the FPI cavity was produced using an ultraviolet (UV) curable resin in between two single mode silica fibers (SMF) as discussed in Chapter 6. This technique leads to the production of FPIs in a simple and straightforward manner, where the sensitivity of the sensor can even be increased (or tuned) by adding additional layers of UV-curable resin (Oliveira et al., 2019). Therefore for the FPI production, a drop UV-curable resin Loctite AA 3936 (Henkel, Germany) was placed in the axial gap between two pieces of SMF-28 (Corning, USA) aligned using a 3D translation stage with micrometer resolution. Thereafter, a second alignment is performed, where axial displacement is performed to define the cavity length by means of monitoring the FSR of the FPI on the interrogator. As the last step on the FPI production, the UV curing lamp is focused on the FPI's cavity with the UV-curable resin for about 40 seconds to promote the curing. The FPI spectrum is shown in the inset of Fig. 10.1(a). The FPI is positioned on the chest band as depicted in Fig. 10.1(a), where the variations on the chest circumferences caused by the respiration cause a curvature change on the FPI, resulting in wavelength shift. It is worth noting that the BR and HR leads to periodic variations on the peak wavelengths of the FPI through chest circumferences variations (mostly related to the breathing) and periodic vibrations (related to the heartbeat). Although the breath-induced curvature variation has much higher amplitude than the ones associated with the heartbeat (Obeid et al., 2011), such variations can be decoupled if the sensor analysis is performed in the frequency domain since both signals have different frequencies (0.1 to 0.8 Hz for breath rate and 1.0 to 3.5 Hz for heart rate). Thus the analysis is performed by means of applying the fast Fourier transform (FFT) on the sensor response. Then the signal is filtered using a Butterworth filter in two different windows, 0.1–0.8 Hz and 1.0–3.5 Hz, which results in the breath and heart rates as depicted in Fig. 10.1(a) inset (Leal-Junior et al., 2019a).

For the breath and heart rate assessment, the FPI-embedded smart textile is positioned on the user's chest and the signal is acquired for about 40 seconds. In order to analyze the sensor at different conditions the FPI-embedded breath and heart rate sensor is analyzed in two conditions, one at which the user is under normal respiration condition, and, the other after an event that resulted in higher respiration and heartbeat rates such as intensive exercises. The heart rate is compared with the one obtained from a reference sensor based on photoplethysmography (PPG) (digiDoc Pulse Oximeter, Norway). Fig. 10.1(b) presents the BR and HR estimations using the FPI-embedded sensor, where the error was higher for the case in which there is a high respiration rate. However, the highest error obtained is below 10% on the worst case, after an exercise where additional movements of the subject can lead to errors on the sensor breath estimation. Thus the proposed sensor system is a feasible option for noninvasive health monitoring with the possibility of remote monitoring using a wireless module in the interrogation device. Similarly, the heart rate assessment (performed after filtering the sensor response on the 1.0 Hz to 3.5 Hz frequency window) also shows good accuracy with the maximum error below 5%.

In general, HR and BR sensors are used in the physiological monitoring in MRI environments and during the sleep (Chen et al., 2014), i.e., in cases in

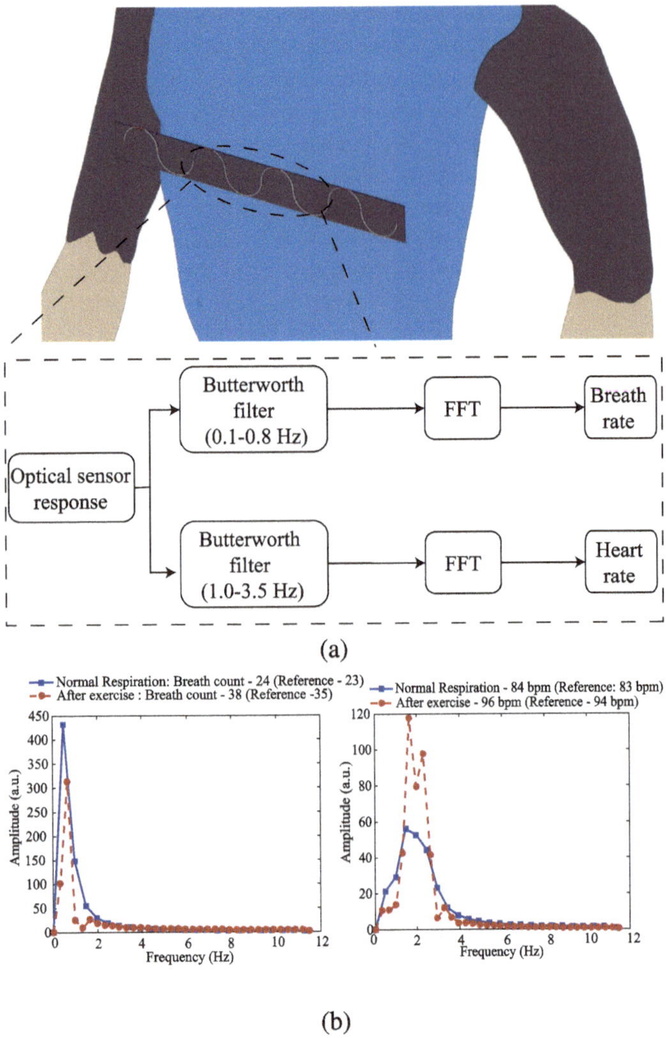

FIGURE 10.1 (a). FPI-embedded smart textile for simultaneous measurement of breath and heart rates. The insets show the block diagram of the technique for BR and HR extraction. (b) Frequency response of the FPI-based sensor for BR and HR assessment.

which there is no dynamic movements of the users. As the sensors are generally based on the correlation between torso movements during the respiration, if the user makes additional movements, the sensor will present errors in the estimation (Koyama et al., 2018), which also can be a source of errors in the tests presented in Fig. 10.1(b). In mitigating the influence of movements in the sensor response, Nishyama et al. (2011) proposed an array with 8 sensors, where each sensor response is compared to reduce the influence of the body movement in the BR assessment, leading to the necessity of a multiplexed sensor system. Another approach is to design a filter for noise reduction that removes some of the non-periodic movements and the signal is only processed when it is stable, i.e., without movements. Although these strategies were able to reduce the influence of the body movement, it was still reported measurement errors due to the body periodic movements.

The frequency-domain analysis can filter out the influence of the nonperiodic movement in the BR or HR estimation. However, the human movement can also have a periodic behavior, where the most common example of human periodic movement is the gait, where the gait cycle is defined as the interval of consecutive periodic events during the gait. The gait periodic behavior can also be quantified through the cadence, which indicates the number of steps per minute (or second) of a person. In normal gait, the cadence is about 100–115 steps per minute (frequency range of 1.6–1.9 Hz), whereas the frequency range for elderly people (or the ones with locomotor impairments) can be as low as 1.4 Hz (Ardestani et al., 2016). Therefore the gait cadence interval is similar to the one of the HR. The gait induces movements in the upper limbs and also periodic displacements in the torso region, which can be detected by the smart textile, leading to inaccuracies on the BR and, especially, on the HR estimations. Fig. 10.2 presents the frequency and amplitude for the periodic movements related to BR, HR, and gait-induced movements, where the amplitude of the chest displacements due to the heartbeat and gait are normalized with respect to the one induced by the breathing. The amplitude induced by the gait cycle is estimated through the spinal angle variation measurement during the gait reported in Schmid et al. (2016). Even though the cadence and HR share the same frequency region, their amplitudes present significant differences as the heartbeat does not lead to major variations in the chest circumference. Thus it is possible to isolate both responses that share the same frequency range, i.e., HR and the gait-induced frequency, by applying a threshold in the signal filtered in the range of 1.0 Hz to 3.5 Hz, which is the HR range. This approach is presented in Fig. 10.2, where the sensor response is filtered in two different windows, 0.1–0.8 Hz and 1.0–3.5 Hz. A FFT is applied in the result of the first window, which results in the BR, whereas the response obtained in the second window has its amplitude limited to 0.04 a.u. to reduce the influence of the cadence and gait-induced periodic movements. Then a FFT is also applied to this signal and the HR is obtained.

It is also worth noting that the BR and HR estimations can be achieved with simpler and lower cost systems than the wavelength-based ones (FBGs and FPIs) previously discussed. In this case, it is possible to employ low-cost optical fiber sensors, generally based on the intensity variation principle, where the optical power attenuation measured in the fiber is directly related to the chest circumference variation during the respiration. In this case, the intensity variation approach for BR and HR monitoring is a curvature sensor with 8 lateral sections, made by removing the fiber cladding and part of its core with a sandpaper with

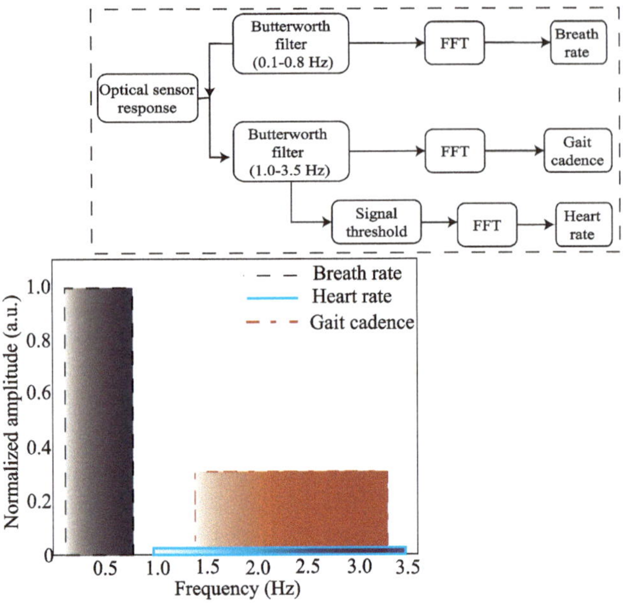

FIGURE 10.2 Frequency and amplitude distribution of the BR, HR, and gait-induced moments on the chest. The figure also shows the signal processing technique to obtain the BR and HR without the influence of periodic movements.

controlled grit size (such as the ones discussed in Chapter 6), that correlates the rate of the curvature differences with the breathing and heart rate. A polymethyl methacrylate (PMMA) POF (HFBR-EUS100Z, Broadcom Limited) with a core diameter of 980 μm, a cladding of fluorinated polymer with 20 μm thickness and a polyethylene coating is used on the sensor fabrication. After its fabrication, the sensor is embedded in an elastic band as shown in Fig. 10.3(a). This sensor works in the transmission mode with a light source (central wavelength at 660 nm and optical power of 3 mW) in one end of the fiber and a photodiode in the other end to perform optical power variation acquisition. The optical fiber-embedded elastic band is fixed on the user's chest in the positions shown in Fig. 10.3. The band stretches, leading to a curvature variation on the sensor, which is also able to detect the body vibrations induced by the heartbeat. The sensor is able of acquiring both BR and HR signals simultaneously, as the sensor response is a combination of the breathing (more evident due to its higher amplitude) and heartbeat waveforms (immersed in the breathing signal due to its much lower amplitude); both curves are shown in Fig. 10.3(b), where the response is amplified in a smaller region to show the heartbeat waveform.

The first set of tests is made with a male subject with the sensor positioned on the regions shown in Fig. 10.3(a) to show if the sensor is capable of operate in different positions without harming its performance, which enable the users to place the sensor in the chest regions that they find more comfortable, increasing the system's usability. It is also noteworthy that the sensor position can indicate the type of breathing, e.g., diaphragmatic, upper costal, and mixed (Krehel et al., 2014). The position 1 is the one closest to the heart, leading to higher influence of the HR in the sensor response and lower influence of the spinal movement during the gait. On the other hand, the opposite occurs in position 3, where there is a lower influence of the HR and higher influence of the gait. Position 2 seems to be the most comfortable one as empirically verified in the experiments.

The waveforms in Fig. 10.3 show the typical ballistocardiography waveform pattern for healthy individuals with the H, I, J, K, L, M, and N waves as thoroughly discussed in Kim et al. (2016), where such waves are divided in divided in preejection (F, G, H), ejection (I, J, K), and diastolic (L, M, N). These waves enable the assessment of the heart conditions in some specific cases. In addition, sensor is capable of measuring the HR and BR in different chest positions, as shown in Fig. 10.4(a), where the signal processing presented in Fig. 10.3 is applied and results in the frequencies responses as well as BR and HR estimations. The results are compared with the reference PPG sensor. It is worth to mention that this tests were made with the same subject (subject 1). Fig. 10.4 results show that the proposed POF-based sensor can operate in different regions of the chest, where the lowest error for the BR was obtained at both positions 1 and 3. The HR errors are higher than the ones obtained in BR due to the higher variability of the heartbeat and the lower signal amplitude when compared with the breathing cycles. However, the errors are lower than 4 bpm in all cases. The results with additional volunteers (four volunteers—two males and two females) are shown in Fig. 10.4(b). In this case, each volunteer was asked to position the sensor in the region that they find most comfortable. Then the HR and BR of each volunteer is recorded for 30 seconds, where the results estimated by the optical fiber sensors are compared with the ones of a photoplethysmography (PPG) sensor. Interestingly, by visual inspection, the males subjects placed the sensor closer to the heart (close to positions 1 and 2 shown in Fig. 10.3), whereas the females placed the sensor in positions 2 or 3, which imply the necessity of sensor operation at different positions. The results in Fig. 10.4(b) show a relative error below 4.5% for the HR measurements and lower than 3.5% for the BR measurements considering all volunteers.

Finally, to demonstrate the sensor feasibility of measuring HR and BR during dynamic movements, Fig. 10.4 shows the results obtained with the tests during gait. The volunteer, with the sensor positioned on his chest, is asked to walk in a straight line (in self-selected velocity) for 30 seconds, the test is performed in positions 2 and 3, which are the ones with higher influence of the gait-induced periodic movements. The steps of the subject were counted at each test to obtain his cadence (in steps per minute). The BR response is presented in Fig. 10.4 by filtering the sensor response in the first window, whereas the response filtered in the second window (1.0 Hz to 3.5 Hz) is presented in without the signal threshold to demonstrate the influence of the gait parameters in the HR measurement. In this case, the peak frequency is on the cadence re-

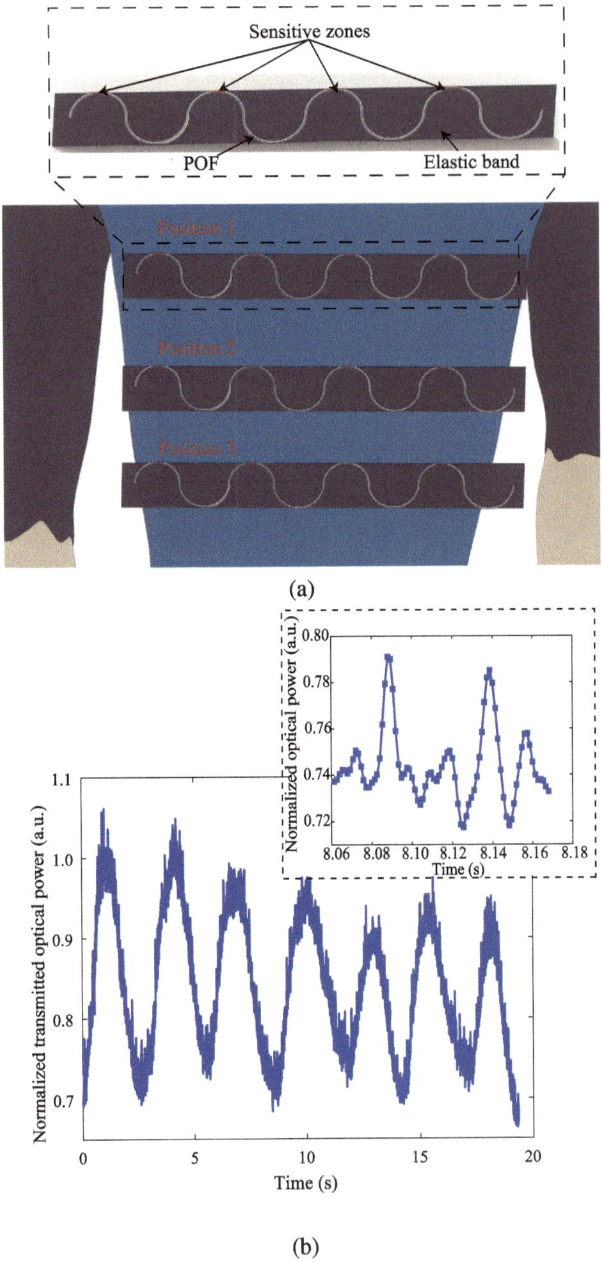

FIGURE 10.3 (a) Schematic representation of POF-based smart textile for HR and BR monitoring. The figure also shows the positions of the sensors on the user's chest region. (b) POF sensor response at position 1, where it can be seen (in the magnified view) a curve that resembles a typical ballistocardiography waveform pattern of the heartbeat, where H, I, J, K, L, M, and N are the typical waves in a ballistocardiography signal.

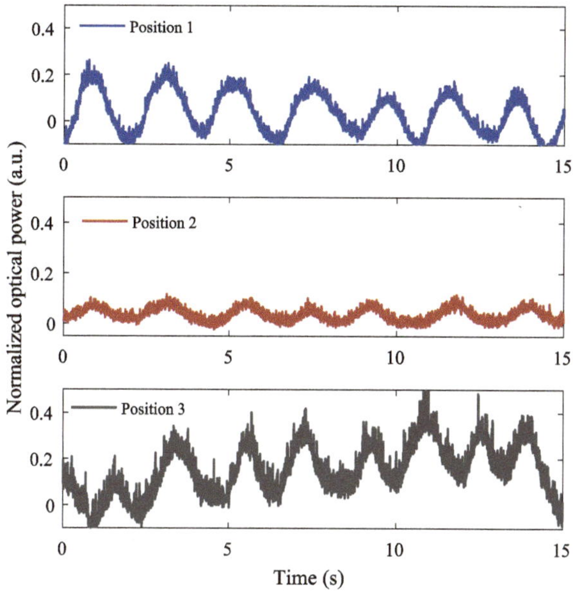

(a)

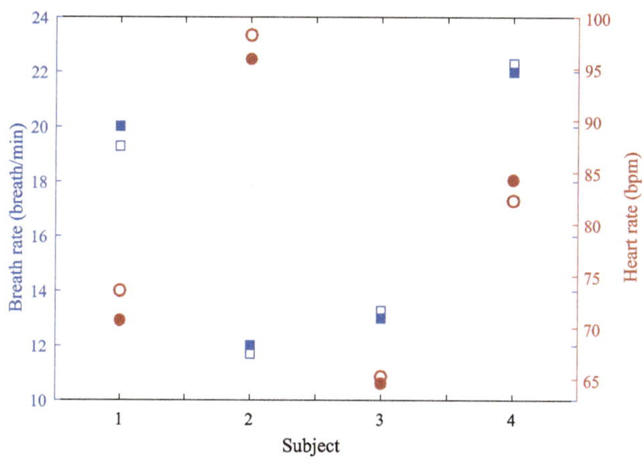

(b)

FIGURE 10.4 (a) Optical fiber-embedded smart textile response in different chest positions for subject 1. (b) BR and HR estimations of all four volunteers.

gion, which has higher amplitude then the heartbeat. Then, applying the signal threshold, it is possible to estimate the HR without influence of the gait cadence. The BR and HR responses presented errors of 1 cpm and 4 bpm, respectively.

These low errors indicate the feasibility of the proposed optical fiber-embedded smart textile for simultaneous measurement of BR and HR, where the sensor has the advantageous feature of performing at different positions in the torso and with insensitivity to user's periodic movements. Thus a simple signal processing technique enables obtaining a lightweight sensor that can be fully embedded in the user's clothes for monitoring physiological parameters even during different activities of daily living.

10.1.2 Body temperature assessment

As many optical fiber sensors are sensitive to temperature variations, it is possible to integrate them in textiles for temperature assessment. The multiplexing capabilities of FBGs make them suitable for textile integration for body temperature assessment. In this case, the temperature can be measured in multiple points distributed in the human body for an accurate assessment of temperature distribution (Li et al., 2012). In addition, it is possible to perform sensor fusion to obtain a more reliable temperature estimation and to investigate the heat transfer mechanisms associated with the environment and human body. As a drawback of this approach, the FBG system needs an interrogation equipment that can present high cost and low portability. Nevertheless, as discussed in Chapter 8 for wearable applications of FBG sensors. An overview of optical fiber-integrated systems for body temperature monitoring is presented in Fig. 10.5.

Another possibility of using quasi-distributed systems for body temperature assessment is the use of the multiplexing technique for intensity variation-based sensors discussed in Chapter 6, which enables the measurement of multiple parameters distributed in the textile. As shown in Fig. 10.5(a), a proof of concept for the multiplexed intensity variation-based sensor system is performed with 3 sensors distributed on the textile. Fig. 10.5(b) shows the optical power variation as a function of the temperature applied on each sensor in the range of 20 °C and 40 °C (within the thermal comfort region and the body temperature). The temperature responses of the sensors show a linear behavior in all analyzed cases, where sensor 1 shows the highest sensitivity, the differences in the sensitivities of all sensors can be related to anisotropy in the fiber material. From the linear regressions obtained in the sensors responses, it is possible to perform a temperature mapping in the textile, as shown in Fig. 10.5(b). In this case, different temperature variations are applied to each sensor, distributed along the optical fiber, which results in distributed and concentrated temperature variations in different regions of the textile, where the quasidistributed array can sense the temperature distribution in the textile. From the sensors responses presented in Fig. 10.5(b), it is possible to observe that the sensors also presented low cross-sensitivity between them, demonstrating the feasibility of the sensing approach. Such results indicate the possibility of using optical fiber sensors embedded in smart textiles for body temperature monitoring as well as in flexible approach for microclimate assessment in objects that are in direct contact with the human skin.

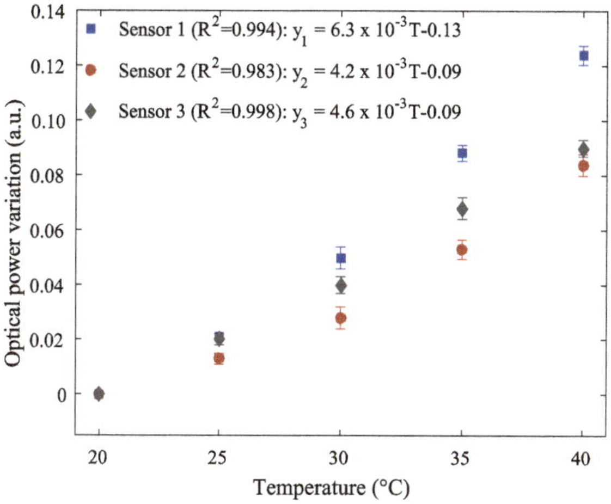

(a)

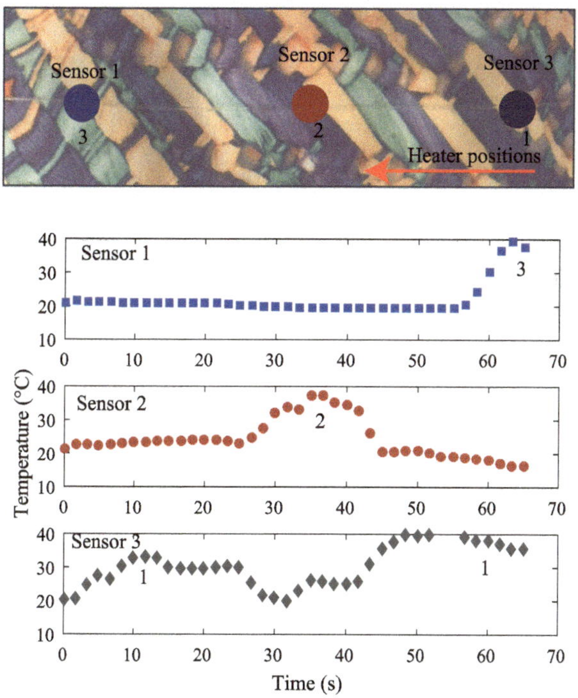

(b)

FIGURE 10.5 (a) Overview of the sensors distribution in a textile for temperature assessment. (b) Schematic representation of a 3-sensors array and temperature responses of each sensor in dynamic and static conditions.

10.2 Smart textile for multiparameter sensing and activities monitoring

The system comprises of special polymer optical fibers with Young's modulus of a few MPa, which close to cotton and other textiles, resulting in a system that does not inhibit or hinder the natural movements of the user. The optical fiber is fabricated using the light polarization spinning method with photopolymerizable resins, as presented in Chapter 4. In addition, a multiplexing technique for intensity variation sensors-based sensors (discussed in Chapter 6) was applied to allow multipoint and multiparameter sensing, which is a common limitation of conventional smart textiles. Such photonic textile also takes advantage of developments in flexible electronics, where light emitting diodes (LEDs) on flexible substrates are used as a light source to further increase the system's flexibility, usability, and transparency on the measurement of multiple parameters with the possibility of applying at different regions of the body.

The system consisting of polymeric optical fiber and flexible LEDs is sewn between two layers of Neoprene fabric. Fig. 10.6 presents the multiplexed photonic textile. The optical fiber connector and the LEDs are connected to a microcontroller and a circuit board with two photodetectors and the batteries. It is also worth to mention that the microcontroller provides the activation sequence of the LEDs as well as the signal acquisition (and storage) of the photodetectors (with acquisition frequency of 100 Hz), which is also transmitted to a gateway through a wireless connection. In addition, the system can be considered a light solution due to its total weight less than 400 g, including the microcontroller, photodetectors, LEDs, and batteries, which is an important parameter for wearable applications. In addition, the proposed smart textile has low energy consumption, since its average consumption is 150 mA for the whole system (i.e., photodetectors, microcontroller, and LEDs), this consumption allows an autonomy up to 10 hours using banks commercially available 10 000 mAh power.

Different transverse forces are applied to each sensor, which can be correlated to the pressure applied considering the area of each sensor. The force characterization is shown in Fig. 10.6(a), where it is possible to notice the high linearity of all sensors and the low standard deviation between the tests. In order to show the low cross sensitivity between the sensors (influence of the response of one sensor on the other), Fig. 10.6(b) shows the responses of the sensors as a function of time under a force of 100 N applied to one sensor at a time (from Sensor 1). It is possible to observe the low cross sensitivity between the sensors, since when the force is applied directly to one sensor, there is no significant variation of the optical signal in the others, showing the feasibility of the proposed LED modulation multiplexing technique. In this way, it is possible to obtain a force map of the interaction between the sensor and the user (or environment) when applying linear regression on the sensors' responses, which indicates the possibility of using the proposed intelligent textile with a sensor mesh to evaluate the pressure interaction between the user and the environment, such as chairs and beds. This assessment is important to provide remote monitoring of user activities and prevent pressure ulcers.

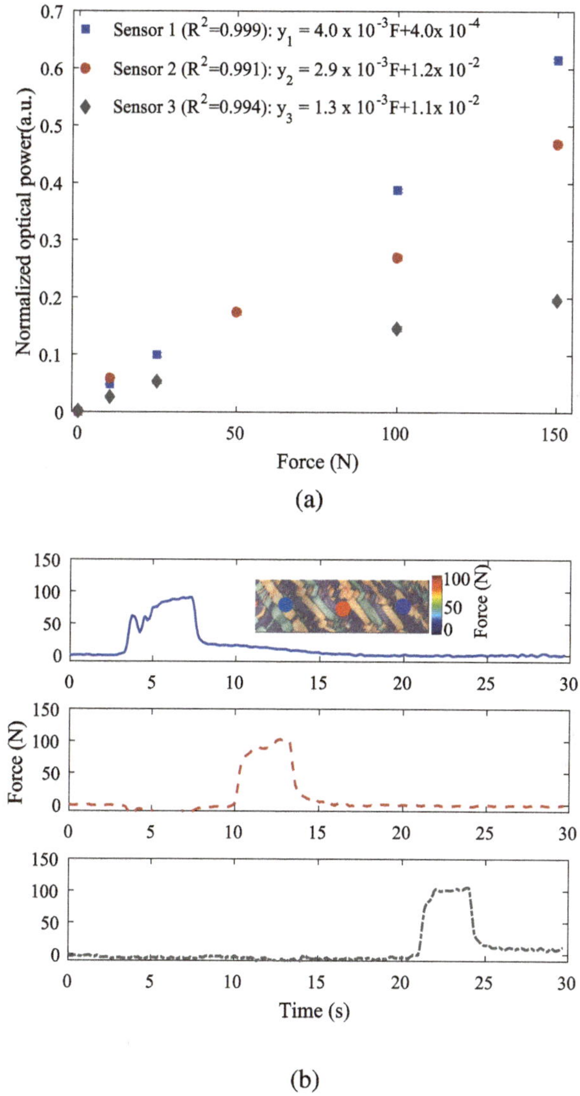

FIGURE 10.6 (a) Transmitted optical power as a function of the time under forces applied on each sensor. (b) Sensors characterization related to the applied force.

The responses of the sensors integrated in the smart textile in relation to the angular displacements applied in different planes are shown in Fig. 10.7. Flexion in different planes and torsions were applied with controlled angles. For flexion, the angles were 0° to 90° and 0° to −90°, while in the evaluation of torsion, angular displacements ranging from 0° to 180° were applied, as shown in Fig. 10.7. When comparing the responses of for each sensor, it is possible to observe differences in their behavior, which can be used to classify each movement in multiple planes. It is also noticeable that Sensors 1 and 2 presented the greatest sensitivity to flexion, in relation to the region where the flexion was applied, resulting in a greater tension in Sensors 1 and 2. In the case of torsion, the higher optical power variation was also obtained in Sensor 1, while again, Sensor 3 showed the smallest signal variation. However, the differences in the sensitivities of the sensors obtained in each case can be used to estimate the displacements applied to the textile, using techniques such as transfer matrix to evaluate three-dimensional displacements.

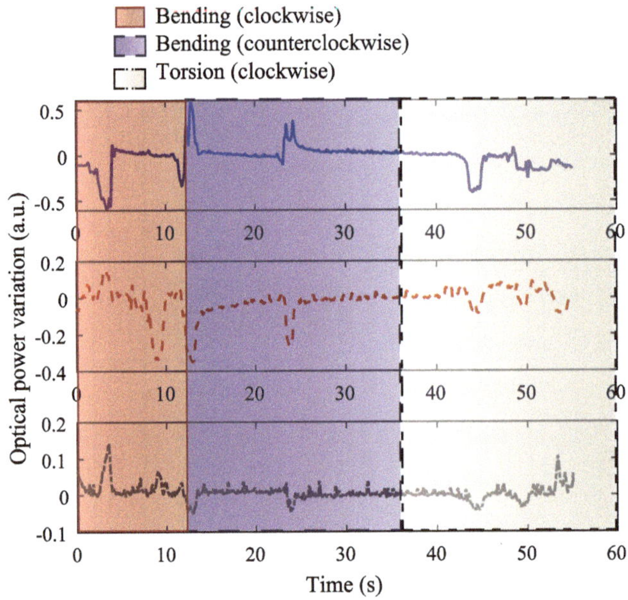

FIGURE 10.7 Sensors' responses as a function of the time for angular displacements in different planes.

The ability to identify the user's activities is assessed by positioning the textile in a healthy volunteer. The textile is positioned on the user's lower back. Its positioning on the lumbar region allows to acquire the signs of the activities characterized by high correlation with the displacement of the user's trunk, since the movements of the lower limbs usually result in trunk variations, the smart textile can be used in the identification of activities related to gait or movement of the lower limbs and trunk. Thus the user was asked to perform three common movements in daily life: walking, squatting, and sitting on a chair. The sensors' responses for each case are shown in Fig. 10.8, in which the sensors are analyzed in the time domain. The response in the time domain shows significant differences in the sensors' responses, mainly in the squat activity, where Sensor 2 presents the highest attenuation of the signal, which is related to its positioning in the center of the user's lumbar region, leading to greater displacements in the fiber when compared to Sensors 1 and 3. Another variation in the sensors' responses was found when the user sat in a chair, where the contact between the human back and the chair resulted in an attenuation in the transmitted optical power. Walking activity leads to the lowest optical power variation in the sensors. To improve the activity detection performance, the PCA technique was applied to the sensors' responses. The PCA represents the data in a new coordinate system using linear transformations and is a technique widely used for dimensionality reduction and as preprocessing for clustering techniques [33]. In this case, PCA was applied to the data in conjunction with k-means to show the possibility of detecting and identifying the activity using the proposed smart textile. Fig. 10.8 shows the scatter plot of the two main components of the sensors' responses obtained in each activity. These two main components represent 98% of all variability, mainly related to the time domain responses of Sensors 2 and 3, which indicates the possibility of reducing dimensionality using only the first two main components (instead of all components). In addition, Fig. 10.8 shows the results in three different groups, which are identified as three clusters, representing each performed activity. It is important to emphasize that a higher number of samples was obtained in the activities of walking and sitting due to their longer duration when compared to the squat activity, as shown in Fig. 10.8. Thus PCA in conjunction with k-means (or another grouping technique) results in a viable option for activities detection using the proposed smart textile. It is also worth mentioning that the proposed system has scalability, since a higher number of sensors can be used, which can lead to the detection of a higher number of activities, depending on the positioning of the system on the user.

The experimental results obtained with the proposed smart textile show the feasibility of this approach in a new system of multiparametric and transparent sensors, which can be used in multiple applications. It is possible to presume that the next generation of textiles will be incorporated with a sensor system to assess the health status and user's activities, where the proposed technology (optical fibers and multiplexing techniques based on LED modulation) may be a key technology for transparent sensor systems in practical applications. The smart textile can also be integrated with IoT modules for remote health monitoring. In addition, due to the scalability of the proposed technique, it is also possible to develop a complete clothing embedded with the proposed sensors for the assessment of multiple parameters, such as heart and breath rates, kinematic parameters of the human, interaction forces, and body temperature. All of these sensors transmit to a gateway for remote health monitoring, which can also be integrated with new classifiers and neural networks to extract all the user's physical and physiological information.

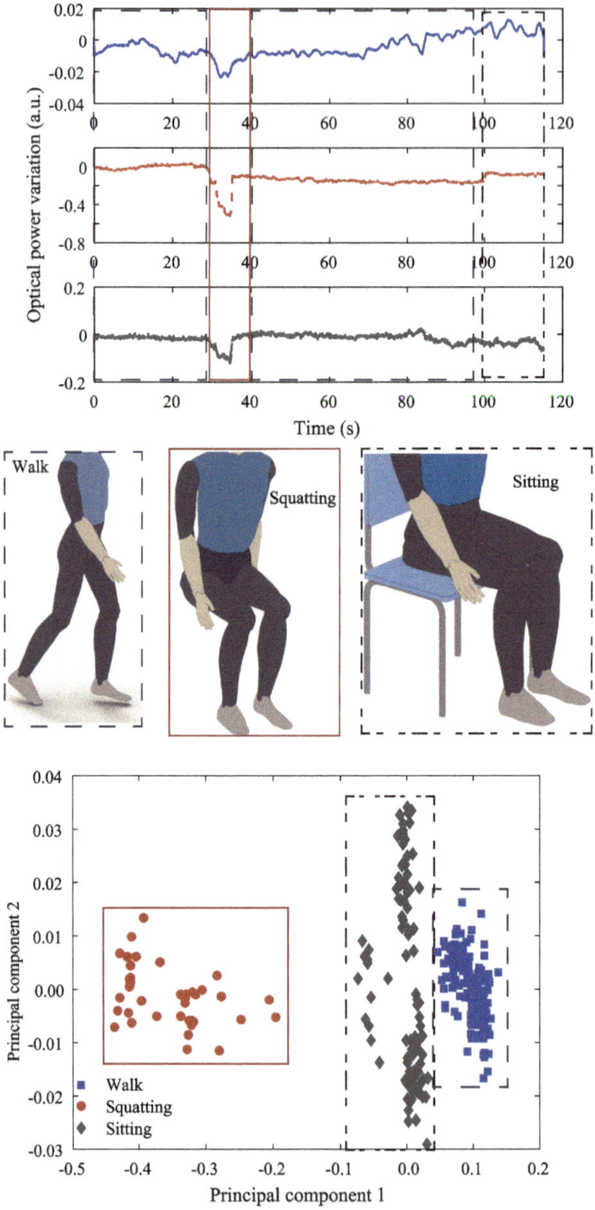

FIGURE 10.8 POF-embedded multiplexed smart textile for activity classification.

10.3 Optical fiber-embedded smart clothing for mechanical perturbation and physical interaction detection

Impact detection and pressure distribution play an important role in different applications, from automotive to structural health monitoring in civil structures as well as mechanical components (Thomas and Khatibi, 2017). In addition, the pressure distribution is an important tool and is commonly used in conjunction with the impact detection for the assessment of structural damage, stress concentration identification due to errors in project or material fatigue (Gandiolle and Fouvry, 2015). The pressure distribution is also important for biomechanical applications such as plantar pressure mapping (Leal-Junior et al., 2019b), human-robot (Leal-Junior et al., 2018), and human-environment (Avellar et al., 2019) interactions.

Investigation on human balance during walking has been an interesting research for many years, in which the assessment of center of mass and center of pressure under different standing and gait conditions are reported (Winter, 1995). In this context, the gait analysis under mechanical perturbations provides an important investigation on human balance control, including the monitoring of spatiotemporal parameters and dynamic stability in the gait (Madehkhaksar et al., 2018). The neuromuscular activity under forward and backward perturbations during gait was also measured, where the magnitude and location of the perturbation can lead to different muscular responses (Mueller et al., 2016). Thus the detection of location and amplitude of the perturbation provide additional information regarding the human natural strategies in the balance control.

The widespread of wearable robots for gait assistance indicates the development of bioinspired devices from the design approaches such as the human-in-the-loop design (Walsh, 2018) to the device control (Tropea et al., 2017), including bioinspired actuation. The investigation of the human balance control can lead to the development of novel robust control strategies for wearable robots with disturbance rejection (Long et al., 2017). In addition, experimental protocols for mechanical disturbances during exoskeleton-assisted gait provide quantitative and qualitative data regarding the exoskeleton control robustness, which also indicate the importance of a wearable impact detection system in biomechanical and health monitoring applications.

Fig. 10.9(a) shows the Smart Garment structure, which consists of two POFs made of polymethyl methacrylate, PMMA (HFBR-EUS100Z, Broadcom Limited) with a core diameter of 980 μm, a cladding of fluorinated polymer with 20 μm thickness and a polyethylene coating that results on a total diameter of 2.2 mm for the fiber considering its coating. The light source used is a light-emitting diode (LED) flexible lamp belt, and to enable the side-coupling of the LED, and at the same time, increase the sensor sensitivity, a lateral section is made on the fiber, where the cladding and part of the core are removed through abrasive removal of material as discussed in Chapter 4. This creates the sensitive zones that act as sensors in this work. In this case, 6 lateral sections were produced on each fiber, resulting in 12 pressure sensors.

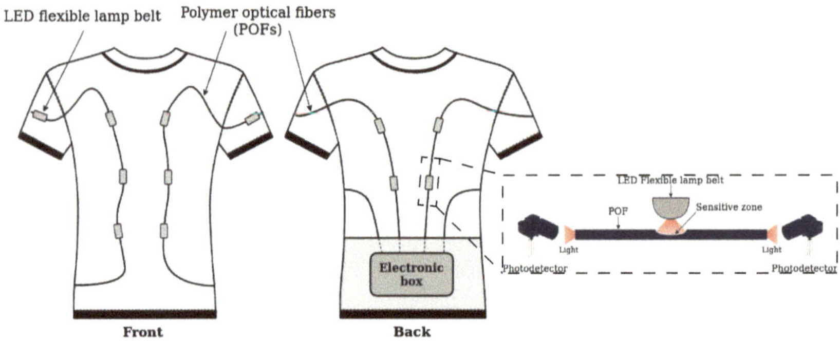

FIGURE 10.9 Smart garment overview.

The results of the proof-of-concept test are presented in Fig. 10.10. Eight perturbations were applied on different sensors while a volunteer was wearing the smart garment. Fig. 10.10(a) shows the perturbation area of each applied pressure and Fig. 10.10(b) shows the normalized sensor responses to pressures applied on each sensor with respective colors representing each perturbation over time. The sensors responses were normalized since in this preliminary assessment the optical power variation of each sensor was used to identify whether there is or not the perturbation, and not to measure the applied force.

The results show that the sensor system is multiplexed and the crosstalk between the sensors is negligible. This makes it possible to identify which sensor was pressed, and hence identify which region of human body the perturbation occurred. The number of sensors was defined to prove the concept, however, it can be increased according to requirement of spatial resolution. In addition, this setup can be employed in other textile applications, which involve impact detection, such as a textile to identify the interaction between exoskeleton and a lower limb or smart gloves.

This section presented a proof-of-concept of intensity variation-based POF pressure sensors embedded in a smart garment for impact detection in perturbation assessment. The proposed smart garment is comprised of two POFs with twelve sensors, which were fabricated by lateral sections made in the fibers and the light coupling of LED flexible lamp belt laterally arranged. A simple pressure test was performed with application of a perturbation on eight different sensors and results showed that the system is capable to detect impact region and the crosstalk between sensors is negligible. The system shows a good alternative to the current electrical textile sensors since this system present low cost and fabrication simplicity, besides allowing the increase of the number of sensors in the same fiber. Future works involve the system application in the real time impact detection, the study of the response latency, and the use of the proposed smart garment in perturbation protocol with patients.

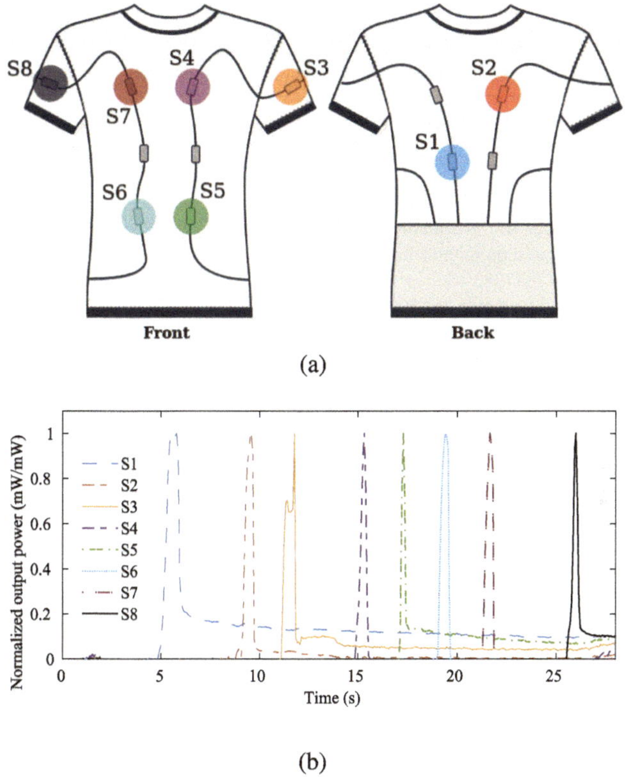

FIGURE 10.10 Results of the proof-of-concept test. (a) Impact area of smart garment. (b) Sensor responses to pressures applied on the smart garment.

References

Ardestani, M.M., Ferrigno, C., Moazen, M., Wimmer, M.A., 2016. From normal to fast walking: impact of cadence and stride length on lower extremity joint moments. Gait and Posture 46, 118–125. https://doi.org/10.1016/j.gaitpost.2016.02.005.

Avellar, L.M., Leal-Junior, A.G., Diaz, C.A.R., Marques, C., Frizera, A., 2019. Pof smart carpet: a multiplexed polymer optical fiber-embedded smart carpet for gait analysis. Sensors 19, 3356. https://doi.org/10.3390/s19153356.

Bohm, M., Reil, J.C., Deedwania, P., Kim, J.B., Borer, J.S., 2015. Resting heart rate: risk indicator and emerging risk factor in cardiovascular disease. American Journal of Medicine 128, 219–228. https://doi.org/10.1016/j.amjmed.2014.09.016.

Chen, Z., Lau, D., Teo, J.T., Ng, S.H., Yang, X., Kei, P.L., 2014. Simultaneous measurement of breathing rate and heart rate using a microbend multimode fiber optic sensor. Journal of Biomedical Optics 19, 057001. https://doi.org/10.1117/1.JBO.19.5.057001.

De Jonckheere, J., et al., 2009. OFSETH: Smart medical textile for continuous monitoring of respiratory motions under magnetic resonance imaging. In: Proc. 31st Annu. Int. Conf. IEEE Eng. Med. Biol. Soc. Eng. Futur. Biomed. EMBC 2009, pp. 1473–1476. https://doi.org/10.1109/IEMBS.2009.5332432.

Gandiolle, C., Fouvry, S., 2015. Fem modeling of crack nucleation and crack propagation fretting fatigue maps: plasticity effect. Wear 330–331, 136–144. https://doi.org/10.1016/j.wear.2015.01.

037.
Kim, C.S., Ober, S.L., McMurtry, M.S., Finegan, B.A., Inan, O.T., Mukkamala, R., Hahn, J.O., 2016. Ballistocardiogram: mechanism and potential for unobtrusive cardiovascular health monitoring. Scientific Reports 6, 1–6. https://doi.org/10.1038/srep31297.
Korhonen, I., Pärkkä, J., Van Gils, M., 2003. Health monitoring in the home of the future. IEEE Engineering in Medicine and Biology Magazine 22, 66–73. https://doi.org/10.1109/MEMB. 2003.1213628. From Duplicate 1 (Health monitoring in the home of the future – Korhonen Ilkka, Pärkkä Juha, Van Gils Mark).
Koyama, Y., Nishiyama, M., Watanabe, K., 2018. Smart textile using hetero-core optical fiber for heartbeat and respiration monitoring. IEEE Sensors Journal 18, 6175–6180. https://doi.org/10. 1109/JSEN.2018.2847333.
Krehel, M., Schmid, M., Rossi, R.M., Boesel, L.F., Bona, G.L., Scherer, L.J., 2014. An optical fibre-based sensor for respiratory monitoring. Sensors (Switzerland) 14, 13088–13101. https:// doi.org/10.3390/s140713088. From Duplicate 1 (An optical fibre-based sensor for respiratory monitoring – Krehel Marek, Schmid Michel, Rossi René M., Boesel Luciano F., Bona Gian Luca, Scherer Lukas J.).
Leal-Junior, A.G.A., Frizera, A., Marques, C., Sánchez, M.R.M., Botelho, T.R.T., Segatto, V.M.M., Pontes, M.M.J., 2018. Polymer optical fiber strain gauge for human-robot interaction forces assessment on an active knee orthosis. Optical Fiber Technology 41, 205–211. https://doi.org/ 10.1016/j.yofte.2018.02.001.
Leal-Junior, A.G., Díaz, C.R., Leitão, C., Pontes, M.J., Marques, C., Frizera, A., 2019a. Polymer optical fiber-based sensor for simultaneous measurement of breath and heart rate under dynamic movements. Optics and Laser Technology 109, 429–436. https://doi.org/10.1016/j.optlastec. 2018.08.036.
Leal-Junior, A.G., Díaz, C.R., Marques, C., Pontes, M.J., Frizera, A., 2019b. 3d-printed pof insole: development and applications of a low-cost, highly customizable device for plantar pressure and ground reaction forces monitoring. Optics and Laser Technology 116, 256–264. https:// doi.org/10.1016/j.optlastec.2019.03.035.
Leitao, C., Antunes, P., Andre, P., Pinto, J.L., Bastos, J.M., 2015. Central arterial pulse waveform acquisition with a portable pen-like optical fiber sensor. Blood Pressure Monitoring 20, 43–46. https://doi.org/10.1097/MBP.0000000000000079.
Leitao, C., Antunes, P., Pinto, J.L., Bastos, J.M., Andre, P., 2017. Carotid distension waves acquired with a fiber sensor as an alternative to tonometry for central arterial systolic pressure assessment in young subjects. Measurement: Journal of the International Measurement Confederation 95, 45–49. https://doi.org/10.1016/j.measurement.2016.09.035.
Li, H., Yang, H., Li, E., Liu, Z., Wei, K., 2012. Wearable sensors in intelligent clothing for measuring human body temperature based on optical fiber Bragg grating. Optics Express 20, 11740. https:// doi.org/10.1364/OE.20.011740.
Long, Y., Du, Z., Cong, L., Wang, W., Zhang, Z., Dong, W., 2017. Active disturbance rejection control based human gait tracking for lower extremity rehabilitation exoskeleton. ISA Transactions, 1–9. https://doi.org/10.1016/j.isatra.2017.01.006.
Madehkhaksar, F., Klenk, J., Sczuka, K., Gordt, K., Melzer, I., Schwenk, M., 2018. The effects of unexpected mechanical perturbations during treadmill walking on spatiotemporal gait parameters, and the dynamic stability measures by which to quantify postural response. PLOS ONE 13, e0195902. https://doi.org/10.1371/journal.pone.0195902.
Majumder, S., Mondal, T., Deen, M., 2017. Wearable sensors for remote health monitoring. Sensors 17, 130. https://doi.org/10.3390/s17010130. From Duplicate 2 (Wearable sensors for remote health monitoring – Majumder Sumit, Mondal Tapas, Deen M).
Mueller, J., Engel, T., Mueller, S., Kopinski, S., Baur, H., Mayer, F., 2016. Neuromuscular response of the trunk to sudden gait disturbances: forward vs. backward perturbation. Journal of Electromyography and Kinesiology 30, 168–176. https://doi.org/10.1016/j.jelekin.2016.07.005.
Nag, A., Mukhopadhyay, S.C., Kosel, J., 2017. Wearable flexible sensors: a review. IEEE Sensors Journal 17, 3949–3960. https://doi.org/10.1109/JSEN.2017.2705700. From Duplicate 1 (Wearable flexible fensors: a review – Nag Anindya, Mukhopadhyay Subhas Chandra, Kosel Jurgen).

Nishyama, M., Miyamoto, M., Watanabe, K., 2011. Respiration and body movement analysis during sleep in bed using hetero-core fiber optic pressure sensors without constraint to human activity. Journal of Biomedical Optics 16, 017002. https://doi.org/10.1117/1.3528008.

Obeid, D., Sadek, S., Zaharia, G., Zein, E.G., 2011. Doppler radar for heartbeat rate and heart rate variability extraction. E-Health and Bioengineering, 24–27.

Oliveira, R.F., Bilro, L., Nogueira, R., Rocha, A.M.M., 2019. Adhesive based Fabry-Pérot hydrostatic pressure sensor with improved and controlled sensitivity. Journal of Lightwave Technology 8724, 1–14. https://doi.org/10.1109/jlt.2019.2894949.

Rothmaier, M., Selm, B., Spichtig, S., Haensse, D., Wolf, M., 2008. Photonic textiles for pulse oximetry. Optics Express 16, 12973. https://doi.org/10.1364/OE.16.012973.

Schmid, S., Studer, D., Hasler, C.C., Romkes, J., Taylor, W.R., Lorenzetti, S., Brunner, R., 2016. Quantifying spinal gait kinematics using an enhanced optical motion capture approach in adolescent idiopathic scoliosis. Gait and Posture 44, 231–237. https://doi.org/10.1016/j.gaitpost.2015.12.036.

Strau, R., Ewig, S., Richter, K., König, T., Heller, G., Bauer, T.T., 2014. The prognostic significance of respiratory rate in patients with pneumonia. Deutsches Aerzteblatt Online 111, 503–508. https://doi.org/10.3238/arztebl.2014.0503.

Thomas, G.R., Khatibi, A.A., 2017. Durability of structural health monitoring systems under impact loading. Procedia Engineering 188, 340–347. https://doi.org/10.1016/j.proeng.2017.04.493.

Tropea, P., Aprigliano, F., Martelli, D., Parri, A., Cortese, M., 2017. An ecologically-controlled exoskeleton can improve balance recovery after slippage. Scientific Reports, 1–10. https://doi.org/10.1038/srep46721.

Walsh, C., 2018. Human-in-the-loop development of soft wearable robots. https://doi.org/10.1038/s41578-018-0011-1. From Duplicate 2 (Human-in-the-loop development of soft wearable robots – Walsh Conor).

Winter, D.A., 1995. Human balance and posture standing and walking control during.

Witt, J., Krebber, K., Demuth, J., Sasek, L., 2011. Fiber optic heart rate sensor for integration into personal protective equipment. In: 2011 International Workshop on Biophotonics, pp. 1–3.

Chapter 11

Smart walker's instrumentation and development with compliant optical fiber sensors

11.1 Smart walkers' technology overview

Walkers are assistive devices that present structural simplicity, low cost, and great potential for maintaining activity and mobility for impaired individuals with residual motor capabilities. Such devices are usually prescribed for patients in need of gait assistance, to increase (static and dynamic) stability and to provide partial body weight support during locomotion (Bateni and Maki, 2005). Although walker-assisted gait offers important benefits, including increased confidence and safety perception during ambulation, there are drawbacks related to the use of the different types (standard frames, two-wheeled, and rollators) of such devices in functional compensation scenarios: standard frames are associated with high energy expenditure (Priebe and Kram, 2011) and increased force levels exerted by the upper limbs (Haubert et al., 2006), while wheeled-based devices present stability issues during ambulation (Van Hook et al., 2003).

In this context, smart walkers (SWs) emerge as a new category of walkers which integrate robotics and sensors technology for compensating the drawbacks of conventional devices while maintaining the benefits of independent ambulation. A great number of additional assistive functionalities are also enabled by SWs: stability and motion support, smart navigation and locomotion, and sensory compensation are just some examples of functions for better adjusting the robotic device to the individual needs of the user (Martins et al., 2015).

Smart walkers (SWs) are usually built upon the structure of conventional three/four-wheeled rollators in which actuators (electric motors) and electronic components are used to provide different levels of assistance (Cifuentes and Frizera, 2016). Autonomous, shared, or manual control strategies are conceived depending on the users' necessities (Martins et al., 2015). For implementing such control strategies, it is important to implement a set of sensor systems that must be embedded or internal to the robotic device, considering that such assistive technology must be designed to operate in nonstructured environments.

https://doi.org/10.1016/B978-0-32-385952-3.00022-6

Typically, SWs control strategies are based on human-robot upper-limb force interactions and/or kinematic information related to the user's gait. Odometry sensors are also used for the actuators' inner control and other technologies such as laser range finders (LRF) (Martins et al., 2015) or different camera-based technologies are used for improving environment interaction and for simultaneous localization and mapping strategies. There is also the possibility of monitoring health or activity-related parameters to improve the assessment of the general condition of the user (Leal et al., 2018).

Considering the limitations of the electronic sensors and aiming at proposing novel instrumentation approaches for robotic walkers, Section 11.2 discusses the design and experimentation of POF-based smart devices for the assessment of physiological parameters of individuals in walker-assisted gait (Leal-Junior et al., 2019). The sensor system consists of two devices. The first one is a smart textile, similar to the one presented in Chapter 10, that is capable of detecting breathing and heart rates as well as the cadence of the user, which is important not only for monitoring the user's activity level but can also be used on control strategies (Cifuentes et al., 2014). The second device consists of a POF-based sensor to be installed in the walker's handles that provides heart rate and oximetry assessment, important information for remote monitoring of patients and for e-health applications.

Additional to the instrumentation advantages of POF-based already mentioned throughout this book, and the optical features of CYTOP-based optical fiber sensors, Section 11.3 presents the development, characterization, and application of a POFBGs array in CYTOP fibers on the instrumentation of a SW. Four different functionalities are addressed. First, the structural health monitoring of the assistive device is presented. Following that, the detection of the user's movement intention and the estimation of the user's gait cadence are addressed. Finally, the possibility of aid on the mapping of structured environment using the frequency response of the FBGs array is discussed. In all of these applications, optical fiber sensors compactness and multiplexing capabilities offer important advantages over the conventional commercial solutions.

In both sections, the UFES Smart Walker (Cifuentes and Frizera, 2016), presented in Fig. 11.1, will be used for the installation of fiber optic sensors and for validating the proposed wearable sensor devices. The SW consists of a custom-built mechanical frame in which different sensors and actuators are installed. Two 3D force sensors (Futek MTA400, USA) are used for upper-limb human-robot interaction and two DC motors control the rotation of the rear traction wheels. Odometry is achieved by means of fusing data from shaft encoders (H1 US Digital, USA) and an IMU (BNO055 Bosch, USA) placed in the SW to measure its orientation and velocity. A LRF (Hokuyo URG-04LX, Japan) is employed to detect the users' legs and to estimate gait cadence during the assisted locomotion.

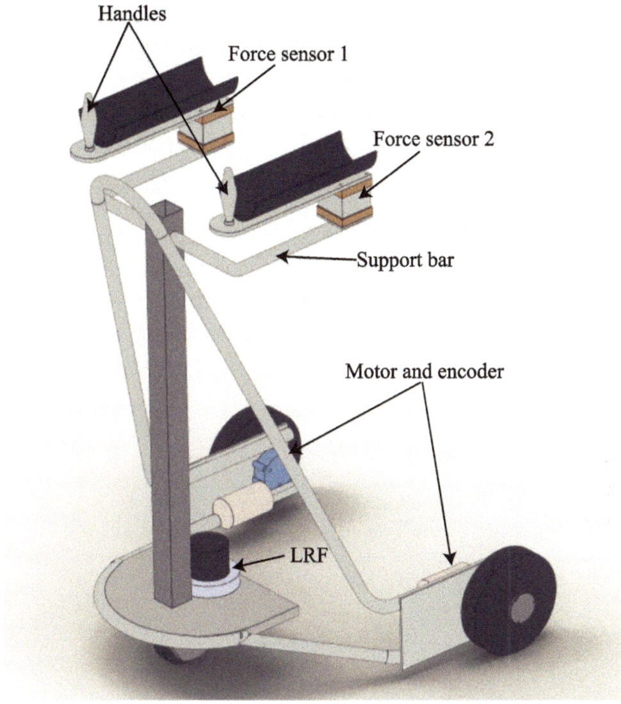

FIGURE 11.1 UFES smart walker.

11.2 Smart walker embedded sensors for physiological parameters assessment

11.2.1 System description

As previously presented in Chapter 10, a POF-based smart textile positioned in the user's chest was used for the assessment of gait cadence and breathing and heart rates. To avoid repetition of similar subjects, the development of such device will not be discussed in detail. The development and validation of the pulse oximetry and heart rate sensors embedded in the SW's handles will be presented below.

Here, it is important to note that oxygen saturation measurement is an important information regarding the user's health and physical conditions. The percentage of hemoglobin with bound oxygen measured in the pulse, known as SpO2, is a key physiological parameter that may be used to assess the evolution of diseases or to monitor intensive physical activities (Berry and Seitz, 2012). Pulse oximetry is based on the blood light absorption properties (Lee et al., 2016): oxygenated hemoglobin (HbO2) has higher absorbance in the near infrared wavelength region than deoxyhemoglobin (Hb), whereas in the red light (about 660 nm) the absorption of the Hb is higher when compared with HbO2

(Cohen et al., 2016). As SpO2 measurements are usually performed on the ear lobe, finger, or toe (Cohen et al., 2016), the proposed SpO2 sensor is embedded in the SW's handles.

Both smart textile and handle sensing devices were used in the configuration described in Fig. 11.2. Considering that gait cadence and heart rate frequencies are within the same range (as it will be shown below), the sensor configuration relies on the premise that the user will need the SW for locomotion, and if the SW is not active, the POF-based smart textile will acquire heart instead of gait cadence. If the SW is being used for locomotion, both smart textile and oximetry sensors will be active and heart rate is obtained with the sensors installed on the walker handles. The smart textile will be used for acquiring gait cadence (and breathing rate).

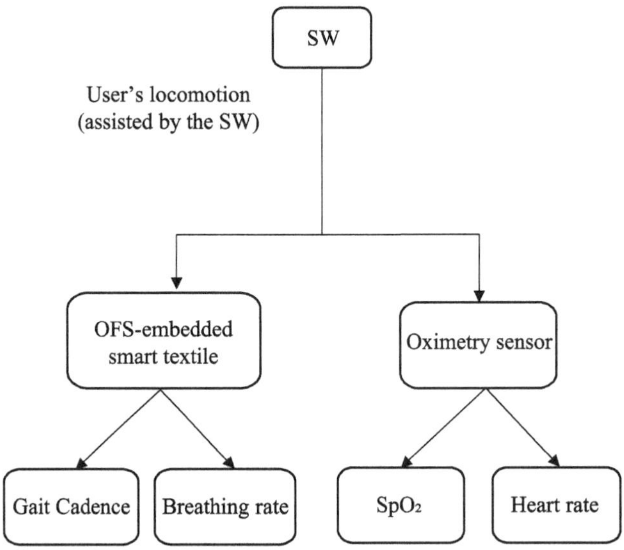

FIGURE 11.2 Block diagram of the POF sensors implementation in the smart walker.

Both sensor devices are shown in Fig. 11.3(a). These sensors combined with those already installed in the robotic device (see Fig. 11.1) offer interesting solutions for e-health (Majumder et al., 2017) and cloud robotics (Kehoe et al., 2015) that could greatly improve the quality of life of walker user's.

Sensor information is sent to the cloud by means of a secure wireless connection to be shared directly with health professionals and hospital database. Control commands for the SW could also be received as presented in different cloud robotics solutions summarized in Kehoe et al. (2015). In this architecture, it is also possible to establish the SW as a gateway for the transmission of the heart and breathing rates signals for remote healthcare when the individual is not directly using the SW, offering a more complete assessment of such parameters throughout the day.

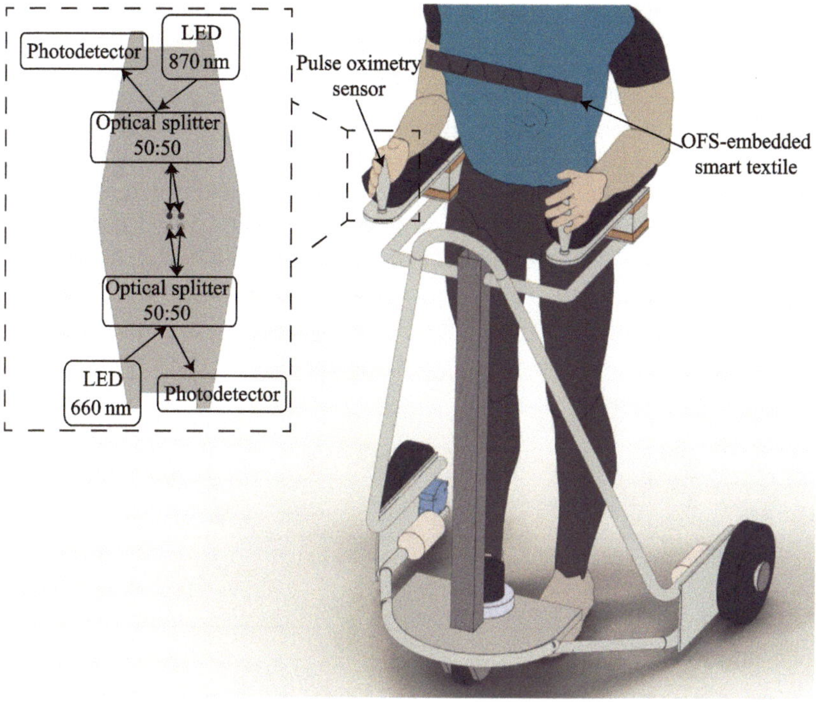

FIGURE 11.3 Schematic representation of the POF-based smart textile and the pulse oximetry sensor.

The sensor installed on the walker's handles is fabricated through additive layer manufacturing (ALM) with embedded POFs (Fig. 11.3). Considering oximetry is based on the blood light absorption properties at different wavelength regions, two LEDs were employed: one with central wavelength at 660 nm (IF-E97, Industrial Fiber Optics, USA) and another (IF-E91D Industrial Fiber Optics, USA) at 870 nm. Optical signals of the light sources are transmitted through two POFs for increasing the measurement region. As the SpO2 is estimated through the light reflected by the individuals's finger (Cohen et al., 2016), two 2x2 light couplers with 50:50 coupling ratio (IF 540 Industrial Fiber Optics, USA) are employed with the configuration also shown in Fig. 11.3.

Light attenuation at each wavelength due to scattering losses are estimated by means of a modified Beer–Lambert law (Krehel et al., 2014), as described in (Delpy et al., 1988). A bandpass filter (0.6 Hz and 3.5 Hz) is used to eliminate movement artifacts and reduce their influence in the SpO2 estimation. It is also interesting to note that a frequency analysis of such signals may also allow the estimation of heart rate (Krehel et al., 2014).

Both sensor devices use PolyMethyl Methacrylate (PMMA) POF with a core diameter of 980 μm and cladding thickness of 10 μm. Photodetectors used in these sensors are phototransistors IF D92 (Industrial Fiber Optics, USA).

11.2.2 Preliminary validation

Four male volunteers (30.5 ± 5.3) participated on the handle sensor (oximetry and heart rate) characterization and validation. Results are compared with a commercial reference photoplethysmography (PPG) sensor (digiDoc Pulse Oximeter, Norway). All volunteers are asked to hold the 3D-printed handle for 30 seconds to 1 minute, where good repeatability was found between measurements with deviations lower than 2%, as shown on Fig. 11.4. Results obtained at 660 nm and 870 nm are normalized and by means of a modified Beer–Lambert law (Krehel et al., 2014) applied with a bandpass filter between 0.6 Hz and 3.5 Hz. The obtained results are shown in Fig. 11.4.

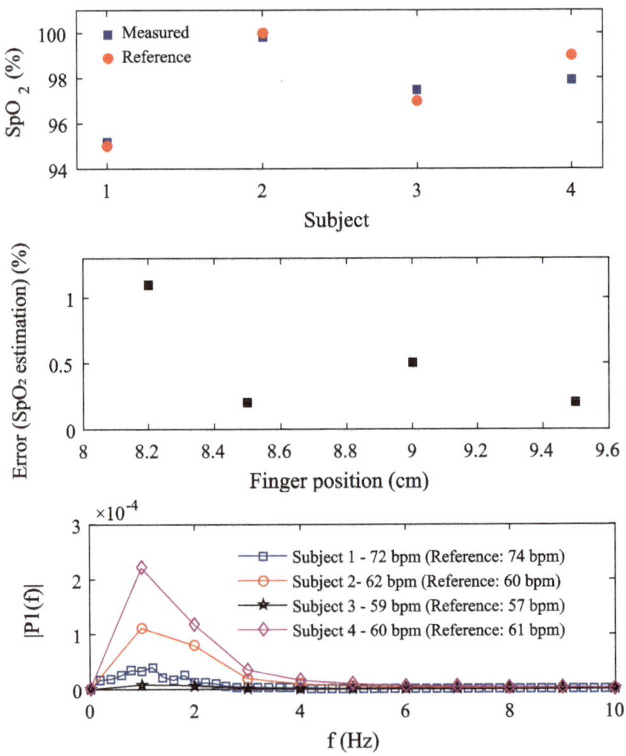

FIGURE 11.4 Oximetry sensor response. SpO2 measurement for each subject, SpO2 measurement error as a function of the finger position and heart rate assessment for each subject in frequency domain.

Since finger placement with respect to the sensing area may lead to deviations in sensor response, finger position of each subject as a function of the measurement errors in shown in Fig. 11.4. Please note that finger position is the measurement from the base of the finger (metacarpal region) to the region of the finger that is placed on the sensing area.

Finally, as previously mentioned, the proposed sensor can also be used for tracking heart rate. Fig. 11.4 shows the sensor response in the frequency domain, where the peak of each response is the heart rate. Since the heart rate is commonly within the range of 0.6 to about 3.5 Hz, a second-order band-pass Butterworth filter is applied. Peak frequencies are then multiplied by 60 for comparison with the reference system that offer such parameter in beats per minute (bpm).

SpO2 measurements using the proposed sensor and reference sensor offered very similar results. Root mean squared errors (RMSE) between the proposed and reference sensor is 0.15%. Nevertheless, the proposed sensor presents a resolution of 0.1% while the reference sensor is 1%. Thus the RMSE is also influenced by the POF system higher resolution. The highest difference between both sensors was found with subject 4, which may be related to the user's specific skin condition or hand positioning in the SW's handles—results indicate that the errors are lower in situations that sensor is positioned closer to the fingertip. Heart rate assessment showed similar positive results when comparing the developed sensor with the reference PPG sensor (digiDoc Pulse Oximeter, Norway): the RMSE between both sensors is 0.25 bpm. RMSE lower than 1 bpm indicate a good comparison between both sensors.

The smart textile was characterized in rest position and during gait, allowing the analysis in both operation conditions. As the breathing frequency interval is within a different range of heart rate and gait cadence, a straightforward way to separate those parameters is to filter the signal in different frequency windows.

A second-order Butterworth filter is applied in the frequency window of 0.1 Hz to 0.6 Hz for the estimation of the breathing rate while another similar filter in the window of 0.8 Hz to 3.5 Hz is used for gait cadence estimation (when the person is walking) or heart rate (when the person is at rest) assessments. The proposed sensor and filtering method for the extraction of the parameters are validated with the same four volunteers that participated in the validation of the sensors installed in the walker's handles.

Fig. 11.5 shows the results for simultaneous assessment of breathing and heart rates using a PPG sensor as reference. POF-based smart textile for heart and breathing rates presented small errors when compared with the reference measuring device. RMSE for heart rate is 1 bpm and for breath count is 0.25 breath counts, showing the applicability of the proposed device for such applications.

A similar analysis was proposed for measuring gait cadence and breathing rate. Reference signals for gait cadence was directly obtained by counting of the user's steps during the test, which lasted 1 minute per subject. The frequency analysis on the sensor response results in an estimated gait cadence of 90 steps per minute and breath rate of 24 breaths per minute, which are equal to the ones obtained with reference systems.

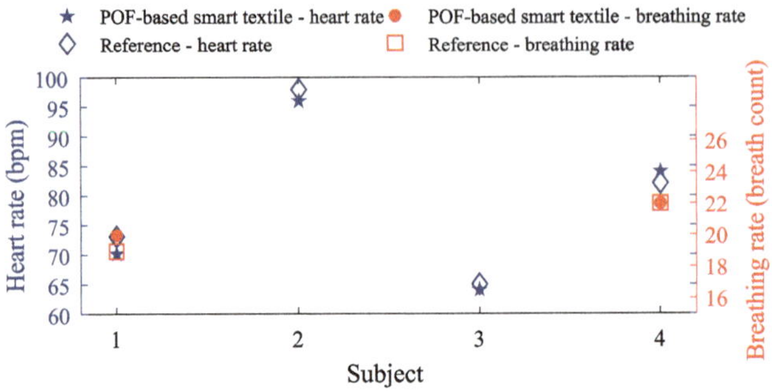

FIGURE 11.5 Heart and breathing rate measurements with the POF-based smart textile.

11.2.3 Experimental validation

Three young healthy volunteers (ages of 30.6 ± 3.1) participated on the validation of this setup. Subjects were asked to walk with the SW in a straight line of about 10 m. Such task was repeated five times for achieving a greater amount of data. Physiological parameters of the users were compared with the reference PPG sensor, whereas the gait cadence is compared with the data obtained from the LRF sensor, which acquires leg movements directly. The gait cadence is also obtained by the SW linear velocity estimated from IMUs and encoders and is compared with the one estimated by the proposed POF sensor.

SpO2 measurements for each test is presented in Fig. 11.6(a), where the mean of the SPO2 is presented. It is possible to observe a slight reduction of the SpO2 for subjects 1 and 2, which can be related to the users' activity during the use of the SW. The maximum reduction is lower than 1% and the SpO2 values are far from the hypoxemia thresholds (90%). A similar analysis is performed for the breathing and heart rates and the results for each subject at each repetition are presented in Fig. 11.6(b) and (c). Gait cadences are presented in Fig. 11.6(d). The results demonstrate the feasibility of the proposed system to estimate all the proposed parameters in walker-assisted locomotion: SpO2, heart rate, and gait cadence.

11.3 Multiparameter quasidistributed sensing in a smart walker structure

11.3.1 Experimental validation

For the second application of fiber optic sensors for smart walker instrumentation by the application of a POFBGs array in CYTOP fibers, five FBGs were inscribed using a femtosecond (fs) laser at 517 nm with 220 fs pulse duration (HighQ laser femtoREGEN). A gradient index multimode CYTOP fiber (Chromis Fiberoptics Inc) with a core diameter of 120 μm, a cladding thickness of 20 μm and a polycarbonate overcladding was used, resulting in a total diam-

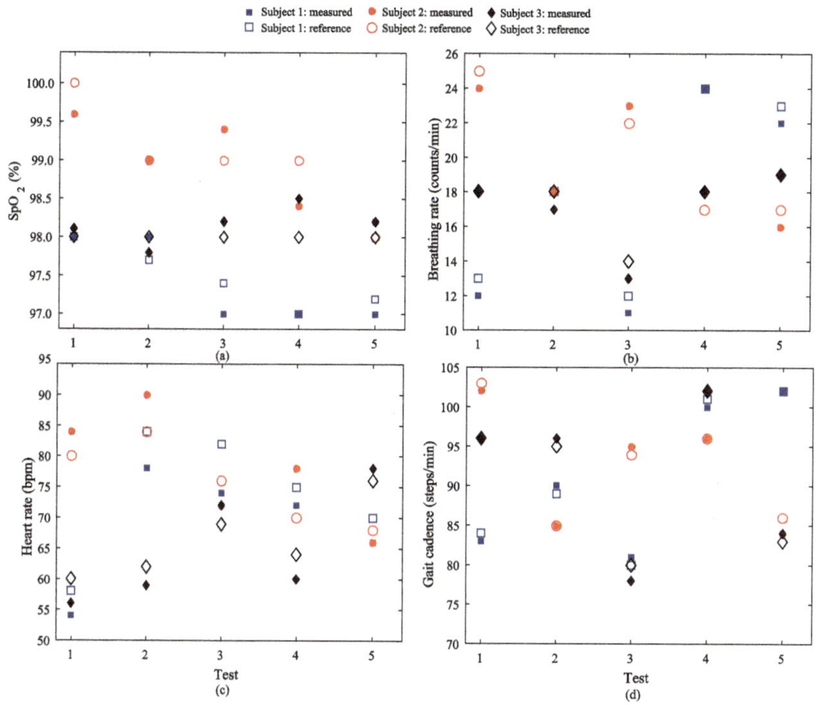

FIGURE 11.6 (a) SpO2 measurements in the SW implementation. (b) Breathing rate measurement. (c) Heart rate measurement. (d) Gait cadence estimation.

eter of 490 μm (Fig. 11.7). The laser beam is focused on the fiber core using a ×50 objective lens to assure uniform modification in the refractive index. Two-dimensional motion at nanometric precision was achieved with an air-bearing translation stage for the direct plane-by-plane inscription.

Temperature experiments were performed using a climatic chamber (1/400 ND, Ethik Technology, Brazil) for the 5-FBG array characterization. Experiments were performed over a range of 25 °C to 55 °C (10 °C steps). Strain characterization was performed with a linear translation stage with micrometer accuracy with strain up to 1000 με in steps of 250 με. Fig. 11.7 shows the 5-FBG array spectrum using a spectrometer I-MON 512 (Ibsen Photonics, Denmark). In this case, parameters (integration time and intensity threshold) are tuned to offer a spectrum with lower spectral distortions, facilitating peak detection using simple algorithms based on a Gaussian fit.

As it will be discussed in the following section, FBGs 1, 2, 4, and 5 will be used for assessing strain while FBG 3 will be used for measuring temperature. The strain response of the FBGs inscribed in CYTOP is presented in Fig. 11.8. FBGs 1, 2, 4, and 5 present a similar response to strain (strain sensitivity of

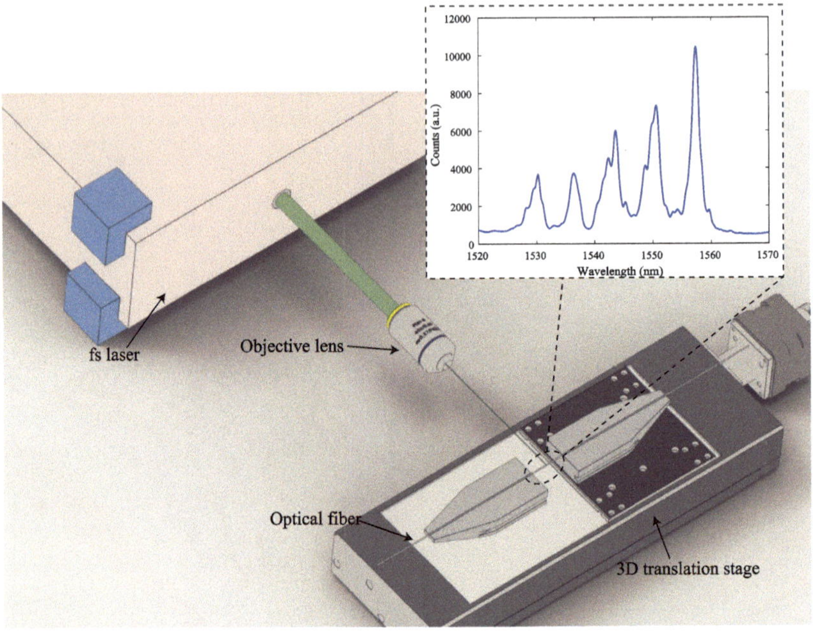

FIGURE 11.7 Schematic representation of setup used for the FBG inscription (inscription parameters were repetition rate of 5 kHz and pulse energy of 80 nJ). Figure inset shows the FBG-array spectrum obtained with an I-MON 512 spectrometer.

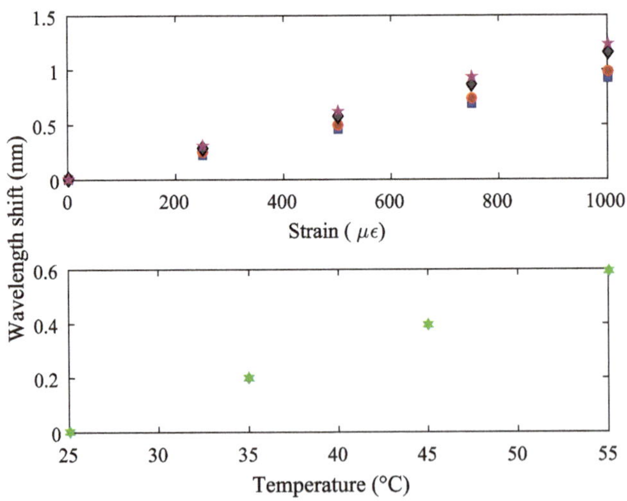

FIGURE 11.8 Wavelengths shift as a function of the strain for FBGs 1, 2, 4, and 5. Temperature response for FBG 3.

1.57±0.15 $pm/\mu\epsilon$) (Mizuno et al., 2019). FBGs 1, 2, 4, and 5 showed linear behavior, where a correlation coefficient (R^2) exceeding 0.99 was obtained in all analyzed cases. Temperature response of FBG 3 (sensitivity of 20.17 $pm/°C$) is shown in Fig. 11.8 (Theodosiou et al., 2018).

FBGs were also characterized under dynamic loadings by applying oscillatory strain cycles with constant strain (750 $\mu\epsilon$) at three different frequencies: 0.5 Hz, 1.5 Hz, and 3.5 Hz (Fig. 11.9). The optical fiber with the inscribed FBG was fixed at one end positioned and the other was attached to a movable (dynamic) stage connected to a DC motor.

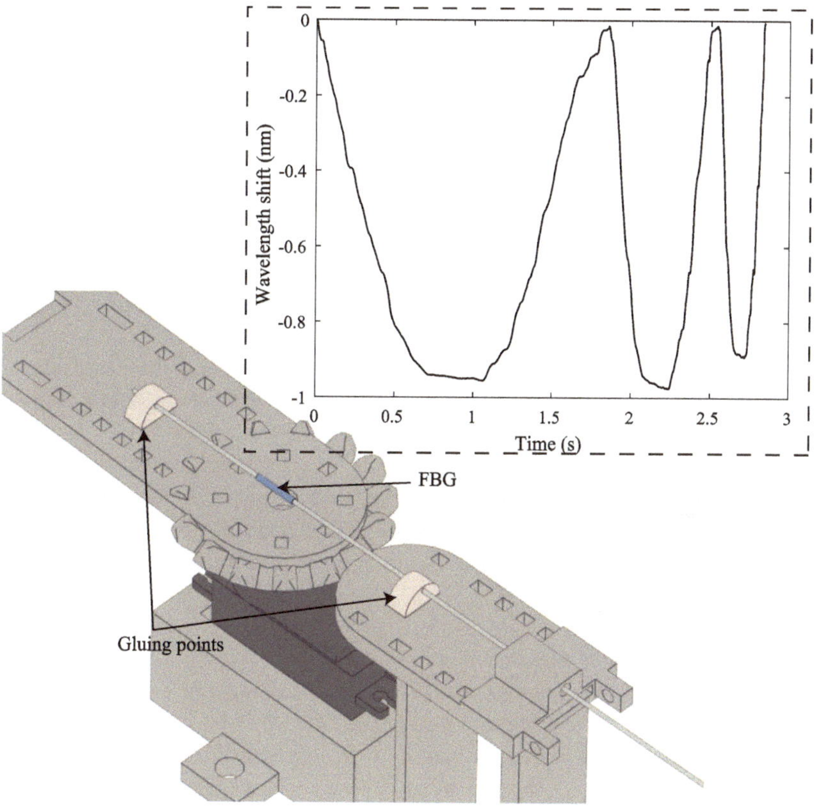

FIGURE 11.9 Schematic representation of the FBG sensors dynamic characterization setup. Figure inset shows the wavelength shift of FBG 1 under oscillatory loadings at 3 different frequencies (0.5 Hz, 1.5 Hz, and 3.5 Hz).

The dynamic response of the FBGs were monitored for three cycles at the same frequency and similar strain amplitude. The wavelength shift for FBG 1 at each frequency is shown in Fig. 11.9. The frequency response of the FBG 1 is also depicted in Fig. 11.9. Tests under dynamic strain at different frequencies show the ability of FBGs to detect frequencies with errors of about 1%, with high repeatability. The deviation between three tests at each frequency shown in the shaded line of Fig. 11.9 is lower than 0.05 Hz for f1 and negligible for f2 and f3.

11.3.2 Experimental validation

After the characterization, the FBG array was installed on the SW: FBGs 1 and 2 were positioned on the SW's right handle, whereas the FBGs 4 and 5 were installed on the SW's left handle. FBG 3 is used for temperature measurement, and thus positioned on a region of SW where there is negligible stress/strain. Temperature assessment is necessary since the Bragg wavelength is temperature dependent. Such parameter is also particularly POFBGs, since the material properties also change with temperature. In this sense, FBG 3 is needed to compensate temperature and improve precision on the measurements obtained by the other FBG-based sensors, as discussed in Chapter 6.

Fig. 11.10 shows the final position of each FBG sensor on the SW's support bar based on the stress/strain distribution on the SW. FBGs 1 and 2 as well as FBGs 4 and 5 are positioned in the regions with the highest strain variations. As force interaction is related to the user's intention to drive the smart walker, the FBGs in these regions will detect different loading conditions on the SW depending on the performed movement (forward motion, turns, and stopping the device). As previously discussed, FBG 3 is used for temperature assessment and is positioned at the center of the bar due to the negligible strain in this area. The distance between the FBGs of the array depicted in Fig. 11.10 is about 20 cm.

Two sets of tests were made. In the first one, the user follows a predetermined path (shown in Fig. 11.11) including right and left turns, in addition to commands for forward movement and to stop the healthcare device. The path along the x and y directions (Fig. 11.11) as a function of time, is depicted in Fig. 11.11. Responses of the FBGs were compared with the forces acquired by the SW's integrated force sensors (see Fig. 11.1).

Fig. 11.12 shows the responses of FBGs 1 and 2 compared with the y-axis of the right handle's force sensor and also presents a similar comparison of FBGs 4 and 5 with the y-axis of the left-handle force sensor. All tested FBGs showed similar responses when compared with the force sensors used as references for the SW's commands. In addition, Fig. 11.12 presents the wavelength variations of FBG 3 during the entire test, showing minor wavelength deviations related to temperature oscillations during the test. Temperature estimated by FBG 3 (27.4 °C) is very similar to the room temperature (27 °C), which was obtained with commercial thermometer.

Data analysis showed a R^2 of about 0.95 when FBGs 1 and 2 are compared with the force sensor in the right handle. Similarly, the R^2 for FBGs 4 and 5 compared with the force sensor in the left handle is about 0.93. This shows good agreement between the force sensors and FBGs array, which proves the

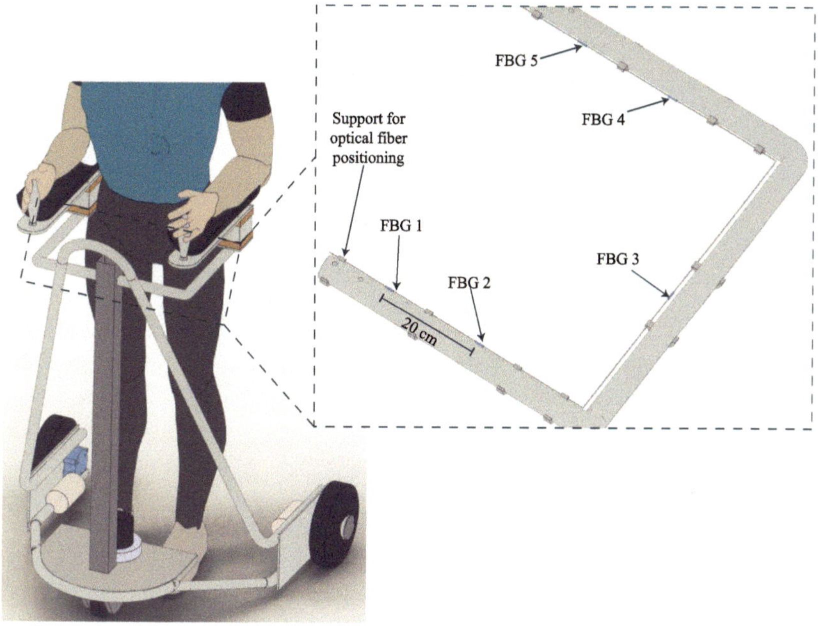

FIGURE 11.10 Positions of the FBGs on the SW structure. The distance between FBGs is 20 cm.

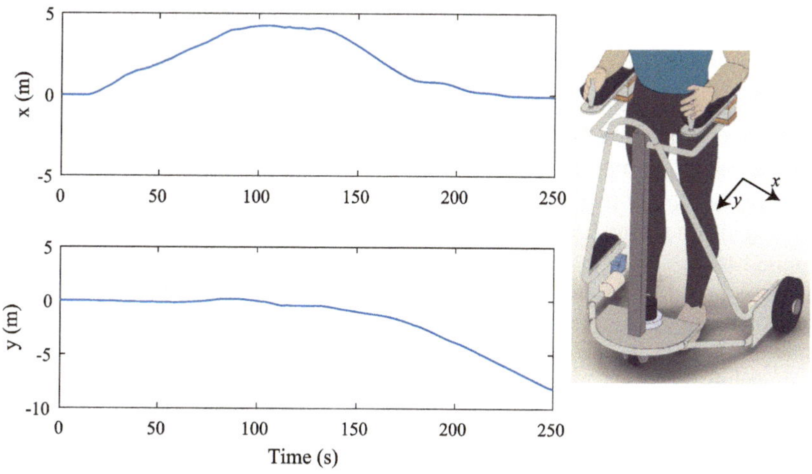

FIGURE 11.11 Temporal variation of the x and y path directions.

feasibility of using such sensors for detecting the users movement intention, commonly used in control strategies for human-robot interaction (Jimenez et al., 2019).

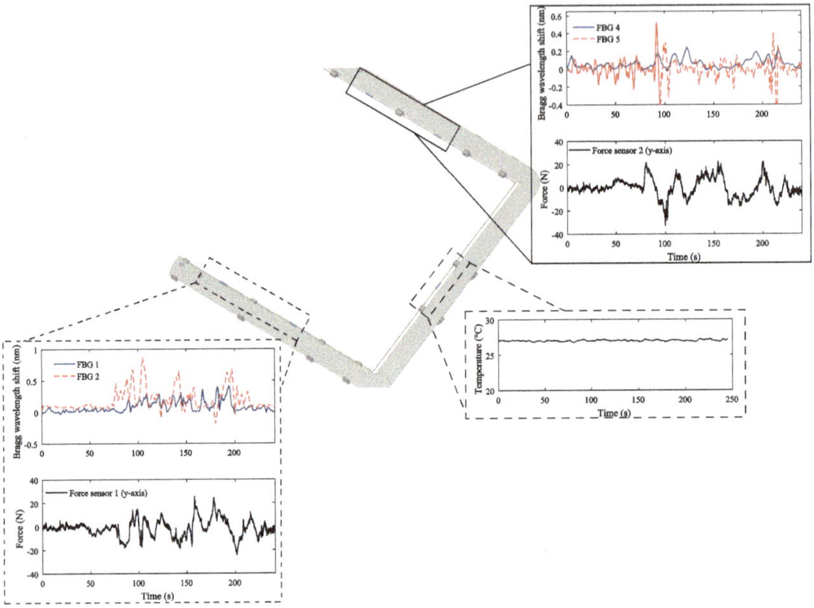

FIGURE 11.12 Comparison between FBGs 1 and 2 with y-axis of the force sensor 1 placed on the SW's right handle. Comparison between FBGs 4 and 5 with y-axis of the force sensor 2 placed on the SW's left handle. Figure also shows the temperature response and wavelength shift of FBG 3 during the test.

For the second test, the user was asked to walk in a straight path over three different floor conditions, where the FBGs array responses in the frequency domain were compared with the LRF for gait cadence assessment and with the accelerometer for measuring floor-induced vibrations analysis during locomotion. Floor conditions vary from a smooth floor, and increasingly rough terrain. A second-order Butterworth low pass filter (cutoff frequency at 1.8 Hz) was applied to the FBG responses for the estimation of gait cadence, whereas a second-order Butterworth high pass filter (cutoff frequency at 2 Hz) was employed to obtain the floor-induced vibrations. Sensors' responses are analyzed in the frequency domain by means of a fast Fourier transform (FFT). The FBG responses have frequency peaks at two regions, accordingly to the proposed filtering architectures: one at frequencies between 0.5 Hz and 1.5 Hz (related to the user's gait cadence), and the other at frequencies exceeding 2 Hz (floor-induced vibrations), which enable to obtain the gait cadence and floor induced vibrations as shown in Fig. 11.13.

Floor induced vibrations can be used for a series of functions, including for assisting simultaneous localization and mapping (SLAM) techniques in structured environments by means of identifying different floor conditions. The determination of floor-induced vibrations and gait cadence also interfere in the estimation of user's intention as components related to such perturbations will

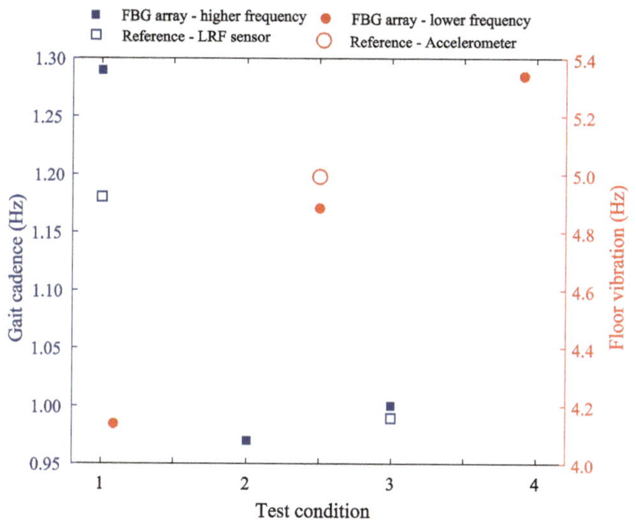

FIGURE 11.13 FBG array, LRF and accelerometer responses in the frequency domain for floor conditions 1, 2, and 3.

appear in the measurements of the interaction forces. Adaptive filtering can be used to filter out such components, allowing a more transparent and safer walker-assisted locomotion (Frizera Neto et al., 2010). The proposed sensor system showed an error of about 5% for the gait cadence estimation and 3.4% for measurement of floor-induced vibrations.

FBG arrays in CYTOP fibers shows promising applications in walker-assisted locomotion especially when compared with conventional instrumentation devices. The proposed sensors system can also be readily employed in healthcare devices for structural health monitoring, for detecting the users' movement intention, estimating gait-related spatiotemporal parameters, and for assisting localization algorithms in structured environments. FBG sensors can also replace traditional force/torque sensors that are traditionally used in admittance control techniques (Jimenez et al., 2019). By replacing traditional sensors with a single technology for multiparametric assessment, FBG arrays can drastically reduce the overall sensor system costs: All the sensors can be integrated in the same fiber and only needs a light source, optical circulator, and spectrometer.

Another important advantage is related to the power consumption, which is a key aspect in mobile/autonomous applications. As the sensor system is based on passive optical components, the use of FBG arrays also results in lower power consumption for the sensor system, where the only elements that require power consumption are the spectrometer and light source.

In this sense, the authors can envisage the next generation of assistive robotics devices with embedded optical fiber sensor arrays instead of or complementing conventional electronic sensors to obtain compact, low cost, and energetically efficient systems.

References

Bateni, H., Maki, B.E., 2005. Assistive devices for balance and mobility: benefits, demands, and adverse consequences. Archives of Physical Medicine and Rehabilitation 86, 134–145. https://doi.org/10.1016/j.apmr.2004.04.023.

Berry, D.C., Seitz, S.R., 2012. Educating the educator: use of pulse oximetry in athletic training. Athletic Training Education 7, 74–80. https://doi.org/10.5608/070274.

Cifuentes, C.A., Frizera, A., 2016. Human-Robot Interaction Strategies for Walker-Assisted Locomotion. Springer Tracts in Advanced Robotics, vol. 115. Springer International Publishing, Cham. From Duplicate 2 (Human-Robot Interaction Strategies for Walker-Assisted Locomotion – Cifuentes Carlos A., Frizera Anselmo).

Cifuentes, C.A., Rodriguez, C., Frizera-neto, A., Bastos-filho, T.F., Carelli, R., Member, S., 2014. Multimodal human – robot interaction for walker-assisted gait. IEEE Systems Journal, 1–11. https://doi.org/10.1109/JSYST.2014.2318698.

Cohen, V.Z.J., Haxha, S., Aggoun, A., 2016. Pulse oximetry optical sensor using oxygen-bound haemoglobin. Optics Express 24, 10115. https://doi.org/10.1364/OE.24.010115.

Delpy, D.T., Cope, M., van der Zee, P., 1988. Estimation of optical path length through tissue from direct time of flight measurement. Physics in Medicine and Biology 33, 1433–1442.

Frizera Neto, A., Gallego, J.A., Rocon, E., Pons, J.L., Ceres, R., 2010. Extraction of user's navigation commands from upper body force interaction in walker assisted gait. BioMedical Engineering Online 9, 1–16. https://doi.org/10.1186/1475-925X-9-37.

Haubert, L.L., Gutierrez, D.D., Newsam, C.J., Gronley, J.A.K., Mulroy, S.J., Perry, J., 2006. A comparison of shoulder joint forces during ambulation with crutches versus a walker in persons with incomplete spinal cord injury. Archives of Physical Medicine and Rehabilitation 87, 63–70. https://doi.org/10.1016/j.apmr.2005.07.311.

Jimenez, M.F., Monllor, M., Frizera, A., Bastos, T., Roberti, F., Carelli, R., 2019. Admittance controller with spatial modulation for assisted locomotion using a smart walker. Journal of Intelligent & Robotic Systems 94, 621–637. https://doi.org/10.1007/s10846-018-0854-0.

Kehoe, B., Patil, S., Abbeel, P., Goldberg, K., 2015. A survey of research on cloud robotics and automation. IEEE Transactions on Automation Science and Engineering 12, 398–409. https://doi.org/10.1109/TASE.2014.2376492.

Krehel, M., Wolf, M., Boesel, L.F., Rossi, R.M., Bona, G.L., Scherer, L.J., 2014. Development of a luminous textile for reflective pulse oximetry measurements. Biomedical Optics Express 5, 2537. https://doi.org/10.1364/BOE.5.002537. From Duplicate 2 (Development of a luminous textile for reflective pulse oximetry measurements – Krehel Marek, Wolf Martin, Boesel Luciano F, Rossi René M, Bona Gian-Luca Scherer Lukas J.).

Leal, A.G., Diaz, C.R., Jimenez, M.F., Leitao, C., Marques, C., Pontes, M.J., Frizera, A., 2018. Polymer optical fiber based sensor system for smart walker instrumentation and health assessment. IEEE Sensors Journal 1748, 567–574. https://doi.org/10.1109/JSEN.2018.2878735.

Leal-Junior, A., Diaz, C., Jimenez, M., Leitao, C., Marques, C., Pontes, M., Frizera, A., 2019. Polymer optical fiber-based sensor system for smart walker instrumentation and health assessment. IEEE Sensors Journal 19. https://doi.org/10.1109/JSEN.2018.2878735.

Lee, H., Ko, H., Lee, J., 2016. Reflectance pulse oximetry: practical issues and limitations. ICT Express 2, 195–198. https://doi.org/10.1016/j.icte.2016.10.004.

Majumder, S., Mondal, T., Deen, M., 2017. Wearable sensors for remote health monitoring. Sensors 17, 130. https://doi.org/10.3390/s17010130. From Duplicate 2 (Wearable sensors for remote health monitoring – Majumder Sumit Mondal Tapas, Deen M.).

Martins, M., Santos, C., Frizera, A., Ceres, R., 2015. A Review of the Functionalities of Smart Walkers. Medical Engineering and Physics 37, 917–928. https://doi.org/10.1016/j.medengphy.

2015.07.006. From Duplicate 1 (A review of the functionalities of smart walkers – Martins Maria, Santos Cristina, Frizera Anselmo, Ceres Ramón).
Mizuno, Y., Ma, T., Ishikawa, R., Lee, H., Theodosiou, A., Kalli, K., Nakamura, K., 2019. Lorentzian demodulation algorithm for multimode polymer optical fiber Bragg gratings. Japanese Journal of Applied Physics 58, 028003. https://doi.org/10.7567/1347-4065/aaf897.
Priebe, J.R., Kram, R., 2011. Why is walker-assisted gait metabolically expensive? Gait and Posture 34, 265–269. https://doi.org/10.1016/j.gaitpost.2011.05.011.
Theodosiou, A., Komodromos, M., Kalli, K., 2018. Carbon cantilever beam health inspection using a polymer fiber Bragg grating array. Journal of Lightwave Technology 36, 986–992. https://doi.org/10.1109/JLT.2017.2768414. From Duplicate 2 (Carbon cantilever beam health inspection using a polymer fiber Bragg grating array – Theodosiou Antreas, Komodromos Michael, Kalli Kyriacos).
Van Hook, F.W., Demonbreun, D., Weiss, B.D., 2003. Ambulatory devices for chronic gait disorders in the elderly. American Family Physician 67, 1717–1724.

Chapter 12

Optical fiber sensors applications for human health☆

12.1 Robotic surgery

The introduction of robots in industry, driven by the third Industrial Revolution, was characterized as a technological investment in manufacturing processes automation to achieve greater efficiency and productivity (Edwards, 1984). A manipulator robot can be defined as an equipment consisting of a set of mechanical, electrical, and electronic components that are programmed to perform repetitive tasks automatically. The adaptability, flexibility, ample working space, and low cost of industrial robots make them promise for application in a lot of processes (Schneider et al., 2014). Initially, their participation in the manufacturing environment consisted of performing functions that could put the integrity of human employees at risk, such as the transport of heavy loads (Sherwani et al., 2020) or functions that required a high degree of precision such as moving objects, painting, and welding (Dumas et al., 2012).

Result of advances focused on continuous evolution, Industry 4.0 represents the fourth stage of industrialization and the most modern in terms of technological processes today. It is characterized by the use of intelligent devices, equipped with sensors, microprocessors, and complete embedded systems that enable a real-time connection of physical and digital systems (Aiman et al., 2016) and make it possible to obtain a productive flow optimized, integrated and automated (Vaidya et al., 2018). Additive Manufacturing, Internet of Things (IoT), Advanced Robotics and Artificial Intelligence can be mentioned as new technologies of this phase (Aiman et al., 2016). This new scenario enabled a change in manipulators performance, which occurred in their own workspace isolated from the human collaborator for a cooperative human-robot approach in many situations. This strategy ensures an environment in which the resource use effectiveness and productivity can be improved by combining robots characteristics with flexibility and dexterity of humans in dealing with unexpected and non-repetitive tasks (Thoben et al., 2017).

The connection between the cyber-physical drove the emergence of smart grids and this fact served as a basis for the appearance of the concept of collaborative robots (cobots). They are characterized by low weight, highly flexible,

☆ This chapter is carried out with the participation of Vitorino Biazi Neto and Carlos Alberto Ferreira Marques.

https://doi.org/10.1016/B978-0-32-385952-3.00023-8

and have a lot of modern control strategies that allow them to operate together with humans in a safe and efficient manner. An example is the impedance control that consists of a technique used to maintain a desired dynamic coupling between robot end-effector position and the contact force (Seraji, 2016). So, it is possible to regulate the robot mechanic impedance during interactions and ensure that the robot will react to contact forces smoothly (Oh et al., 2014).

Many of these cobots are preconfigured so that they can be programmed automatically when guided in a specific trajectory. So, the user has the possibility of not having advanced technical knowledge in robotic programming. Due to this, many improvements, such as high precision in robotic surgery, have been achieved in various research fields (Sherwani et al., 2020). The first concepts of robotic surgery appeared before the need to perform surgeries without the physical presence of the doctor in the operating room. This scenario is characteristic of critical situations in which access to a properly equipped surgical center is difficult, thus limiting the surgeon's performance capabilities. Some of these situations are battlefields and space stations (Mack, 2001).

Technological advances of the 20th century also had an impact on the medical field. The surgical procedures practice has changed from an invasive approach, in which the patient has a painful postoperative consequence of traumatic access to the affected area, to a minimally invasive approach, which results in lower complications and surgical mortality (Mack, 2001). The MIS proposal is the application of high definition cameras combined with endoscopic technology and precise instrumentation to access regions of the body through small incisions. In this way, it is possible for the surgeon to perform the operation without directly seeing or touching the affected organs (Diana and Marescaux, 2015). Some advantages for opting for minimally invasive surgery (MIS) over conventional open surgery is the reduction of anesthesia time, incision size, intraoperative blood loss, postoperative infections, and hospitalization time (Okamura, 2009).

Including collaborative robots in this scenario brought even more benefits in terms of surgical procedures as they compensate the dexterity and precision limitations of the humans in manipulating the microinstruments (Okamura, 2009). In addition, the surgeon exposure to ionizing radiation and physical exhaustion during very long procedures can be reduced (Schostek et al., 2009).

Zeus (Computer Motion, Goleta, California, USA) is a commercial product with three robotic arms monitored and controlled remotely by the surgeon. One of the robotic arms handles a camera and can be positioned as desired for a stable image by using voice commands. The other two arms have a haptic interface, to be remotely teleoperated, in which the doctor movements are transferred to the robots during surgeries. Da Vinci (Intuitive Surgical, Sunnyvale, California, USA) can be considered an evolution of Zeus. The main difference between them is the fact that this new system is equipped with a stereoscopic 3D camera, which collects images expanded up to 10 times and can be controlled by the surgeon for a more accurate and safe navigation (Diana and Marescaux, 2015).

Despite the advantages of MIS, this type of surgery has some peculiarities that can affect the quality of the results such as instruments long length and high stiffness. These attributes end up limiting the reception of contact force information and reduce the tactile sensitivity of the medics (Song et al., 2011). Because of that, they lose their haptic perception and cannot have an accurate control of hands movements (Jaschinski et al., 2011). Some research show that the lack of contact force feedback causes an increase of about 50% in the average value and doubles the peak value of the force on human tissues (Wagner et al., 2002). In the case of minimally invasive robotic surgery (MIRS), as the procedure is performed by robots, force feedback is eliminated. Recovering this sensitivity through tactile sensors has been an alternative found by research groups.

Optical fiber sensors are electrically passive and have electromagnetic immunity. These characteristics make them superior to electrical-based sensors (piezoelectric, piezoresistive, and capacitive sensors) for procedures applying magnetic resonance imaging (MRI) and for brain and heart surgery because they are unaffected by neural and cardiac activities. In addition, some equipment sterilization methods are based on high heat, pressure, and humidity, which can damage the sensors electronic circuits and the optical sensors can withstand these effects. Besides that, polymers optical fibers also have good performance and compatibility during interactions with bio-environment and so they do not offer risks of contamination and body reaction to the patients. Besides that, they are compact, lightweight, present chemical stability and multiplexing capabilities being suitable for general surgery.

The optical fiber sensors are based on three working principles: wavelength (WM), phase (PM), and light intensity modulation (Hooshiar et al., 2020). The working configuration consists of a light source coupled to one tip of the fiber and a receptor coupled to the other end. The light is transmitted through the fiber and it is received by the detector on the other side (Bandari et al., 2020).

Yip et al. (2010) presented a proposal of an optical fiber sensor based on the light intensity modulation to be used in cardiac MIS. That strategy is important because the heart chambers pressurization due to the heartbeats can limit the surgeon's sensory feedback. The sensor consists in a miniature uniaxial force sensor. This application was mainly concerned with preventing potential damage to the electronic circuits due to cardiac electrical interference. These sensor characteristics are waterproof to evade the heart chamber blood from entering the sensors circuits, electrical passivity to allow them to avoid electrical activity in the heart, and miniaturization to prevent them to be affected by friction and ancillary forces that form the incision.

The system consists of three optical fiber cables, each containing a transmitting fiber and a receiving fiber connected to an LED and a phototransistor circuit, respectively. A circular plate is positioned on the end of the cables and it reflects the light coming from the transmission fibers. This light is then received by the phototransistors and its intensity will depend of the distance between the plate and the cables extremity. Tip forces approximate the reflection plate to the fibers and increase the received light power. The cables are positioned in a triangle formation to minimize sensitivity to rotations of the plate.

Due to the dynamic pressure and heart movements, this type of equipment requires robust insulation. The set insulation is a silicone seal between the sensor and the instrument shaft. Besides that, an elastic element is placed between the fibers and the plate to convert force to displacement.

The system needs a nonlinear calibration due to the viscoelastic properties of the silicone rubber and quadratic fit proved to favorable. A damping term was inserted in the equation to deal with hysteresis effect. The sensor presented 0.126 N RMS error (<3.2% of force range), force range of 0–4 N and frequency range of 0–6 Hz. The relation between the applied force and the phototransistor voltages connected to the three optical fiber can be seen in Eq. (12.1):

$$F = K_1 \cdot \begin{bmatrix} V_{fiber1}^2 \\ V_{fiber2}^2 \\ V_{fiber3}^2 \end{bmatrix} + K_2 \cdot \begin{bmatrix} V_{fiber1} \\ V_{fiber2} \\ V_{fiber3} \end{bmatrix} + K_3 \cdot \frac{d}{dt} \begin{bmatrix} V_{fiber1} \\ V_{fiber2} \\ V_{fiber3} \end{bmatrix}. \quad (12.1)$$

Noh et al. (2016) also make a sensor system based in intensity modulation. It consists of a multiaxis force sensor for catheter tip. The set configuration is composed by four optical fibers, a mechanical structure, a light source, a reflector and a CCD camera, and all of these components are protected by a surrounding jacket. Fig. 12.1 shows the schematic representation of the components.

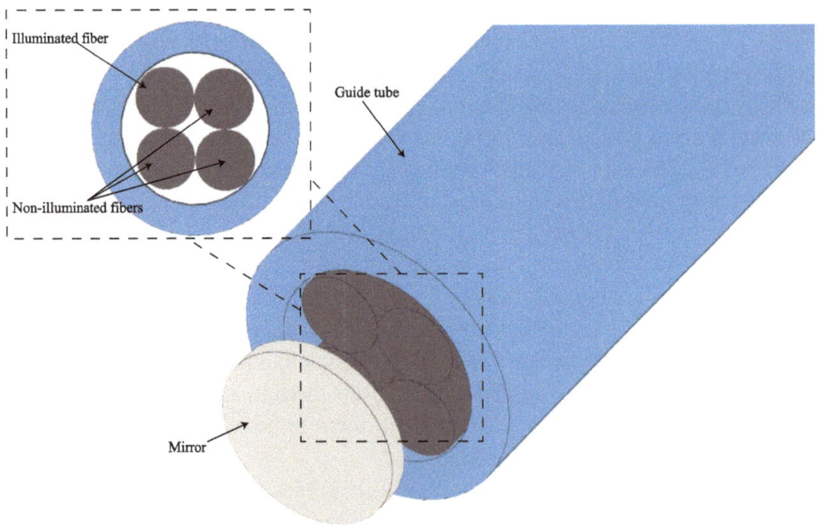

FIGURE 12.1 Schematic representation of the sensor arrangement.

The four fibers are positioned side by side and the light source transmits constant light through one of them. The light hits a mirror on the mechanical structure and the other three optical fibers receive and send the reflected light to the CCD camera. An image processing software was developed to estimate the light coming from the fibers. The mirror is mounted over three cantilever beams and when a force is applied to the sensor tip the beams on the structure suffer a deflection and the mirror has its inclination changed.

The light intensity collected by the CCD camera is proportional to the beams deflection and because of that it can be linearly related to the external force. The equipment was calibrated using a multilinear regression and it showed a linear relationship between the light intensity values and the force applied to the sensor. Eq. (12.2) shows the linear relation between analog samples of the light intensity and the force components:

$$\begin{bmatrix} F_x \\ F_y \\ F_z \end{bmatrix} = K \cdot \begin{bmatrix} a_1 \\ a_2 \\ a_3 \end{bmatrix} \quad (12.2)$$

Song et al. (2011) produced a robotic manipulator with 7 DoF for tests involving MIS sensors. The tactile sensing system which they proposed is based in wavelength modulation with Fiber Bragg Gratings (FBGs) and it was tested on their manipulator. The FBGs were chosen because they are independent of the light fluctuations and immune to sources power and connector loses. That occurs thanks to its sensing principle attached to Bragg wavelength shift. Other positive advantage is that the system requires less wires to data transmission due to its multiplexing capabilities. Tests carried out with the sensors show that they can detect forces slightly above 10N. They also have a resolution of 0.05N and a maximum error of 0.1N in the two directions.

There are two forces to be measured in the surgical procedure: one is the surgical tool-tip force, which must be controlled to prevent tissue damage, and the other is the grip force on the forceps, which measures organs hardness. Four flexible beams were positioned in line with the longitudinal axis of the chosen shaft and four FBGs were inserted on these beams to measure the strains on them. Other four FBGs are placed away from the beam ones to compensate temperature effect. Fig. 12.2 shows the beams arrange and positions over the robotic arm.

The strains that occurs in the optical fiber due to applied stress provokes a Bragg wavelength shift. Light emitted by the source is sent to the FBGs from a 1x4 coupler via a circulator and the Bragg wavelength light returns via circulator to a tunable FFP filter, which is converted to wavelength by a control voltage. The wavelength shift passes through temperature compensation with the four reference FBGs and it is finally translated into strains.

Using the measured strains, it is possible to obtain the force applied to the sensor module acting on the Z axis and the moments acting on the X- and Y-axis using the Euler's beam theory (Eq. (12.3)). This methodology considers the

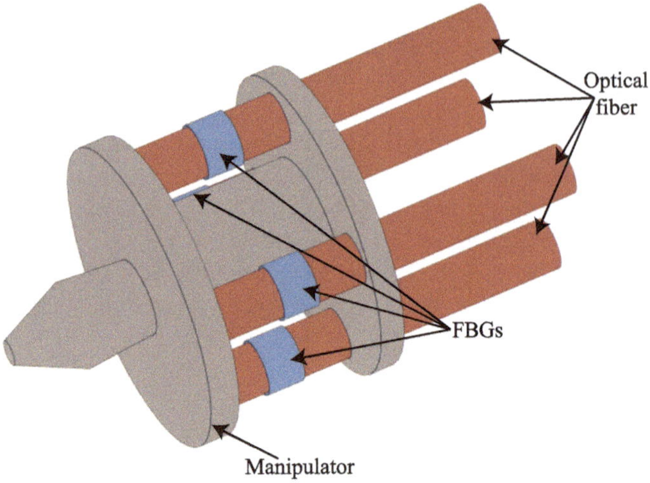

FIGURE 12.2 Beams with the FBGs arranged in the manipulator (Song et al., 2011).

others forces and moments components negligible:

$$\begin{bmatrix} \epsilon_1 \\ \epsilon_2 \\ \epsilon_3 \\ \epsilon_4 \end{bmatrix} = \frac{1}{E} \cdot \begin{bmatrix} 1/A & C/I & 0 \\ 1/A & -C/I & 0 \\ 1/A & 0 & 0 \\ 1/A & 0 & 0 \end{bmatrix} \cdot \begin{bmatrix} F_z \\ M_x \\ M_y \end{bmatrix}. \tag{12.3}$$

E, A, C, and I are Young's modulus, the area of sensor's cross-section, the largest distance from the neutral surface and the moment of inertia, respectively.

He et al. (2012) developed a tactile measurement system applying FBGs to monitor the forces acting on forceps tip during vitreoretinal surgeries. The equipment has a resolution of 0.25 mN, a sampling rate of 100 Hz, and a temperature compensation scheme. The used forceps instrument is divided into two parts: the tip and the handle. Microforceps used to manipulate human tissue is positioned at the end of a shaft. The researchers were concerned with a tool design that would allow the FBGs to be positioned on the surface of the microforceps shaft. For that, an orientation template has been designed to align the FBG sensors along the shaft. An optical interrogator was used to monitor the FBGs and had an interface with a computer for recording and acquiring data.

Due to the misalignment between the shaft and the FBGs, an offset in the value of the measured force comes up when the clamp is open. It is desired to obtain the clamping force on the forceps when the surgeon is performing the surgical tasks and because of that the equipment calibration is done with the tool closed. The calibration procedure was made based on the work of Iordachita et al. (2009). For an increment of 25 mg in weight applied on the tool-tip shaft measured by an analytical balance, the wavelength position of the FBGs is also collected with the optical interrogator. Eq. (12.4) shows the calibration equation that linearly relates the wavelength shift in the FBGs and the surgery force components in x and y directions:

$$\begin{bmatrix} \Delta S_1 \\ \Delta S_2 \\ \Delta S_3 \end{bmatrix} \begin{bmatrix} K_{1,x} & K_{1,y} \\ K_{2,x} & K_{2,y} \\ K_{3,x} & K_{3,y} \end{bmatrix} \cdot \begin{bmatrix} F_x \\ F_y \end{bmatrix}. \tag{12.4}$$

The force components then can be obtained using the sensor readings and the Moore–Penrose inverse of the calibration matrix.

The research of Gonenc et al. (2017) was the development of a 3-DoF force system for a microforceps tool used in vitreoretinal surgery. The principal objective of the equipment is to permit the measurement of the interaction forces between the tool and tissue during membrane peeling. Four FBG sensors were embedded on the tool shaft to maximize the coupling between transverse and axial forces. A motorized unit was applied to perform the grasp function without any mechanically coupled handle mechanism. That ensures highly repeatable behavior in FBG sensor outputs during operation. The system was designed as a modular unit so that it can be adapted to robotic surgery

The FBG sensors were chosen because they can be fitted through a small incision on the sclera; they are sterilizable and biocompatible and can provide mN accuracy order. Three lateral FBGs were fixed around guide tube using medical epoxy adhesive to capture the transverse force (Fx and Fy) at the tool tip. A fourth FBG sensor is used to detect the tensile forces (Fz) along the tool axis that showed up due to retina manipulation. Experiments were carried out by the researchers with objective of prevent the sensor to be affected by the tool actuation. The axial FBG was positioned so that the sensitive zone is located out of the guided tube close to the tip. The sensor segment of the prototype is showed in Fig. 12.3.

During the experiments, an optical sensing interrogator was used to collect Bragg wavelength shift of each FBG sensor. The roll angle and the pitch angle of the microforceps were adjusted with the aid of two rotary stage. The angle's manipulation is important to apply pure transversal and axial loadings or to combine them. The tool tip is coupled to a wire hook hanging aluminum washers that were previously weighed using a precision scale. This strategy permits the application of a lot of forces combinations on the forceps. The system calibration was made using linear and second-order Bernstein polynomials fits. The second method provided better accuracy for the equipment.

12.2 Biosensors

12.2.1 Introduction to biosensing

The most optical sensing mechanisms for biosensing include fluorescence, absorption, surface plasmon resonance (SPR) among others (Yin et al., 2018). The

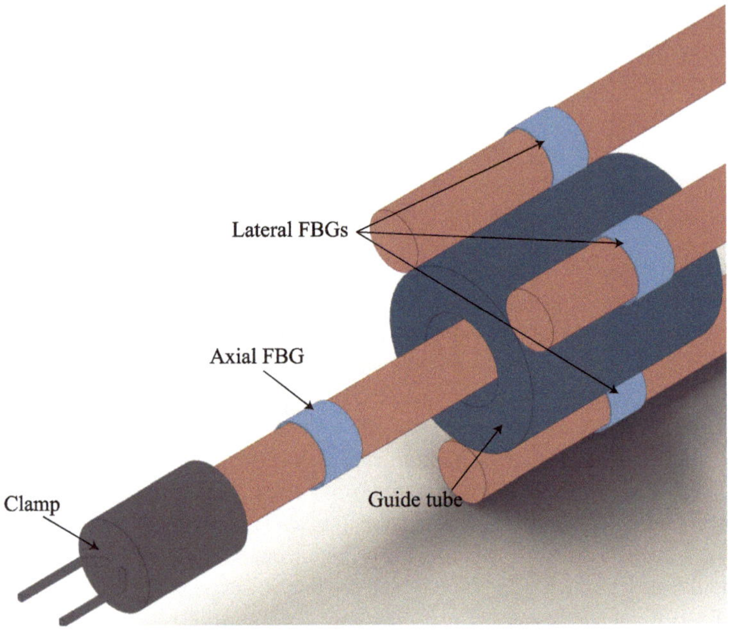

FIGURE 12.3 Prototype of the sensor segment (Gonenc et al., 2017).

first two occur when no coatings or metal layers are used for the detection of analytes. Consequently, detection is accomplished through the direct interaction of the evanescent wave (EW) with the molecules of interest. SPR mechanism is the principle used in plasmonic fiber optic sensors. In this mechanism, photons from a light source interact with surface electrons on a metal layer and lead them to oscillate collectively. The excitation of these surface plasmons (SPs) can result in the propagation of surface waves at the intersection of a metal and dielectric nanofilms (surface plasmon polariton, SPP). When a strongly localized oscillation of surface electrons takes place in nanoparticles and nanostructures, it is designated localized surface plasmon resonance (LSPR). The first SPR sensors were fabricated based on the Kretschmann configuration using prism structures. However, the use of prism-based SPR biosensors has a main disadvantage due to the bulkiness of the setup, which reduces its applicability in-situ biosensing (Monfared, 2020a). The first fiber optic sensor based on SPR was proposed by Jorgenson and Yee in 1993 (Zhao et al., 2019). From that time on, SPR fiber optic biosensors attracted considerable attention and achieved satisfactory results (Zhao et al., 2019). Therefore plasmonic fiber optic biosensors have been used in the most diverse areas of research since they can solve issues associated with conventional SPR biosensors and add the following potential characteristics: harsh environment tolerance, remote, and distributed sensing (Monfared, 2020a).

In general, biosensors can be divided into catalytic and affinity biosensors, depending on the recognition properties of the biological components. The first category is also known as metabolism sensors, which include kinetic devices based on the achievement of a steady-state concentration of a transducer-detectable species. In this case, an isolated enzyme, a microorganism, a subcellular organelle, or a tissue slice can be the biocatalyst. As for the second category, the receptor molecule binds to the analyte irreversibly and noncatalytically. In this instance, a physicochemical change can be measured resulting from the binding event between the target molecule and the bioreceptor. The bioreceptor can include an antibody, a nucleic acid, or a hormone receptor (Marazuela and Moreno-bondi, 2002; Cristea et al., 2015). The biosensors that employ antibodies as the bioreceptor molecules are called as immunosensors. These are one of the main classes of biosensors owing to the highly sensitive and selective immunoreaction that takes place once an antibody recognizes and binds to a specific antigen (Cristea et al., 2015; Alzawa, 1994).

The quality of the results obtained by a biosensor is validated by determining its analytical performance. This is achieved by assessing parameters known as figures of merit (FOM), which include sensitivity, selectivity, limit of detection (LOD), accuracy, signal-to-noise-ratio (SNR), repeatability, and reproducibility. Sensitivity corresponds to the ratio between the change in output (for instance, wavelength and intensity) in response to variations of the measured analyte (e.g., RI and concentration). Selectivity denotes the ability to discern the response of the analyte of interest from every other interference. The LOD indicates the lowest analyte concentration, which is not related to noise, detectable by the sensor with acceptable degree of certainty. Accuracy refers to the proximity between the sensor output value and the actual value of the measurand. SNR is determined by the ratio between the analytical signal and the background noise. Finally, repeatability and reproducibility designate the sensor capability to obtain a similar response over several repetitions using the same measured sample, with the former considered when performing under the same conditions (operators, equipment, laboratories, intervals of time, etc.), and the latter under different conditions (Caucheteur et al., 2015; Justino et al., 2016).

12.2.2 Background on optical fiber biosensing working principles

12.2.2.1 Evanescent wave

The light propagating in the fiber can be divided into the guided field in the core (core modes) and the evanescent field in the cladding that decays exponentially to zero before reaching the external medium (Yin et al., 2018). Moreover, the short distance that the EW travels beyond the core-cladding interface is called the penetration depth and corresponds to the distance where the evanescent field decreases to 1/e of its initial value at the interface (Leung et al., 2007). The detection of analytes of interest can be accomplished in the penetration depth, through the interaction of the EW with those molecules.

When the EW interacts directly with molecules, a net flow of energy is created across the cladding in the surrounding medium to maintain the evanescent field. Due to this transfer of energy, the reflectance is attenuated. This phenomenon can be utilized to produce absorption sensors based on EWs, more specifically, attenuated total reflection (ATR) sensors. In the present case, the variation of the absorbance properties of the sensitive layer occurs during the interaction with the analyte, due to the concentrations of the species analyzed (Marazuela and Moreno-bondi, 2002). In brief, absorption-based optical fiber sensors rely on the fact that the selected target or transducer must modify the intensity of the light propagated through the optical fiber core as a function of the selected measurand (Hernaez et al., 2017).

12.2.2.2 SPR and LSPR

The SPR principle was introduced to the field of biosensing by Liedberg et al. in 1983 (Liedberg et al., 1983) and is currently the mechanism behind of most biosensors in the field of biomedicine. This optical detection technique has been extensively used as it allows label-free monitoring of biomolecular interactions in real-time with fast response, high sensitivity, and very low LOD (Gandhi et al., 2019; Chiu et al., 2018). SPs are a quantum of collective oscillations of free electrons taking place at the interface between two materials of different dielectric constant, usually a metal and a dielectric. SPs that propagate along the transverse magnetic (TM) direction (P-polarized) are considered surface plasmon waves (SPWs), light waves whose magnetic vector is perpendicular to the plane of incidence. When light traveling through the dielectric medium is incident on the metal, if under the resonance condition, its photons can excite the SPs and cause the resonant oscillation of the conduction electrons at the interface between the mediums, a phenomenon known as SPR (Yin et al., 2018). The resonance or phase-matching condition represents the situation when the propagation constants of the incident light and of the SPs are equal and, therefore, their momentums match and resonance ensue. For this reason, SPR is very sensitive to RI changes, providing it the ability to monitor biochemical reactions occurring at the metal-dielectric interface.

Several designs have been proposed to promote SPR excitation, each resorting to different coupling devices such as prism coupler, waveguide coupler, and grating coupler (Zhao et al., 2019). Among the existing methods, the Kretschmann–Raether configuration remains the most popularly used in commercial systems nowadays. In this approach, light is injected through a high RI glass prism towards its metal-coated base (usually with gold or silver), with an incidence angle greater than the critical angle so that TIR occurs. This generates an EW that propagates along the interface separating the metal film and the surrounding dielectric medium (Prabowo et al., 2018). If the propagation constant of the EW matches that of the SP, thereby fulfilling the resonance condition, the wave can transfer energy to the SP (Caucheteur et al., 2015), leading to a decrease of the reflection at a given angle. Hence, monitoring the reflectance of the SPW can be used to monitor changes of the surrounding environment. Nevertheless, these prism-based sensors are expensive and bulky, which hinders in-situ biosensing, particularly in locations of difficult access. For these reasons, vast research was devoted to optical fiber technology as it provides a highly-integrated and miniaturized structure compared to their bulky counterparts, with greater flexibility in terms of design, materials, and performance, along with the capacity for in-vivo measurement (Monfared, 2020a). In these sensors, the light guided in the fiber core must be exposed to the surrounding medium so that it can interact with metal coatings or particles placed on the fiber in order to elicit SPR. While the first reported optical fiber plasmonic sensors resorted to removal of the cladding, numerous other techniques and structures have emerged since then, as will be described in the following sections. In these structures, spectral and intensity interrogation techniques are generally used, by measuring wavelength shifts and optical power changes in the transmission or reflection spectra (Caucheteur et al., 2015).

In addition, LSPR-based optical fiber sensors have also been the focus of extensive research, having demonstrated promising performance results (Qi et al., 2020). In this optical detection technique, the incident light waves are confined within conductive nanoparticles or nanostructures, whose dimensions are inferior to the wavelength of the light. The excitation of localized SPs (LSPs) arises from the interaction between the incident light and the electrons in the conduction band of the metal, inducing strong localized oscillations of surface electrons (Caucheteur et al., 2015). Despite having the same excitation mechanism, contrary to SPR, which involves lossy propagation along the intersection between metal and dielectric nanofilm, LSPs are nonpropagating and oscillate locally in the nanostructure as a result of particle size restriction (Qi et al., 2020). For LSPR, detection relies on monitoring the increase in absorption, at a given wavelength, of light waves propagating through nanoparticles dispersed in liquids or deposited on solid substrates. Once changes occur in the area surrounding the nanoparticles, e.g., RI changes; this resonant wavelength is shifted, revealing the LSPR. Furthermore, the resonant frequency is highly dependent on the dielectric environment, distance between particles and intrinsic properties of the metal nanoparticles, namely composition, size, shape, geometry, and aspect ratio (Khalaf et al., 2017). Therefore tuning these factors can optimize the detection performance toward a specific target.

Different plasmonic materials have been deposited onto optical fiber biosensors for both SPR and LSPR, with gold and silver being the most common (Monfared, 2020a). These two noble metals are frequently used for nanofilms and nanoparticles, showing LSPR in the visible region (Caucheteur et al., 2015). Between the two, gold demonstrates a larger shift of the resonance peak, whereas silver evidences a narrower peak width and fewer optical losses. However, silver oxidizes easily once exposed to aqueous environment, unlike gold that is chemically inert and, therefore, much more stable. The deposition of a thin bimetallic layer onto the silver surface, prior to application in biological systems, can inhibit oxidation but leads to a decrease in sensor performance (Rifat et al., 2017). The performance of plasmonic biosensors is highly influenced by the shape of the deposited metal particles and the characteristics of the coating, namely morphology, thickness (usually 30–70 nm) and rugosity. For example, in a sensor coated with gold nanofilm, the greatest surface sensitivity is achieved with a thickness of 50 nm, corresponding to the narrowest and deepest SPR attenuation.

Over the years, graphene and its derivates, namely graphene oxide (GO) and reduced GO (RGO) started being included in plasmonic optical fiber sensors, either as a substitute for metallic materials or applied in addition to them (Hernaez et al., 2017). Graphene is a mechanically strong and chemically inert two-dimensional (2D) carbon allotrope, with zero band gap and a hexagonal lattice structure that prevents oxidation (Rifat et al., 2017). Several studies have proved that the addition of a graphene layer to a metal surface improves sensitivity and resolution due to $\pi - \pi$ stacking and its high surface to volume ratio, which enhances the absorption of analytes (Rifat et al., 2017). As an example, in 2017, Jiang et al. developed a LSPR biosensor using a U-shape POF covered with a silver film, a layer of polyvinyl alcohol (PVA), graphene and silver nanoparticles. Their studies revealed that graphene improved sensitivity, reaching 700.3 nm/RIU, while also delaying the oxidation process of the silver nanoparticles, maintaining sensor stability (Jiang et al., 2017). Plasmonic biosensors can also be based on GO as it provides several benefits including high biocompatibility, colloidal stability, and easy chemical modification of the surface oxide. GO comprises a large amount of oxygen-containing functional groups that aid in biomolecule immobilization through covalent bonding. Other alternative materials are emerging in this field, including, molybdenum disulfide (MoS2) (Wang and Wang, 2018c; Wang et al., 2020a), zinc oxide (ZnO) (Kim et al., 2019), titanium dioxide (Tahhan et al., 2017), titanium nitride (Monfared, 2020b), yet further research still needs to be conducted.

12.2.2.3 Gratings-assisted sensors

An optical fiber grating is a kind of fiber device with a permanent period change of RI on the fiber core and was demonstrated by Ken Hill in 1978 (Hill et al., 1978). These fall into two general classifications based on the grating period: short period grating and long period grating (LPG). Short period fiber grating corresponds to gratings with period less than 1 μm, and includes uniform fiber Bragg gratings (FBGs), discussed in Chapter 6 and tilted fiber Bragg gratings (TFBGs). The information provided by the optical fiber grating is encoded in wavelength. As the wavelength is an absolute parameter, the optical signal can be processed in a way that the information remains immune to energy fluctuations along the optical path (Majumder et al., 2008).

As mentioned before, TFBGs are short period (500 nm) gratings with a periodic pattern uniformly tilted by an angle θ (< 45°) with respect to the fiber axis, leading to a more complex coupling. In addition to the self-backward coupling of the core mode at the Bragg wavelength, TFBGs also redirect some light to the cladding, which allows coupling to circularly and noncircularly symmetric copropagating or counter propagating cladding modes, with the direction of the coupled light depending on the tilt angle of the TFBG (Chiavaioli et al., 2017). For this reason, the transmitted spectrum shows multiple narrow attenuation bands that are related to the coupling with different cladding modes. In this case, the Bragg condition is slightly modified, because it takes into account the resonance wavelengths of the various cladding modes $\lambda_{res(m)}^{TFBG}$ given by

$$\lambda_{res(m)}^{TFBG} = (n_{eff}^{core} - n_{eff}^{clad(m)}) \times \frac{\Lambda}{\cos(\theta)'} \tag{12.5}$$

where n_{eff}^{core} and $n_{eff}^{clad(m)}$ are the effective RIs of the core at the mth cladding mode, respectively. Conforming to phase matching condition, when the effective RI of the corresponding cladding modes is identical to the surrounding RI, maximum sensitivity is achieved. Besides that, as a consequence of phase matching, the spectral position of a resonance depends on the effective RI of the corresponding cladding mode, which in turn depends on the optical properties of the medium close to the cladding surface. Thus when the surrounding RI increases, the resonance wavelengths of the corresponding cladding modes shift to longer wavelengths (Chiavaioli et al., 2017).

On the other hand, in an LPG the grating period typically ranges from 100 μm to 700 μm and coupling occurs between the fundamental core mode and copropagating cladding modes. For this reason, multiple attenuation bands at discrete resonance wavelengths are present in the transmission spectrum, with each band related to coupling to a specific cladding mode. These resonance wavelengths, $\lambda_{res(m)}^{LPG}$, satisfy the phase-matching condition expressed by

$$\lambda_{res(m)}^{TFBG} = (n_{eff}^{core} - n_{eff}^{clad(m)}) \times \Lambda \tag{12.6}$$

where n_{eff}^{core} and $n_{eff}^{clad(m)}$ are the effective RIs of the core at the mth cladding mode, respectively. Considering that the cladding modes propagate near the interface between cladding and surrounding medium, they are sensitive to changes in the RI of the latter. As such, once the surrounding RI increases, the resonance wavelengths shift to shorter values (Chiavaioli et al., 2017). Also, LPGs are sensitive to bending, axial strain, and temperature, which can cause unwanted spectral fluctuations (Wang et al., 2020b).

12.2.2.4 Other fibers

In order to overcome the limitations of conventional optical fibers, special optical fibers have been proposed and investigated. There are two main limitations when using conventional fibers: due to the limited geometry of the fiber, RI deviations of the core, and coating are limited, and because of the phenomenon of TIR, the RI of the core cannot be lower than that of the coating. For these reasons, microstructured optical fibers (MOFs), structures that have air holes or micro-capillaries inside the fiber, have been used owing to their higher degree of integration, longer interaction distance between light and analyte, which occurs along the entire length of the fiber, and improved robustness. Furthermore, in MOFs, the fabrication of the sensing region happens inside of the fiber instead of on the exterior, which allows effective interaction between the analyte of interest and the evanescent field of the core mode (Zuber et al., 2019).

In recent years, a variety of MOF structures have been proposed, such as hexagonal MOFs, semicircular channel MOFs, exposed core MOFs among others. Moreover, when air holes in the MOFs present periodic formation (periodic microstructure), usually in the region of the coating and along the entire length of the fiber, they are known as photonic crystal fibers (PCFs) (Berghmans et al., 2014). These can be divided into solid core PCFs, that operate based on TIR mechanism, and hollow-core PCFs, in which the guiding mechanism depends on the material that is placed at the core site (Monfared, 2020a). PCFs can further be classified according to the external structure in, e.g., D-shape (Monfared, 2020b) and H-shape (Liu et al., 2017c). In most MOFs, the main objective is to facilitate simple analyte infiltration and provide large interaction area.

12.2.3 Biofunctionalization strategies for fiber immunosensors

The development of an optical fiber immunosensor requires the immobilization of antibodies on the fiber surface, a step that has a strong impact on sensor performance. The main aspects to be considered include the orientation and density of antibodies per unit of nanofilm surface area, along with the preservation of the biological activity of the antibody. The orientation determines the ability of the epitope to be recognized by the antigens, while the density regulates the efficiency and prevents nonspecific interactions with undesirable biomolecules (Lara and Perez-Potti, 2018).

12.2.3.1 Bare silica optical fiber

Vast research has been realized regarding the immobilization of antibodies on silica fibers, with oriented or site-directed immobilizations garnering increasing attention (Liu and Yu, 2016). In this case, a common approach consists in, first, cleaning the glass surface using a piranha solution, introducing reactive hydroxyl groups on the fiber. This is followed by a silanization process that can be achieved using different reagents according to the protocol used, for instance, 3-aminopropyl-trimethoxysilane (APTMS) (Loyez et al., 2018), 3-aminopropyl-triethoxysilane (APTES) (Sun et al., 2017), 3-mercaptopropyl-trimethoxysilane (MTS) (Tang et al., 2018), 3-(Triethoxysilyl) propylsuccinicanhydride (TESPSA) (Janczuk et al., 2020), 3-aminopropyl diethoxymethylsilane (ADMS) (Xu et al., 2020) solutions. This silanization process produces a free amine group (NH2-), which in a covalent immobilization method, allows the fiber to further react with aldehyde groups (-CHO) of glutaraldehyde (GA) to form imines, followed by immobilization of the antibodies. Considering a physical adsorption method via electrostatic immobilization, the silanized optical fiber is directly immersed in the antibody solution (Sun et al., 2017). An alternative electrostatic adsorption approach involves treating the fiber surface with NaOH solution, providing it a negative charge, and subsequently adding Poly(allylamine) (PAA) solution to produce a self-assembled monolayer (SAM) of PAA by electrostatic interaction as PAA is positively charged (Jiang et al., 2017). Another strategy consists in binding the antibodies through an intermediate protein that is directly immobilized on the silanized glass surface. Fig. 12.4 shows a general schematic of each common biofunctionalization mechanism.

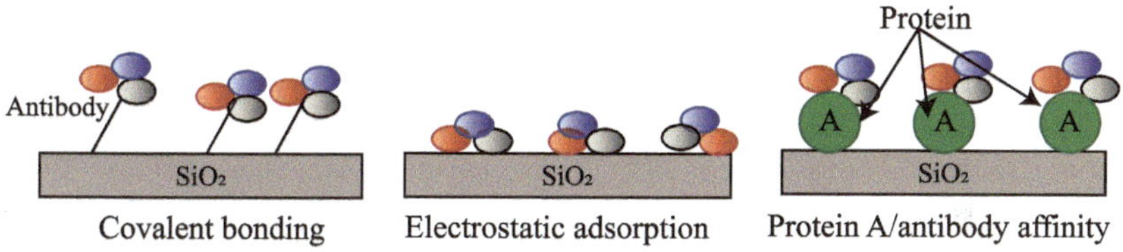

FIGURE 12.4 Biofunctionalization mechanisms.

12.2.3.2 *Polymer optical fiber*

The functionalization process in a POF whose surface is of pure PMMA (unclad POF) can be achieved by reacting PMMA with hexamethylenediamine (HMDA), yielding an aminated surface (Lopes et al., 2018). First, the sensors are treated with sulfuric acid (H2SO4) to produce carboxylic groups on the POF surface and reduce methyl ester groups through acid hydrolysis. Afterwards, the probes are incubated in HMDA in order to form amine groups on the hydrolyzed surface (George et al., 2021) becoming more hydrophilic. Then they are incubated in GA solution, followed by incubation in a solution of Staphylococcal Protein A (SPA), which is used to immobilize the antibodies (Lopes et al., 2018). Alternatively, in Lopes et al. (2018), the antibodies are linked to the aldehyde functionalized POF surface through their amine groups by Schiff's base links.

12.2.3.3 *Metal coated fibers*

Different metal coated optical fiber immunosensors have been developed in a large scale, in order to allow SPR or LSPR (in the presence of nanoparticles) phenomenon to detect a variety of analytes as explained before. Detection in these types of sensors is accomplished due to the presence of antibodies. Thereby, it is crucial to know how to immobilize them on the surface of the sensor to capture the target antigen.

In many studies and investigations, gold is the chosen metal to trigger SPR as well as gold nanoparticles to create LSPR. Cysteamine is one of the cross-linking agents used to immobilize the antibody on the gold surface. Therefore one approach consists in, first immersing the sensors in a cysteamine solution to bind the thiol (-SH) functional groups of the cysteamine molecule to the gold surface through strong ion-exchange interactions. Second, the antibodies are immobilized by binding covalently their carboxylic acid (-COOH) functional groups to the cysteamine -NH2 functional groups. For that, it is used a solution of 1-ethyl-3-[3-dimethylaminopropyl]carbodiimide hydrochloride (EDC) and N-Hydroxysuccinimide (NHS) with an antibody concentration to activate the carboxylic acid groups, which the sensors are immersed for antibody immobilization. Lastly, in order to avoid nonspecific interactions with cysteamine amine functional groups that not reacted, it is applied a solution of bovine serum albumin (BSA) (da S Arcas et al., 2018).

Another approach utilizes protein A instead of cysteamine, since it cannot only specifically bind to the Fc region of the antibodies, but also can capture antibodies with a high degree of orientation. The first step of this method consists in the modification of gold with -COOH groups and its activation through EDC and NHS chemistry. The following step is the immobilization of the protein A on the sensing surface, and after the antibody immobilization onto the protein A surface. Lastly, BSA is used as already described (Wang and Wang, 2018a).

In addition, some researches have been using dopamine as the cross-linking agent instead of cysteamine and protein A. In this way, the antibody of the respective antigen to be detected is immobilized on the surface of dopamine by covalent binding (Wang et al., 2017).

12.2.3.4 Carbon-based materials as fiber coating

Among carbon-based materials, GO is widely used to coat optical fibers with the purpose of developing immunosensors with enhanced sensitivity. The structure of this material consists in a two-dimensional honeycomb lattice with carbon atoms, containing functional groups such as hydroxyl, epoxy, carboxyl on its basal plane and sheet edges (Esposito et al., 2021). As a result, GO can covalently bind to antibodies due to the abundant presence of functional groups on its surface (Wang and Wang, 2018b). Therefore it is used EDC and NHS to activate, by cross-linking chemistry, the -COOH functional groups that appear at the edges of GO flakes (Sridevi et al., 2015). More specifically, EDC activates the -COOH groups, whereas NHS increases the coupling efficiency of EDC. After the activation, the -NH2 groups of the antibodies bind covalently with those -COOH groups and generate an amide bond by releasing water molecules (Wang and Wang, 2018b).

Another functionalization method, uses protein A before the antibody immobilization. By using this protein, proper antibody orientation is achieved when they are immobilized, since this protein specifically binds to the Fc region of the antibodies (Wang and Wang, 2018b). As a result, masking of the antibody-binding sites does not occur. In both methods, at the end, BSA is used to passivate the surface.

12.2.3.5 Oxide semiconductors

Some oxide semiconductors have already been used to coat sensors, but their use is still under investigation. Recent studies have shown that these materials can improve sensor sensitivity owing to the enhancement of the SPR signal characteristics.

MoS2 is one of the oxide semiconductors that has been investigated and used in immunosensors. The properties of this oxide make it suitable for the functionalization of sensors, since it facilitates a simple and rapid functionalization process. MoS2 not only provides higher binding sites for antibody immobilization, but also avoids the use of cross-linkers for the biofunctionalization process. In this instance, antibodies bind to MoS2 through physisorption via hydrophobic interactions (Kaushik et al., 2019). For surface passivation, it is used BSA as described above (Viter et al., 2018).

ZnO is another oxide semiconductor that has been used to produce immunosensors. The functionalization of this oxide begins with its signaling with APTES, and then it is activated with GA. After that process, protein A is deposited over the surface of the sensor, followed by the antibody immobilization and surface passivation with BSA (Maya et al., 2018).

12.2.4 Immunosensing applications in medical biomarkers detection

For every disease, early screening and diagnosis significantly improves patient care and leads to higher survival rates. This can be achieved by quantifying disease-specific biomarkers resorting to portable and user-friendly optical fiber immunosensors, which enable POC testing near the patient and in real time (Nie et al., 2020). Over the years, there has been an increase of reported diagnostic devices for biomarker detection, with cancer being one of the most common targets.

12.2.4.1 Cancer biomarkers

Loyez et al. described a sandwich optical fiber assay to detect HER2 breast cancer biomarker based on SPR using an unclad end-face reflected biosensor. In this research, the cladding was removed from the fiber tip (1 cm length), which was then coated with a gold film via sputtering to enhance the sensitivity toward RI changes. The performance of direct (label-free) and indirect approaches for detecting HER2 proteins was evaluated. In both cases, thiolated anti-HER2 aptamers were directly linked to the gold film surface, however, the indirect approach involved a signal amplification by further immobilization of anti-HER2 antibodies in a sandwich layout (aptamers-HER2-antibodies). By monitoring wavelength shifts of the SPR minimum in the spectral response, the biosensor demonstrated a sensitivity of 0.17 nm/nM. Also, the sandwich assay proved to lower the LOD of HER2 biomarkers from 0.6 μg/mL to 9.3 ng/mL compared to label-free method (Loyez et al., 2021). Sun et al. (2017) also proposed another HER2 biomarker immunosensor, using a fiber optic taper interferometer cascaded with an FBG. The FBG, which is insensitivity to RI changes, was used as a temperature thermometer, whereas the tapered fiber optic interferometer showed a RI sensitivity of 2333 nm/RIU, determined in sucrose solution. The obtained results evidenced a LOD of 2 ng/mL. In 2017, Ribaut et al. (2017) reported an innovative gold coated-TFBG immunosensor for the detection of CK17, a biomarker of interest for lung cancer diagnosis. The gold-coated TFBGs were immersed in S2-PEG6-COOH to create a SAM. Next, through EDC/NHS chemistry, AbCK17 was immobilized on the surface of the TFBGs. Finally, BSA was used to passivate the surface. With the purpose of developing a sensing platform that can assess nonliquid environments, detections were made on CK17 encapsulated in gel matrix that were mimicking tissue samples. The sensor could selectively detect CK17 with a LOD of 0.1 ng/mL. Then the fiber sensor was packed to be utilized for ex-vivo testing of a human lung biopsy and successfully distinguished samples of healthy tissue from tissue with tumors. The LOD of CK17 antigen could reach 0.4 nM, based on the same sensing platform. A sensitive PCF-based immunosensor was proposed by Liu et al. (2017b), for the detection of alpha fetoprotein (AFP), an oncofetal glycoprotein and a biomarker for hepatocellular carcinoma. The surface of the silica PCF was pretreated with ADMS and GA, followed by covalent immobilization of the antibody, which was used to capture the AFP human antigen, that in turn captured the Alexa Fluor 488-labeled AFP secondary antibody. The results demonstrated a linear dynamic range from 0.1 ng/mL to150 ng/mL and a LOD of 0.1 ng/mL, a value that was lower than that of ELISA (3.5 ng/mL) and below the accepted decision threshold for AFP in serum to establish disease diagnosis (10 ng/mL). In 2019, Luo et al. (2019) resorted to ex-TFBG coated with gold nanoshells for label-free detection of soluble programmed death ligand-1 (sPD-L1). In this research, anti-sPD-L1 monoclonal antibodies were attached to the LSPR-based ex-TFBG surface employing the SPA method. The immunosensor evidenced a LOD of 1 pg/mL in buffer solutions and of 5 pg/mL in complex serum media, such as fetal bovine serum.

12.2.4.2 Cardiac biomarkers

Cardiovascular diseases, including acute myocardial infarction (AMI) and heart failure (HF), are a major cause of morbidity and mortality across the world (Zhou et al., 2018). Hence, several cardiac biomarkers have been researched through the years for early screening of these diseases, including cardiac troponin I (cTnI), an indicator of AMI, and Human N-terminal pro-B-type natriuretic peptide (NT-proBNP), a HF biomarker. In 2018, Liu et al. (2018) proposed a cTnI label-free immunosensor deploying a phase shifted microfiber Bragg grating (PS-mFBG). The π-phase-shift in modulation granted a narrow band notch signal in the reflection spectrum of the mFBG, drastically enhancing sensor resolution. The sensing mechanism was based on monitoring the wavelength shift of the notch in the reflective spectrum, caused by changes of the

RI surface because of the antigen-antibody binding. This biosensor obtained a log-linear sensing range from 0.1 ng/mL to 10 ng/mL, a LOD of 0.03 ng/mL, and a RI sensitivity, measured in a mixture of ethanol and DI water (RI range 1.333–1.365), of 32.57 nm/RIU. In 2017, Luo et al. (2017) developed a highly specific immunosensor platform using ex-TFBGs (tilt angle > 80°) for fast and label-free detection of NT-proBNP. In blank PBS, the ex-TFBGs showed a RI sensitivity around 156.03 nm/RIU and a FWHM of, approximately, 2.3 nm, that was concluded to be narrower than that of conventional LPGs, which is around or over 20 nm. The immunoassays were performed in vitro in human serum samples by monitoring the shift of the resonance wavelength, with a fiber optic grating demodulation system, once in the presence of NT-proBNP antigens. The attained results indicated a LOD around 0.5 ng/mL and an average sensitivity of 45.967 pm/(ng/mL) at a concentration range of 0.0–1.0 ng/mL.

12.2.4.3 Cortisol biomarker

Stress, when it is persistent and uncontrollable is considered pathological, which can trigger diseases like depression and cardiovascular diseases (Pinto et al., 2017). Therefore the development of technology capable of monitoring stress is essential. Stress involves a large number of neuronal circuits, and when it is promoted, leads to glucocorticoids release, in particular cortisol (Holsboer and Ising, 2010). The substantial variation of this hormone occurs due to exposure to psychological, environmental, or emotional stress (Usha et al., 2017). As a result, cortisol is one of the most important stress biomarkers.

Nowadays, cortisol levels quantification is still realized through conventional laboratory techniques, which take a long time to give the results, cannot be performed at POC, among other disadvantages. Consequently, immunosensors have been produced to overcome these drawbacks. In 2020, Sharma et al. (2020) simulated a SPR fiber optic immunosensor for salivary cortisol detection at the wavelength of 830 nm. The sensor presented a silver layer with 2D materials, conventional (graphene, tungsten disulfide (WS2), and MoS2) and transition metal carbides (MXenes: Ti3C2, Ti3C2O2, Ti3C2F2) considered one at a time, which operated in two modes ("AIM" and "IIM"). The sensor that showed a superior balanced set of performance parameters under both modes was the Ti3C2O2-based probe. Through the simulation, this probe was able to achieve a LOD of 15.7 fg/m. Recently, Leitão et al. (2021b) reported a SPR unclad POF immunosensor coated with gold/palladium (AuPd) alloy. This sensor was modified with anticortisol antibody for cortisol detection. The detection mechanism relied on the shift of the SPR wavelength achieved through the RI variation on the AuPd surface, due to the antibody-antigen binding reaction. For a range from 0.005 to 10 ng/mL of cortisol concentration, the proposed sensor had a 15 nm wavelength shift, which allowed to conclude a high sensitivity on the part of this sensor. The attained sensitivity and LOD were 3.56 ± 0.20 nm/log(ng/mL) and 1 pg/mL, respectively. In this research, selectivity tests were also performed in a sensor functionalized with antibodies for human chorionic gonadotropin (anti-hCG antibodies) in which the variance of the resonance wavelength was only 1 nm, much lower than the sensor modified with anticortisol antibodies. In another 2021 study, Leitão et al (2021a) developed and tested an ultrasensitive gold coated TFBG immunosensor based on SPR, functionalized with anticortisol antibodies using cysteamine for cortisol detection. In this investigation, it was practiced as an alternative interrogation method. Hence, the local maximum of the plasmonic signature of the lower envelope of the spectra was signalized to monitor the SPR mode. This sensor was tested for a linear cortisol concentration range of 0.1–10 ng/mL, obtaining a total wavelength shift of 3 nm and a sensitivity of 0.275 ± 0.028 nm/ng.mL-1.

12.2.4.4 Cortisol biomarker

The detection of inflammatory biomarkers is crucial for obtaining early disease diagnosis, screening diseases, and even for monitoring treatment efficacy (Nie et al., 2020). Immunosensors produced for this purpose can detect these biomarkers almost without any difficulties and overcome the problems associated with laboratory techniques, as they presented are less time-consuming and can be performed in POC. The development of diagnostic devices for biomarkers detection has been growing recently, including for inflammatory biomarkers.

Liu et al. (2017a) demonstrated a cytokine interleukin-6 (IL-6) detection device in 2016. This device consisted on a sandwich immunoassay scheme with a silica optical fiber coated with gold nanoparticles and, in turn, functionalized with IL-6 antibodies. The attained LOD of this sensor was 1 pg/mL for a linear detection range of 1–400 pg/mL.

More recently, in 2020, Nie et al. (2020) reported portable pencil-like immunosensor (PPS) platform, in other words, a portable and versatile chemiluminescence-based optical fiber immunosensor for the detection of IL-6, procalcitonin (PCT), and C-reactive protein (CRP) in human serum samples. This PPS platform consisted of a unique pencil-like optical fiber-based sensor, a reagent strip consisting of a series of pencil-cap-like wells, and a battery-powered photon counting detector for recording chemiluminescence. Hence, this platform combined an immunosensor with immunoassay process, chemiluminescence detection, and data analysis in a portable suitcase-like device. This sensor presented a LOD for IL-6, PCT, and CRP of 1.05 pg/mL, 10.64 pg/mL, and 29.40 ng/mL, respectively, and an excellent linear relationship from 5 to 10000 pg/mL for IL-6, from 0.05 to 200 ng/mL for PCT, and from 0.1 to 80 μg/mL for CRP. After 14 days of storage at room temperature, this sensor still maintains 90% of response, which is a good characteristic for field assays.

References

Aiman, M., Bahrin, K., Othman, F., Hayati, N., Azli, N., Talib, F., 2016. Journal teknologi full paper Industry 4.0: a review on industrial automation and robotic. Journal Teknologi 78, 2180–3722. www.jurnalteknologi.utm.my.

Alzawa, M., 1994. Immunosensors for clinical analysis, vol. 31.

Bandari, N., Dargahi, J., Packirisamy, M., 2020. Tactile sensors for minimally invasive surgery: a review of the state-of-the-art, applications, and perspectives. IEEE Access 8, 7682–7708. https://doi.org/10.1109/ACCESS.2019.2962636.

Berghmans, F., Geernaert, T., Baghdasaryan, T., Thienpont, H., 2014. Challenges in the fabrication of fibre Bragg gratings in silica and polymer microstructured optical fibres. Laser & Photonics Reviews 8, 27–52.

Caucheteur, C., Guo, T., Albert, J., 2015. Review of plasmonic fiber optic biochemical sensors: improving the limit of detection. Analytical and Bioanalytical Chemistry 407, 3883–3897.

Chiavaioli, F., Baldini, F., Tombelli, S., Trono, C., Giannetti, A., 2017. Biosensing with optical fiber gratings. Nanophotonics 6, 663–679.

Chiu, N.F., Yang, C.D., Chen, C.C., Lin, T.L., Kuo, C.T., 2018. Functionalization of Graphene and Graphene Oxide for Plasmonic and Biosensing Applications. Elsevier, Inc.

Cristea, C., Florea, A., Tertiş, M., Sndulescu, R., 2015. Immunosensors

da S Arcas, A., da S Dutra, F., Allil, R.C.S.B., Werneck, M.M., 2018. Surface plasmon resonance and bending loss-based u-shaped plastic optical fiber biosensors. Sensors (Switzerland) 18, 1–16.

Diana, M., Marescaux, J., 2015. Robotic surgery. British Journal of Surgery 102, 15–28. https://doi.org/10.1002/bjs.9711.

Dumas, C., Caro, S., Cherif, M., Garnier, S., Furet, B., 2012. Joint stiffness identification of industrial serial robots. Robotica 30, 649–659. https://doi.org/10.1017/S0263574711000932.

Edwards, M., 1984. Robots in industry: an overview. Applied Ergonomics 15, 45–53. https://doi.org/10.1016/S0003-6870(84)90121-2.

Esposito, E., et al., 2021. Long period grating in double cladding fiber coated with graphene oxide as high-performance optical platform for biosensing. Biosensors & Bioelectronics 172.

Gandhi, M.S.A., Chu, S., Senthilnathan, K., Babu, P.R., Nakkeeran, K., Li, Q., 2019. Recent advances in plasmonic sensor-based fiber optic probes for biological applications. Applied Science 9, 5.

George, A., Amrutha, M.S., Srivastava, P., Sunil, S., Sai, V.V.R., Srinivasan, R., 2021. Development of a u-bent plastic optical fiber biosensor with plasmonic labels for the detection of Chikungunya non-structural protein 3. Analyst 146, 244–252.

Gonenc, B., Chamani, A., Handa, J., Gehlbach, P., Taylor, R.H., Iordachita, I., 2017. 3-DOF force-sensing motorized micro-forceps for robot-assisted vitreoretinal surgery. IEEE Sensors Journal 17, 3526–3541. https://doi.org/10.1109/JSEN.2017.2694965.

He, X., Balicki, M.A., Kang, J.U., Gehlbach, P.L., Handa, J.T., Taylor, R.H., Iordachita, I.I., 2012. Force sensing micro-forceps with integrated fiber Bragg grating for vitreoretinal surgery. Optical Fibers and Sensors for Medical Diagnostics and Treatment Applications XII 8218, 82180W. https://doi.org/10.1117/12.909602.

Hernaez, M., Zamarre no, C.R., Melendi-Espina, S., Bird, L.R., Mayes, A.G., Arregui, F.J., 2017. Optical fibre sensors using graphene-based materials: a review. Sensors (Switzerland) 17, 1–24.

Hill, K.O., Fujii, Y., Johnson, D.C., Kawasaki, B.S., 1978. Photosensitivity in optical fiber waveguides: application to reflection filter fabrication. Applied Physics 32, 647–649.

Holsboer, F., Ising, M., 2010. Stress hormone regulation: biological role and translation into therapy. Annual Review 61, 81–109.

Hooshiar, A., Najarian, S., Dargahi, J., 2020. Haptic Telerobotic Cardiovascular Intervention: A Review of Approaches, Methods, and Future Perspectives. IEEE Reviews in Biomedical Engineering, vol. 13, pp. 32–50.

Iordachita, I., Sun, Z., Balicki, M., Kang, J.U., Phee, S.J., Handa, J., Gehlbach, P., Taylor, R., 2009. A sub-millimetric, 0.25 mN resolution fully integrated fiber-optic force-sensing tool for retinal microsurgery. International Journal of Computer Assisted Radiology and Surgery 4, 383–390. https://doi.org/10.1007/s11548-009-0301-6.

Janczuk, M.R., et al., 2020. Immunosensor based on long-period fiber gratings for detection of viruses causing gastroenteritis. Sensors (Switzerland) 20, 1–11.

Jaschinski, Thomas, Mosch, Christoph G., Eikermann, Michaela, Am Neugebauer, Edmund, Sauerland, Stefan, 2011. Laparoscopic versus open surgery for suspected appendicitis – update of an existing cochrane review. Surgical Endoscopy and Other Interventional Techniques 25, S25. https://doi.org/10.1002/14651858.CD001546.pub4.

Jiang, Shouzhen, et al., 2017. A novel U-bent plastic optical fibre local surface plasmon resonance sensor based on a graphene and silver nanoparticle hybrid structure. Journal of Physics 50, 165105.

Justino, C.I.L., Duarte, A.C., Rocha-santos, T.A.P., 2016. Critical overview on the application of sensors and biosensors for clinical analysis. TrAC Trends in Analytical Chemistry 85.

Kaushik, S., Tiwari, U.K., Deep, A., Sinha, R.K., 2019. Two-dimensional transition metal dichalcogenides assisted biofunctionalized optical fiber spr biosensor for efficient and rapid detection of bovine serum albumin. Scientific Reports 9, 1–11.

Khalaf, A.L., et al., 2017. Modified plastic optical fiber with CNT and graphene oxide nanostructured coatings for ethanol liquid sensing. Optics Express 25, 5.

Kim, H.M., Park, J.H., Lee, S.K., 2019. Fiber optic sensor based on zno nanowires decorated by Au nanoparticles for improved plasmonic biosensor. Scientific Reports 9, 1–9.

Lara, S., Perez-Potti, A., 2018. Applications of nanomaterials for immunosensing. Biosensors 8, 4.

Leitão, C., et al., 2021a. Cortisol in-fiber ultrasensitive plasmonic immunosensing. IEEE Sensors Journal 21, 3028–3034.

Leitão, C.L., 2021b. Cortisol AuPd plasmonic unclad POF biosensor. Biotechnology Reports 29.

Leung, A., Shankar, P.M., Mutharasan, R., 2007. A review of fiber-optic biosensors. Sensors and Actuators. B, Chemical 125, 688–703.

Liedberg, B., Nylander, C., Lunström, I., 1983. Surface plasmon resonance for gas detection and biosensing. Sensors and Actuators 4, 299–304.

Liu, Y., Yu, J., 2016. Oriented immobilization of proteins on solid supports for use in biosensors and biochips: a review. Mikrochimica Acta 183, 1–19.

Liu, G., Zhang, K., Nadort, A., Hutchinson, M.R., Goldys, E.M., 2017a. Sensitive cytokine assay based on optical fiber allowing localized and spatially resolved detection of interleukin-6. ACS Sensors 2, 218–226.

Liu, X., Song, X., Dong, Z., Meng, X., Chen, Y., Yang, L., 2017b. Photonic crystal fiber-based immunosensor for high-performance detection of alpha fetoprotein. Biosensors & Bioelectronics 91, 431–435.

Liu, Z., Tam, H.Y., Htein, L., Tse, M.L.V., Lu, C., 2017c. Microstructured optical fiber sensors. Journal of Light 35, 3425–3439.

Liu, T., et al., 2018. A label-free cardiac biomarker immunosensor based on phase-shifted micro fiber Bragg grating. Biosensors & Bioelectronics 100, 155–160.

Lopes, R.N., Rodrigues, D.M.C., Allil, R.C.S.B., Werneck, M.M., 2018. Plastic optical fiber immunosensor for fast detection of sulfate-reducing bacteria. Measurement 125, 377–385.

Loyez, M., Albert, J., Caucheteur, C., Wattiez, R., 2018. Cytokeratins biosensing using tilted fiber gratings. Biosensors 8, 3.

Loyez, M., Lobry, M., Hassan, E.M., DeRosa, M.C., Caucheteur, C., Wattiez, R., 2021. Her2 breast cancer biomarker detection using a sandwich optical fiber assay. Talanta 221.

Luo, B., et al., 2017. Human heart failure biomarker immunosensor based on excessively tilted fiber gratings. Biomedical Optics Express 8 (1), 57. https://doi.org/10.1364/boe.8.000057.

Luo, B., et al., 2019. Label-free and specific detection of soluble programmed death ligand-1 using a localized surface plasmon resonance biosensor based on excessively tilted fiber gratings. Biomedical Optics Express 10 (10), 5136–5148.

Mack, M.J., 2001. Minimally invasive and robotic surgery. Journal of the American Medical Association 285, 568–572. https://doi.org/10.1001/jama.285.5.568.

Majumder, M., Gangopadhyay, T.K., Chakraborty, A.K., Dasgupta, K., Bhattacharya, D.K., 2008. Fibre Bragg gratings in structural health monitoring-present status and applications. Sensors and Actuators. A, Physical 147, 150–164.

Marazuela, M.D., Moreno-bondi, M.C., 2002. Fiber-Optic Biosensors – an Overview, pp. 664–682.

Maya, Y.C., Villar, I.D., Socorro, A.B., Corres, J.M., Botero-Cadavid, J.F., 2018. Optical fiber immunosensors optimized with cladding etching and ito nanodeposition 2018. IEEE Photonics Conference (IPC) 31, 4–5. https://doi.org/10.1109/IPCon.2018.8527306.

Monfared, Y.E., 2020a. Overview of recent advances in the design of plasmonic fiber-optic biosensors. Biosensors 10, 7.

Monfared, Y.E., 2020b. Refractive index sensor based on surface plasmon resonance excitation in a d-shaped photonic crystal fiber coated by titanium nitride. Plasmonics 15, 535–542.

Nie, R., Huang, J., Xu, X., Yang, L., 2020. A portable pencil-like immunosensor for point-of-care testing of inflammatory biomarkers. Analytical and Bioanalytical Chemistry 412, 3231–3239.

Noh, Y., Liu, H., Sareh, S., Chathuranga, D.S., Wurdemann, H., Rhode, K., Althoefer, K., 2016. Image-based optical miniaturized three-axis force sensor for cardiac catheterization. IEEE Sensors Journal 16, 7924–7932. https://doi.org/10.1109/JSEN.2016.2600671.

Oh, S., Woo, H., Kong, K., 2014. Frequency-shaped impedance control for safe human-robot interaction in reference tracking application. IEEE/ASME Transactions on Mechatronics 19, 1907–1916. https://doi.org/10.1109/TMECH.2014.2309118.

Okamura, A.M., 2009. Haptic feedback in robot-assisted minimally invasive surgery. Current Opinion in Neurology 19, 102–107. https://doi.org/10.1097/MOU.0b013e32831a478c.

Pinto, V., Sousa, P., Catarino, S.O., Correia-Neves, M., Minas, G., 2017. Microfluidic immunosensor for rapid and highly-sensitive salivary cortisol quantification. Biosensors & Bioelectronics 90, 308–313.

Prabowo, B.A., Purwidyantri, A., Liu, K.C., 2018. Surface plasmon resonance optical sensor: a review on light source technology. Biosensors 8, 3.

Qi, M., Zhang, N.M.Y., Li, K., Tjin, S.C., Wei, L., 2020. Hybrid plasmonic fiber-optic sensors. Sensors, 1–27.

Ribaut, C., et al., 2017. Cancer biomarker sensing using packaged plasmonic optical fiber gratings: towards in vivo diagnosis. Biosensors & Bioelectronics 92, 449–456.

Rifat, A.A., et al., 2017. Photonic crystal fiber based plasmonic sensors. Sensors and Actuators. B, Chemical 243, 311–325.

Schneider, U., Momeni-K, M., Ansaloni, M., Verl, A., 2014. Stiffness modeling of industrial robots for deformation compensation in machining. In: IEEE International Conference on Intelligent Robots and Systems, pp. 4464–4469.

Schostek, S., Schurr, M.O., Buess, G.F., 2009. Review on aspects of artificial tactile feedback in laparoscopic surgery. Medical Engineering and Physics 31, 887–898. https://doi.org/10.1016/j.medengphy.2009.06.003.

Seraji, H., 2016. Force Tracking in Impedance Control.pdf, pp. 97–117.

Sharma, A.K., Kaur, B., Marques, C., 2020. Simulation and analysis of 2d material/metal carbide based fiber optic spr probe for ultrasensitive cortisol detection. Optik (Stuttg) 218.

Sherwani, F., Asad, M.M., Ibrahim, B.S., 2020. Collaborative robots and industrial revolution 4.0 (IR 4.0). In: 2020 International Conference on Emerging Trends in Smart Technologies, ICETST 2020.

Song, H., Kim, K., Lee, J., 2011. Development of optical fiber Bragg grating force-reflection sensor system of medical application for safe minimally invasive robotic surgery. Review of Scientific Instruments 82. https://doi.org/10.1063/1.3606502.

Sridevi, S., Vasu, K.S., Asokan, S., Sood, A.K., 2015. Sensitive detection of C-reactive protein using optical fiber Bragg gratings. Biosensors & Bioelectronics 65, 251–256.

Sun, D., Ran, Y., Wang, G., 2017. Label-free detection of cancer biomarkers using an in-line taper fiber-optic interferometer and a fiber Bragg grating. Sensors (Switzerland) 17, 11.

Tahhan, S.R., Chen, R.Z., Huang, S., Hajim, K.I., Chen, K.P., 2017. Fabrication of fiber Bragg grating coating with tio2 nanostructured metal oxide for refractive index sensor. Journal of Nanotechnology 2017.

Tang, Min, et al., 2018. Development of an optical fiber immunosensor for the rapid and sensitive detection of phthalate esters. Sensors and Actuators. B, Chemical 258, 304–312.

Thoben, K.D., Wiesner, S.A., Wuest, T., 2017. "Industrie 4.0" and smart manufacturing-a review of research issues and application examples. International Journal of Automation Technology 11, 4–16. https://doi.org/10.20965/ijat.2017.p0004.

Usha, S.P., Shrivastav, A.M., Gupta, B.D., 2017. A contemporary approach for design and characterization of fiber-optic-cortisol sensor tailoring lmr and zno/ppy molecularly imprinted film. Biosensors & Bioelectronics 87, 178–186.

Vaidya, S., Ambad, P., Bhosle, S., 2018. Industry 4.0 – a glimpse. Procedia Manufacturing 20, 233–238. https://doi.org/10.1016/j.promfg.2018.02.034.

Viter, R., et al., 2018. Analytical, thermodynamical and kinetic characteristics of photoluminescence immunosensor for the determination of Ochratoxin A. Biosensors & Bioelectronics 99, 237–243.

Wagner, C.R., Stylopoulos, N., Howe, R.D., 2002. The role of force feedback in surgery: analysis of blunt dissection. In: Proceedings – 10th Symposium on Haptic Interfaces for Virtual Environment and Teleoperator Systems, HAPTICS, pp. 68–74, 2002.

Wang, W., et al., 2017. A label-free fiber optic SPR biosensor for specific detection of C-reactive protein. Scientific Reports 7, 1–8.

Wang, B.T., Wang, Q., 2018a. Sensitivity-enhanced optical fiber biosensor based on coupling effect between spr and lspr. IEEE Sensors Journal 18, 8303–8310.

Wang, Q., Wang, B., 2018b. Sensitivity enhanced spr immunosensor based on graphene oxide and spa co-modified photonic crystal fiber. Optics & Laser Technology 107, 210–215.

Wang, Q., Wang, B.T., 2018c. Surface plasmon resonance biosensor based on graphene oxide/silver coated polymer cladding silica fiber. Sensors and Actuators. B, Chemical 275, 332–338.

Wang, Q., Jiang, X., Niu, L., Fan, X., 2020a. Enhanced sensitivity of bimetallic optical fiber spr sensor based on mos2 nanosheets. Optics & Laser Technology 128.

Wang, R., Ren, Z., Kong, D., Hu, B., He, Z., 2020b. Highly sensitive label-free biosensor based on graphene-oxide functionalized micro-tapered long period fiber grating. Optical Materials 109.

Xu, X., Nie, R., Huang, J., Yang, L., 2020. Chemiluminescent optical fiber immunosensor combining surface modification and signal amplification for ultrasensitive determination of hepatitis b antigen. Sensors 20.

Yin, M., Gu, B., An, Q.F., Yang, C., Guan, Y.L., Yong, K.T., 2018. Recent development of fiber-optic chemical sensors and biosensors: mechanisms, materials, micro/nano-fabrications and applications. Coordination Chemistry Reviews 376, 348–392.

Yip, M.C., Yuen, S.G., Howe, R.D., 2010. A robust uniaxial force sensor for minimally invasive surgery. IEEE Transactions on Biomedical Engineering 57, 1008–1011. https://doi.org/10.1109/TBME.2009.2039570.

Zhao, Y., Tong, R. jie, Xia, F., Peng, Y., 2019. Current status of optical fiber biosensor based on surface plasmon resonance. Biosensors & Bioelectronics 142.

Zhou, W., et al., 2018. Ultrasensitive label-free optical microfiber coupler biosensor for detection of cardiac troponin i based on interference turning point effect. Biosensors & Bioelectronics 106, 99–104.

Zuber, A.A., Klantsataya, E., Bachhuka, A., 2019. Biosensing. Comprehensive Nanoscience and Nanotechnology 1, 105–126.

Chapter 13

Conclusions and outlook

13.1 Summary

The goal of this book is to provide novel instrumentation approaches for healthcare devices, especially gait assistance devices, using optical fiber sensors. The motivation of this work (as described in Chapter 1) is some of the limitations of conventional electronic sensors, which include electromagnetic sensitivity, necessity of precise assemblies (increasing the system's complexity and weight), and in most cases, the inability of being embedded or positioned in flexible structures. These disadvantages are undesirable for the rigid robots as well as in wearable sensors and specially undesirable for soft robotics, where the robots' structures and actuators are made with flexible materials, which is a trend on wearable systems and assistive devices, as discussed in Chapter 2. Thus the proposed sensors are compared with the electronic ones in their best operation conditions (precisely assembled and aligned in rigid structures), where the electronic sensors present high accuracy and can be used as a reference for the comparison with the optical fiber sensors. In this way, another contribution of this book is to show that the optical fiber sensors present low errors when compared with electronic sensors in their best conditions for operation. Nevertheless, as mentioned above, optical fiber sensors can also be applied in flexible structures in which potentiometers and encoders cannot be employed with the same precision. Furthermore, the high flexibility of optical fibers, combined with their small dimensions make them suitable for wearable sensors that can be applied on the movement analysis, activity monitoring and physiological parameters assessment. As locomotion plays an important role in society and consists of different phases and types. Chapter 3 describes the human gait and shows an overview of its features.

In the second part of this book, the fundamentals of optical fiber sensors are depicted. The fundamental aspects of optical fibers systems, which include their operation principles, types, and fabrication. In addition, the commonly used components such as optical circulators and optical couplers/splitters, light sources (including light emitting diodes and lasers) are discussed in Chapter 4. This chapter also presents the photodetectors operation principles and its variants, which also include the description of commonly used systems and approaches for optical spectrum analyzers and spectrometers. As a critical aspect of optical fiber sensors systems, especially when polymer optical fibers are used, the connectorization between optical fibers and their connectors are also discussed in Chapter 4.

https://doi.org/10.1016/B978-0-32-385952-3.00024-X

Another important parameter on optical fiber sensors applications, especially for thermomechanical parameters sensing is the material properties. The materials used in optical fiber fabrications as well as their mechanical, thermal, and optical properties are depicted in Chapter 5. For polymer optical fibers, the viscoelastic response is presented and characterized for different optical fiber materials, which leads to the assessment of the temperature and frequency dependency of the viscoelastic properties in polymer materials. In addition, the thermal and chemical treatments in optical fibers are discussed and characterized in Chapter 5. The dynamic characterization of POF materials is presented, where the materials characterized are PMMA (for intensity variation-based sensors) and CYTOP (for FBG sensors). This chapter is the cornerstone of the works and is presented in the remainder of the book. In this case, creep response, temperature, and frequency dependency as well as humidity dependency of PMMA complex modulus are thoroughly discussed. Moreover, the assumptions of material-related hysteresis in POF sensors are also confirmed by means of sequential stress-strain cycles. Another contribution of this chapter is on the characterization of the CYTOP fibers with and without FBGs, which not only serves as basis for the development of the fiber Bragg grating (FBG) sensors discussed in Chapter 6.

Then Chapter 6 closes the second part of this book. In this case, the optical fiber sensors approaches are discussed. In spite of many sensors approaches and variants, Chapter 6 focuses on intensity variation-based sensors (including its multiplexing technique), Fabry–Perot interferometers (FPI) and FBGs, where its principles, models, fabrication, and interrogation/signal acquisition are discussed. In addition, temperature cross-sensitivity compensation techniques are discussed since the temperature can influence the sensors responses, which can be undesirable for some applications, especially the ones related to movement analysis, robots and actuators instrumentation discussed in this book.

The third part of this book includes the optical fiber sensors applications in healthcare, assistive devices, and compliant actuators instrumentation. In Chapter 7, the exoskeleton and orthosis instrumentation are discussed. In this case, intensity variation-based POF sensors were used due to: (i) their low cost (in some cases, even lower than the ones of electronic sensors for the same performance), (ii) ease of implementation, since the sensor system only needs low cost LEDs and photodetectors, and (iii) simplicity on the signal processing, since the only acquired signals are the analog response of the sensors. Furthermore, as the advances of grating inscription as well as novel portable and low relative cost FBG interrogators occur, FBG sensors were also explored taking advantages of the high multiplexing capabilities and wavelength-encoded data. The later offers a crucial advantage (especially when compared with intensity variation-based sensors) as the data is an absolute quantity and immune to light source power deviations. Regarding the applications, it was also proposed the complete instrumentation of two exoskeletons with sensors for human-robot interaction forces assessment, joint angle measurements and even microclimate sensors.

Thereafter, Chapter 8 presents the optical fiber sensors development for movement analysis as wearable sensors incorporated in textiles or insoles. The intensity variation-based sensors were used on the development of a portable and low cost system for joint angle analysis. Similarly, an FBG-based knee sensor is developed, where an FPI is used as an optical filter for a portable and low cost acquisition of the FBG responses (without the need of optical spectrum analyzers). The multiplexing capabilities of the FBGs also enable the development of a plantar pressure mapping system using FBGs embedded in an insole, where the pressure distribution is acquired in 6 points distributed on the foot. After the positive feedback of the multiplexing technique (discussed in Chapter 6), a 3D-printed insole was presented with 15 independent sensors. The proposed insole not only is a low-cost and highly customizable device (due to the materials and methods employed), but also is the insole with optical fiber sensing technology with the highest number of sensors, where the number of sensors on the proposed insole is more than two times higher than the ones of the FBG-based insole proposed.

In Chapter 9, the development of optical fiber sensors in compliant (and soft) actuators are discussed. First, using the background provided by the second part of this book, compensation techniques for of POF sensors are proposed and validated. It is noteworthy that the background and fundamentals provided also enable the development of temperature and humidity sensors based on stress-optic effects on the fiber under torsion and a similar approach was made to propose a novel torque sensor, which was applied in a series elastic actuator (SEA). In addition, FBG arrays were applied in SEA's spring for multipoint assessment of the strain. In all analyzed scenarios, the application of FBG arrays resulted in either higher precision of the sensor system, especially when combined with sensor fusion techniques, with the additional advantage over intensity variation-based sensors regarding the wavelength-encoded data, which is insensitive to light source power fluctuations.

In the last part of the book, case studies and additional applications were depicted. The first group of applications were discussed in Chapter 10, where the multifunctional smart textiles were presented. The textiles comprised in optical fiber-integrated fabrics, where the interaction of such textiles with optical signals resulted in the possibility of measuring heartbeat and breath rate as well as body temperature in configurations transparent for the user, portable and with lower cost. In this case, the FPIs and intensity variation sensors were described in detail, but additional sensors such as other interferometers configurations and FBGs were also proposed in these applications and were briefly described. The multiplexing technique for intensity variation-based sensors also resulted in the possibility of multiparameter assessment using the photonic textiles (optical fiber-embedded textiles), where the human-environment interaction can be estimated from the pressure mapping in the textiles and, if combined with the displacement measurement, can also result in the activity monitoring. In this case, three activities were monitoring: (i) walking, (ii) squatting, and (iii) sitting, where the sensors were able to identify each activity. Another case study was presented in Chapter 11, where the instrumentation of a smart walker (SW) was presented using intensity variation-based sensors for oximetry, gait cadence, heartbeat, and breath rates to evaluate the user's health condition during the SW usage. In addition, an FBG array was employed in the SW for multiparameter measurements using a single optical fiber cable, resulting in the increase of compactness and reduction of the cost per sensor in a healthcare device, where the array was able to identify the user's movement intention, gait cadence, floor-induced vibration, and has the additional possibility of performing the structural health monitoring of the assistive device. Thereafter, Chapter 12 includes two different applications of optical fiber sensors: one is related to the instrumentation of robotic manipulators and devices for robotic surgery, an ever growing field in robotics for medical applications. In this case, different approaches for the instrumentation of the manipulator end-effector (or tool) are presented, including 3D sensors for force feedback on the robot and/or its operator. In another large field of applications, the biosensors using optical fiber systems are described where multiple sensing schemes as well as biofunctionalization are described. Applications on biomarkers detection are presented as an important development of optical fiber sensors field on biomedical applications, which are in accordance with the necessities of remote healthcare applications as well as the diseases diagnosis.

13.2 Final remarks and outlook

This book paved the way for a multitude of applications in healthcare devices as well as novel sensing approaches taking into account the material mechanical properties. For these reasons, many research fields and future works are available on each of the three keystones of the book (material characterization, optical fiber sensors techniques, and healthcare applications). Many of these works are already under investigation in different research groups around the world.

The future works perspectives for the material characterizations are listed as follows:

- Mechanical, optical, and thermal characterizations of optical fibers with different materials;
- Characterization of novel 3D-printed optical fibers and optical fiber based on thermosetting resins;
- Development of new polymer optical fibers using environmental-friendly techniques;
- Investigation of optical fiber material properties after the FBG inscription using UV nanosecond pulsed and CW lasers;
- Investigation of the UV radiation influence on the material response of different optical fibers;

- Analysis of different dopants for POFs and their influence on the material properties under different conditions of UV radiation;
- Analysis of the POF fabrication parameters such as pulling force and extrusion temperature and their influence on POF viscoelastic response;
- Characterization of photoelastic constant and thermooptic coefficients in different optical fibers.

The future works and research directions involving optical sensors (emphasis on intensity variation sensors and FBGs) are listed below:

- Analysis of intensity variation-based sensors under higher frequency regime;
- Intensity variation-based sensors in acoustic and ultrasound applications;
- Low cost biochemical sensors based on POF material features;
- Development of 3D sensors for movement analysis;
- Investigations of the spectral response of FBGs under different loading conditions for innovative sensing solutions;
- Novel cost-effective techniques for grating inscription;
- Application of the compensation approaches in nonuniform gratings;
- Development of novel portable and low cost FBG interrogators;
- Smartphone integration with FBG sensors.

Finally, regarding perspectives and outlook of future works using optical fiber sensors for healthcare applications include:

- Applications of the 3D FBG-based curvature sensor in robotics and movement analysis;
- Shape reconstruction in robotic elements using optical fiber sensors;
- Novel applications in healthcare devices and novel smart devices with embedded optical sensors;
- Applications in control systems of soft robotics;
- Tendon-driven actuators using POFs as structural and sensing element;
- Further applications of the proposed multiplexing technique for POF intensity variation-based sensors.
- Optical fiber-integrated clothing and accessories.

Index

I

P

Kolibri 360 GmbH
Ferdinand-Jühlke-Straße 7
99095 Erfurt, Deutschland
produktsicherheit@kolibri360.de